The Physics of Microdroplets

Scrivener Publishing
100 Cummings Center, Suite 41J
Beverly, MA 01915-6106

Scrivener Publishing Collections Editors

James E. R. Couper	Ken Dragoon
Richard Erdlac	Rafiq Islam
Norman Lieberman	Peter Martin
W. Kent Muhlbauer	Andrew Y. C. Nee
S. A. Sherif	James G. Speight

Publishers at Scrivener
Martin Scrivener (martin@scrivenerpublishing.com)
Phillip Carmical (pcarmical@scrivenerpublishing.com)

The Physics of Microdroplets

Jean Berthier
CEA/LETI, University of Grenoble, France

and

Kenneth A. Brakke
Department of Mathematical Sciences,
Susquehanna University, USA

Scrivener

Co-published by John Wiley & Sons, Inc. Hoboken, New Jersey, and Scrivener Publishing LLC, Salem, Massachusetts.
Published simultaneously in Canada.

For general information on our other products and services or for technical support, please contact our Customer Care Department within the United States at (800) 762-2974, outside the United States at (317) 572-3993 or fax (317) 572-4002.

Wiley also publishes its books in a variety of electronic formats. Some content that appears in print may not be available in electronic formats. For more information about Wiley products, visit our web site at www.wiley.com.

For more information about Scrivener products please visit www.scrivenerpublishing.com.

Cover design by Russell Richardson.

The cover shows a Surface Evolver simulation of a liquid plug in an enclosed rectangular hydrophobic channel, with square pillars added to constrict the channel (top). The pinched regions between the pillars have high curvature, hence high pressure, so liquid flows into the low-curvature, low-pressure larger blobs (middle), until the neck pinches off to form two stable blobs (bottom).

Library of Congress Cataloging-in-Publication Data:

ISBN 978-0-470-93880-5

Printed in the United States of America

10 9 8 7 6 5 4 3 2 1

Contents

Acknowledgements

J. Berthier

I would like to thank all my colleagues who have contributed with photographs and sketches: R. Berthier, N. Sarrut, P. Dalle, J. Wagh, P. Clementz, P. Pouteau, F. Baleras, N. Chiarutinni, M. Bestehorn, and G. Morris.

I am grateful to E. Berthier and his colleagues B. Casavant and D.J. Beebe at University of Wisconsin for their discussion on "Suspended microfluidics" and sketches for chapter 8 "Open microfluidics".

Many thanks to J.C. Mourrat (EPFL) for the fruitful discussion on the analytical model for the twist mode of chapter 11 "Capillary Self-assembly for 3D Microelectronics". I also am indebt to L. Di Cioccio, S. Mermoz and C. Frétigny for their help for this same chapter.

I am also grateful to D. Peyrade for the interesting discussions on capillary alignment of gold nanospheres in chapter 9 "Droplets, Particles and Interfaces".

I would like to thank my company for having given me encouragements and support for this project, especially L. Malier director of the Leti, and C. Peponnet my group leader.

Finally I thank my children Erwin, Linda and Rosanne for constantly encouraging me during this work.

K. Brakke

I would like to thank my Ph.D. advisor Fred Almgren for introducing me to the mathematics of soap films and liquid surfaces and his support and encouragement during the early development of my Surface Evolver. I would also like to thank the Geometry Center at the University of Minnesota for numerous summer visits and a sabbatical, which greatly contributed to Evolver development.

Preface

Capillary phenomena are intriguing. During the many years I (Jean) have spent with my colleagues working on microsystems for biotechnology, I have observed the difficulty to predict – and sometimes understand – the behavior of droplets and interfaces at the micro scale. First, optical observation is not straightforward – it is not easy to locate an interface seen from above in the microscope. Second, the analysis of the observed phenomena is complicated. In my personal experience, that was the case for pancreatic cells encapsulation in micro-flow-focusing devices, liquid-liquid extraction systems, digital microfluidics, capillary valves, spontaneous capillary flows in closed and open channels, in cracks, and between fibers.

And the difficulty is even more important for the conception of new microsystems. Questions such as "where is the interface going to anchor?" or "will the particles cross the interface?" or "will the interface de-pin when the capsule arrives?" or "will the capillary force be sufficient?" are repeatedly being asked. Although illustrious pioneers such as P-G. de Gennes, D. Quéré, G.M. Whitesides, and others have contributed to the knowledge of interface behaviors on a theoretical standpoint, much is left to understand for the engineer having to design a microchip or the student behind his computer or the biologist at his lab bench.

In this book, Ken and I have attempted to give the reader the tools for solving these capillary and surface tension problems, present theoretical tools derived from previous works of colleagues and our personal experience, as well as provide calculation tools through the Surface Evolver numerical program.

I first heard about Evolver at a Nanotech Conference in 2004 and its potential for two-phase microflows and droplets behavior. Although it cannot treat the dynamics of a flow, it can be used to predict the stable shape and location of droplets and interfaces. A typical example is that of a capillary valve where the bulging out of the interface directly depends on the applied pressure. Besides, useful information can be gained by considering that an interface or a droplet has not reached its equilibrium position: this is for example the case of spontaneous capillary flows or droplets moving up a step or a slope. Finally, at the microscale, interfaces are restored nearly immediately by capillary and surface tension forces, which frequently dominate the other forces like weight, viscosity, and inertia. This applies for example to self-alignment problems.

I started to work with Evolver for predicting the behavior of droplets in digital microfluidic systems. Because the electrowetting effect can often be translated into a capillary effect (capillary equivalence), Evolver is well suited to treat such problems. I had the fortune that the author of Surface Evolver, Kenneth Brakke, agreed to assist me with the handling of the numerical program and our cooperation was extremely fruitful. After a few years of working on this topic, as well on the theoretical, numerical and experimental aspects, I had the opportunity to write the book Microdrops and Digital Microfluidics in 2008.

But many capillary problems were still to be tackled outside the domain of digital microfluidics. I continued to use Evolver, again with Ken's help. When our Evolver tool box was sufficient, we thought that it could be useful to make it available to the scientific community and decided to write this new book with my publisher Martin Scrivener. The Evolver files corresponding to the examples and problems of this book are available for the reader at the internet address http://www.susqu.edu/brakke/physicsofmicrodrops.

We hope that our work will be useful to boost the developments of microfluidic systems and that this book will find an echo in the micro and nanotechnology world.

Jean Berthier, Grenoble, February 22, 2012
Kenneth A. Brakke, Susquehana University, February 22, 2012

Introduction

From Conventional Single-phase Microfluidics to Droplets and Digital Microfluidics

Starting in the year 1980, microfluidics was at first a mere downscaling of macrofluidics. Its development was triggered by the emergence of biotechnology and materials science, imagined by visionary pioneers like Feynman [1], deGennes [2], Whitesides [3] and others. In particular, biotechnology was as a new science at the boundary of physics and biology. The goal was to give biological, medical and pharmaceutical research new automation tools to boost the development of new drugs, fabricate new body implants and increase the potentialities of fundamental research. In reality, this plan imagined by these first researchers has been extremely effective and produced even more discoveries than what was first expected. In a way, biotechnology developments bloomed according to Feynman's words: "The best way to predict the future is to invent it." The foreseen goals have required the downscaling of fluidic systems to the "convenient" size to work at the proper scale characteristic of a population of biologic targets. At the same time, it was found that the downscaling brought economy in costly materials, fluids, and devices; that sensitivity was increased and operating times were greatly reduced by the integration of many functions on the same microchip. Gradually, as microsystems based on microflows become conventionally used, new approaches were investigated that required even less volume of sample fluids. This trend to downscaling has promoted the development of new microfluidic approaches such as droplet and digital microfluidics. Reduction of the liquid vessel containing the biological targets was found to be possible by the use of microdroplets. New systems based on the confinement of biologic targets in extremely small vessels like microdrops are emerging. In such approaches the liquid volumes are reduced to a few picoliters.

Domains of Application

Historically, genomics and proteomics were the first beneficiaries of the development of biotechnology, and now it is the turn of cellomics. Also, these developments have spread beyond the domain of biotechnology and created a "cloud" of new applications in other domains such as bioinformatics, bioengineering, tissue engineering, etc. At the same time, microfluidic techniques reached other domains, such as materials science, microelectronics and mechatronics. It has been quickly demonstrated that biochemical reactions such as PCR for the recognition of DNA can be performed with the same efficiency in droplets, with a lesser amount of replicas [4-6]. Proteins can be crystallized in droplets, resulting in a greater ability to investigate their structures by X-ray crystallography [7]. In biology, single cell research has become feasible, after encapsulating the cell in a droplet or a gelled (polymerized) droplet [8-10] or manipulating

1

cells on a digital microfluidic chip [11]. Chemical reactions can also be performed with very small amounts of chemical species inside droplets [12-14]. The use of droplet and digital microfluidics soon extended beyond the limits of biotechnology. Electrowetting droplets are now commercially used in optics as tunable lenses [15] and screen displays [16]. In mechatronics, electrowetting switches (or CFA, for "capillary force actuators") have been shown to be much more effective than electrostatic switches of the same size [17]. Self-assembly techniques using capillary forces produced by a droplet surface are currently used in materials science for manipulating gold nano-spheres for coating applications [18]. Self-alignment using capillary forces is also a promising approach to 3D-microelectronics, which is required to circumvent the present limitations of 2D assembly [19-21]. The examples are many showing the interest in microdrops.

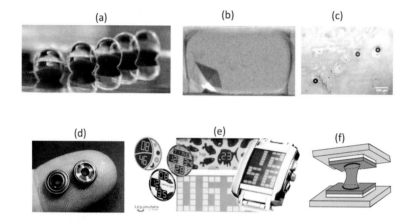

Figure 1 Different applications using microdroplets: (a) droplets moved with magnetic beads for PCR application [6]; (b) protein crystallization in a microdrop (from [7], ©Wiley-VCH Verlag GmbH & Co. KGaA. Reproduced with permission); (c) encapsulated cells in a polymerized alginate matrix (photo courtesy CEA-LETI); (d) tunable lenses by Varioptics (from [15], courtesy Varioptics); (e) screen displays by Liquavista, (from [16], courtesy Liquavista); (f) schematic of a capillary force actuator (not to scale) [17].

Organization of the Book

This book is dedicated to the study of droplets and interfaces principally in a steady or quasi-steady state, although some dynamic considerations have been added when it was judged useful. The first chapter presents the general considerations leading to the concepts of surface tension and capillary forces, associated to the notions of surface energy and contact angle. Young's and Laplace's laws, which are the two "pillars" of any capillary approach, are described, commented and exemplified. The second chapter presents the theory of liquid surfaces in space, including some ways to prove certain surfaces are minimums of energy. Chapter 3 is devoted to the determination of the shape, surface area, and volume of droplets. In chapter 4, the shape and behavior of sessile droplets (droplets place on a solid surface) is investigated for many different configurations of chemical and geometrical surface inhomogeneities: drops at the boundaries between hydrophilic and hydrophobic substrates, or on geometrical inhomogeneities such as steps or grooves or corners. The fifth chapter concerns the behavior of droplets in asymmetric geometries; in a first part, the Hauksbee problem is treated and an extension to hydrophobic surfaces is given. In a second part, the Concus-Finn relations are presented. In chapter 6, the behavior of droplets in microwells and closed microchannels is investigated. The cases of wetting and non-wetting plugs are treated as well as that of trains of droplets. Chapter 7 is dedicated to the phenomena of capillary rise, capillary pumping and capillary valving. In the first two parts, we analyze how capillary forces can contribute to moving a liquid in horizontal or vertical tubes. In the third part, we analyze the opposite: how to find a geometry that can stop a capillary flow. The focus of chapter 8 is open microfluidics, i.e. microflows partially guided by a solid wall, but also in contact with air or another liquid, which is becoming a very important issue in biotechnology; this type of microflow rely mainly on capillary forces and if necessary on electrowetting forces to move the fluid. Chapter 9 deals with the contact and potential engulfment of droplets and particles by interfaces. Examples pertaining to encapsulation of polymerized droplets and capillary assembly are presented. Chapter 10 is on digital microfluidics, a convenient way to manipulate droplets on a planar, or locally planar surface, which has seen many developments lately. We present the state of the art and new developments in this technique. In chapter 11, we treat an example of the use of capillary forces: the ongoing approach to 3D-microelectronics by assembling stacks of chips on a wafer. A promising approach to achieve chip positioning and alignment is that of capillary self-assembly.

References

[1] R. Feynman, Chap 6 in *Building biotechnology* by Y.E. Friedman, third edition, Logos press, 2008.

[2] P-G de Gennes, F. Brochart-Wyart, D. Quéré. *Capillary and wetting phenomena: drops, bubbles, pearls, waves.* Springer, 2002.

[3] G.M. Whitesides, Chap 9 in *Biotechnology and Materials Science – Chemistry for the future*, by L.M. Good, ACS publications, 1988.

[4] P.-A. Auroux,Y. Koc,A. deMello, A. Manz and P. J. R. Day, *Miniaturised nucleic acid analysis*, Lab Chip **4**, pp. 534–546, 2004.

[5] E. Wulff-Burchfield, W.A. Schell, A.E. Eckhardt, M. G. Pollack, Zhishan Hua, J. L. Rouse, V. K. Pamula, Vijay Srinivasan, J. L. Benton, B. D. Alexander, D. A. Wilfret, M. Kraft, C. Cairns, J. R. Perfect, and T. G. Mitchell, "Microfluidic Platform versus Conventional Real-time PCR for the Detection of Mycoplasma pneumoniae in Respiratory Specimens," *Diagnostic microbiology and infectious disease* **67**(1), pp. 22–29, 2010.

[6] http://www.quantalife.com/technology/ddpcr

[7] Bo Zheng, L. Spencer Roach, and R. F. Ismagilov, "Screening of Protein Crystallization Conditions on a Microfluidic Chip Using Nanoliter-Size Droplets," *JACS* **125**, pp. 11170-11171, 2003.

[8] T. Thorsen, R. W. Roberts, F. H. Arnold, S. R. Quake, "Dynamic pattern formation in a vesicle-generating microfluidic device," *Phys. Rev. Lett.* **86**, pp. 4163–4166, 2001.

[9] S.L. Anna, N. Bontoux, and H.A. Stone, "Formation dispersions using flow focusing in microchannels," *Appl. Phys. Lett.* **82**(3), pp. 364–366, 2003.

[10] J.F. Edd, D. Di Carlo, K.J. Humphry, S. Köster, D. Irimia, D.A. Weitz, M. Toner, "Controlled encapsulation of single cells into monodispersed picoliter drops," *Lab Chip* **8**(8), pp. 1262–1264, 2008.

[11] D. Witters, N. Vergauwe, S. Vermeir, F. Ceyssens, S. Liekens, R. Puers and J. Lammertyn, "Biofunctionalization of electrowetting-on-dielectric digital microfluidic chips for

miniaturized cell-based applications," *Lab Chip* **11**, pp. 2790–2794, 2011.

[12] H. Song, J. D. Tice, R. F. Ismagilov, "A microfluidic system for controlling reaction networks in time," *Angew. Chem.* **42**, pp. 767–771, 2003.

[13] A. Gnther, K.F. Jensen, "Multiphase microfluidics: from flow characteristics to chemical and material synthesis," *Lab. Chip* **6**, pp. 1487–1503, 2006.

[14] J. Atencia, D.J. Beebe, "Controlled microfluidic interfaces," *Nature* **437**, pp. 648–655, 2005.

[15] Variotics^{TM}: http://www.varioptic.com/en/tech/technology01.php

[16] Liquavista^{TM}: http://www.liquavista.com/files/LQV060828XYR-15.pdf

[17] C. R. Knospe and S.A. Nezamoddini, "Capillary force actuation," *J. Micro-Nano Mech.* **5** p. 5768, 2009.

[18] O. Lecarme, T. Pinedo-Rivera, K. Berton, J. Berthier, D Peyrade, "Plasmonic coupling in nondipolar gold collidal dimers," *Applied Physics Letters* **98**, 083122, 2011.

[19] T. Fukushima, T. Tanaka, M. Koyanagi. "3D System Integration Technology and 3D Systems," *Advanced Metallization Conference Proceedings*, pp. 479–485, 2009.

[20] K. Sato, T. Seki, S. Hata, A. Shimokohbe, "Self-alignment of microparts using liquid surface tension – behavior of micropart and alignment characteristics," *Precision Engineering* **27**, pp. 42–50, 2003.

[21] J. Berthier, K. Brakke, F. Grossi, L. Sanchez and L. Di Cioccio, "Self-alignment of silicon chips on wafers: A capillary approach," *JAP* **108**, 054905, 2010.

1

Fundamentals of Capillarity

1.1 Abstract

In this first chapter, the fundamentals of capillarity are presented. We follow a conventional approach [1], first presenting surface tension of an interface, which is the fundamental notion in capillarity theory; this notion leads naturally to that of wetting, then to Laplace's law, and to the introduction of Young contact angles and capillary forces. Next, different applications of capillary forces are shown, and the problem of the measurement of surface tensions is presented.

1.2 Interfaces and Surface Tension

1.2.1 The Notion of Interface

Mathematically speaking, an interface is the geometrical surface that delimits two fluid domains. This definition implies that an interface has no thickness and is smooth (i.e. has no roughness). As practical as it is, this definition is in reality a schematic concept. The reality is more complex, the boundary between two immiscible liquids is somewhat blurred and the separation of the two fluids (water/air, water/oil, etc.) depends on molecular interactions between the molecules of each fluid [2] and on Brownian diffusion (thermal agitation). A microscopic view of the interface between two fluids looks more like the scheme of figure 1.1. However,

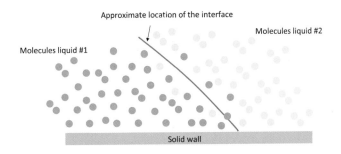

Figure 1.1 Schematic view of an interface at the molecular size.

in engineering applications, it is the macroscopic behavior of the interface that is the focus of attention, and the mathematical concept regains its utility. At a macroscopic size, the picture of figure 1.1 can be replaced by that of figure 1.2, where the interface is a mathematical surface without thickness and the contact angle θ is uniquely defined by the tangent to the surface at the contact line.

In a condensed state, molecules attract each other. Molecules located in the bulk of a liquid have interactions with neighboring molecules on all sides; these interactions are mostly van der Waals attractive interactions for organic liquids and hydrogen bonds for polar liquids like water [2]. On the other hand, molecules at an interface have interactions in a half space with molecules of the same liquid, and in the other half space interactions with molecules of the other fluid or gas (figure 1.3).

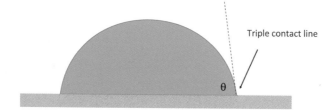

Figure 1.2 Macroscopic view of the interface of a drop.

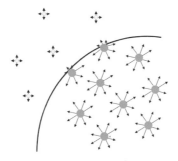

Figure 1.3 Simplified scheme of molecules near an air/water interface. In the bulk, molecules have interaction forces with all the neighboring molecules. At the interface, half of the interactions have disappeared.

Consider an interface between a liquid and a gas. In the bulk of the liquid, a molecule is in contact with 4 to 12 other molecules depending on the liquid (4 for water and 12 for simple molecules); at the interface this number is divided by two. Of course, a molecule is also in contact with gas molecules, but, due to the low densities of gases, there are fewer interactions and less attraction than on the liquid side. The result is that there is locally a dissymmetry in the interactions, which results in an excess of surface energy. At the macroscopic scale, a physical quantity called "surface tension" has been introduced in order to take into account this molecular effect. The surface tension has the dimensions of energy per unit area, and in the International System it is expressed in J/m^2 or N/m (sometimes, it is more practical to use mN/m as a unit for surface tension). An estimate of the surface tension can be found by considering the molecules' cohesive energy. If U is the total cohesive energy per molecule, a rough estimate of the energy excess of a molecule at the interface is $U/2$. Surface tension is a direct measure of this energy excess, and if δ is a characteristic molecular dimension and δ^2 the associated molecular surface area, then the surface tension is approximately

$$\gamma \approx \frac{U}{\delta^2}. \tag{1.1}$$

This relation shows that surface tension is important for liquids with large cohesive energy and small molecular dimension. This is why mercury has a large surface tension whereas oil and organic liquids have small surface tensions. Another consequence of this analysis is the fact that a fluid system will always act to minimize surface area: the larger the surface area, the larger the number of molecules at the interface and the larger the cohesive energy imbalance. Molecules at the interface always look for other molecules to equilibrate their interactions. As a result, in the absence of other forces, interfaces tend to adopt a flat profile, and when it is not possible due to boundary constraints or volume constraints, they take a rounded shape, often that of a sphere. Another consequence is that it is energetically costly to expand or create an interface: we will come back on this problem in Chapter 10 when dividing a droplet into two "daughter" droplets by electrowetting actuation. The same reasoning applies to the interface between two liquids, except that the interactions with the other liquid will usually be more attractive than a gas and the resulting dissymmetry will be less. For example, the contact energy (surface tension) between water and air is 72 mN/m, whereas it is only 50 mN/m between water and oil (table 1.1). Interfacial tension between two liquids may be zero: fluids with zero interfacial tension are said to be miscible. For example, there is no surface tension between fresh and salt water: salt molecules will diffuse freely across a boundary between fresh and saltwater.

The same principle applies for a liquid at the contact of a solid. The interface is just the solid surface at the contact of the liquid. Molecules in the liquid are attracted towards the interface by van der Waals forces. If the attractions to the solid are strong, the liquid-solid interface has negative surface energy, and the solid is said to be wetting or hydrophilic (or lyophilic for non-water liquids, but we will use the term hydrophilic for all liquids). If the attractions are weak, the interface energy is positive, and the solid is nonwetting or hydrophobic (or lyophobic).

Usually surface tension is denoted by the Greek letter γ, with subscripts referring to the two components on each side of the interface, for example γ_{LG} at a Liquid/Gas interface. Sometimes, if the contact is with air, or if no confusion can be made, the subscripts can be omitted. It is frequent to speak of "surface tension" for a liquid in contact with a gas, and "interfacial tension" for a liquid in contact with another liquid. According to the definition of surface tension, for a homogeneous interface (same molecules at the interface all along the interface), the total energy of a surface is

$$E = \gamma S, \tag{1.2}$$

where S is the interfacial surface area.

In the literature or on the Internet there exist tables for surface tension values [3,4]. Typical values of surface tensions are given in table 1.1. Note that surface tension increases as the intermolecular attraction increases and the molecular size decreases. For most oils, the value of the surface tension is in the range $\gamma \approx 20 - 30$ mN/m, while for water, $\gamma \approx 70$ mN/m. The highest surface tensions are for liquid metals; for example, liquid mercury has a surface tension $\gamma \approx 500$ mN/m.

Table 1.1 Values of surface tension of different liquids in contact with air at a temperature of 20 °C (middle column, mN/m) and thermal coefficient α (right column, mN/m/°C).

liquid	γ_0	α
Acetone	25.2	−0.112
Benzene	28.9	−0.129
Benzylbenzoate	45.95	−0.107
Bromoform	41.5	−0.131
Chloroform	27.5	−0.1295
Cyclohexane	24.95	−0.121
Cyclohexanol	34.4	−0.097
Decalin	31.5	−0.103
Dichloroethane	33.3	−0.143
Dichloromethane	26.5	−0.128
Ethanol	22.1	−0.0832
Ethylbenzene	29.2	−0.109
Ethylene-Glycol	47.7	−0.089
Isopropanol	23.0	−0.079
Iodobenzene	39.7	−0.112
Glycerol	64.0	−0.060
Mercury	425.4	−0.205
Methanol	22.7	−0.077
Nitrobenzene	43.9	−0.118
Perfluorooctane	14.0	−0.090
Polyethylen-glycol	43.5	−0.117
PDMS	19.0	−0.036
Pyrrol	36.0	−0.110
Toluene	28.4	−0.119
Water	72.8	−0.1514

1.2.2   The Effect of Temperature on Surface Tension

The value of the surface tension depends on the temperature. The first empirical equation for the surface tension dependence on temperature was given by Eötvös in 1886 [5]. Observing that the surface tension goes to zero when the temperature tends to the critical temperature T_C, Eötvös proposed the semi-empirical relation

$$\gamma = \left(\frac{1}{v_L}\right)^{\frac{2}{3}} (T - T_C),$$ (1.3)

where v_L is the molar volume. Katayama (1915) and later Guggenheim (1945) [6] have improved Eötvš's relation to obtain

$$\gamma = \gamma^* \left(1 - \frac{T}{T_C}\right)^n,$$ (1.4)

where γ^* is a constant for each liquid and n is an empirical factor, whose value is 11/9 for organic liquids. Equation (1.4) produces very good results for organic liquids. If temperature variation is not very important, and taking into account that the exponent n is close to 1, a good approximation of the Guggenheim-Katayama formula is the linear approximation

$$\gamma = \gamma^* (1 + \alpha T).$$ (1.5)

It is often easier and more practical to use a measured reference value (γ_0, T_0) and consider a linear change of the surface tension with the temperature,

$$\gamma = \gamma_0 (1 + \beta(T - T_0)).$$ (1.6)

Comparison between (1.4) and (1.6) for $\gamma = 0$ at $T = T_C$ requires

$$\beta = -\frac{1}{T_C - T_0}.$$ (1.7)

Relations (1.5) and (1.6) are shown in figure 1.4. The value of the reference surface tension γ_0 is linked to γ^* by the relation

$$\gamma^* = \gamma_0 \frac{T_C - T_0}{T_0}.$$ (1.8)

Typical values of surface tensions and their temperature coefficients α are given in table 1.1.

The coefficient α being always negative, the value of the surface tension decreases with temperature. This property is at the origin of a phenomenon which is called either Marangoni convection or thermocapillary instability (figure 1.5). If an interface is locally heated by any heat source (such as radiation, convection or conduction), the surface tension is reduced on the heated area according to equations (1.5) or (1.6) . A gradient of surface tension is then induced at the interface between the cooler interface and the warmer interface. We will show in section 1.3.7 that surface tensions can be viewed as forces; as a consequence, there is an imbalance of tangential forces on the interface, creating a fluid motion starting from the warm region (smaller value of the surface tension) towards the cooler region (larger value of the surface tension). This surface motion propagates to the bulk under the influence of viscosity. If the temperature source is temporary, the motion of the fluid tends to homogenize the temperature and the motion gradually stops. If a difference of temperature is maintained on the interface,

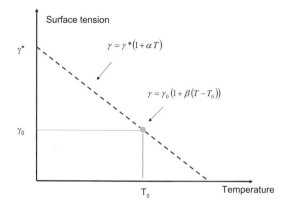

Figure 1.4 Representation of the relations (1.5) and (1.6).

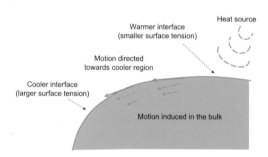

Figure 1.5 Sketch of interface motion induced by a thermal gradient between two regions of the surface. The motion of the interface propagates into the bulk under the action of viscous forces.

the motion of the fluid is permanent; this is the case of a film of liquid spread on a warm solid. Depending on the contrast of temperature between the solid surface and the liquid surface, the motion of the liquid in the film has the morphology of convective rolls, hexagons or squares. Figure 1.6 shows hexagonal patterns of Marangoni convection in a film of liquid heated from below [7]. The white streamlines in the left image show the trajectories of the liquid molecules.

1.2.3   The Effect of Surfactants

"Surfactant" is the short term for "surface active agent". Surfactants are long molecules characterized by a hydrophilic head and a hydrophobic tail, and are for this reason called amphiphilic molecules. Very often surfactants are added to biological samples in order to prevent the formation of aggregates and to prevent target molecules from adhering to the solid walls of the microsystem (remember that microsystems have extremely large ratios between the wall areas and the liquid volumes). Surfactants diffuse in the liquid, and when reaching the interface they are captured because their amphiphilic nature prevents them from escaping easily from the interface. As a consequence, they gather on the interface, as is sketched in figure 1.7, lowering the surface tension of the liquid.

As the concentration of surfactants increases, the surface concentration increases also.

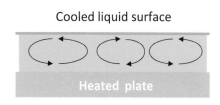

Figure 1.6 Marangoni convection, due to thermocapillary instabilities, makes hexagonal patterns in a thin film of liquid. Reprinted with permission from [7], ©AIP 2005.

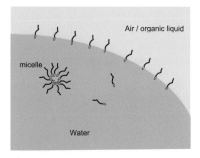

Figure 1.7 Schematic view of surfactants in a liquid drop.

Above a critical value of the concentration, called CMC for Critical Micelle Concentration, the interface is saturated with surfactants, and surfactant molecules in the bulk of the fluid group together to form micelles. The evolution of the value of the surface tension as a function of the concentration in surfactants is shown in figure 1.8. At very low concentration, the slope is nearly linear; when the concentration approaches the CMC, the value of the surface tension drops sharply; above CMC, the value of the surface tension is nearly constant [8]. For example, pure water has a surface tension 72 mN/m, and water with Tween 80 at a concentration above the CMC has a surface tension of only 30 mN/m.

In the limit of small surfactant concentration ($c \ll$ CMC), the surface tension can be expressed as a linear function of the concentration

$$\gamma = \gamma_0(1 + \beta(c - c_0)). \tag{1.9}$$

Equation (1.9) is similar to equation (1.6) (different β, of course). We have seen how a temperature gradient results in a gradient of surface tension leading to Marangoni type of convection. Similarly, a concentration gradient results in a gradient of surface tension, and consequently to a Marangoni convection, as in figure 1.9. Note that the direction of the motion is always towards the largest value of surface tension. Spreading of surfactant molecules on an interface can be easily seen experimentally: an instructive example is that of a thin paper boat with a cavity at the rear (figure 1.10). When the boat is placed gently on the surface of water, it rests on the surface of water suspended by surface tension forces. Upon putting a drop of soap solution/detergent in the notch, boat accelerates rapidly. Soap molecules try to spread over the surface of water.

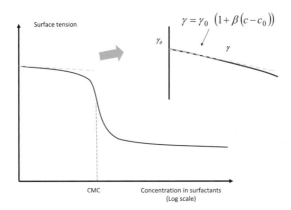

Figure 1.8 Evolution of the value of the surface tension as a function of the surfactant concentration.

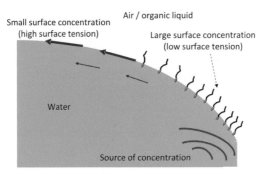

Figure 1.9 Schematic Marangoni convection induced by a gradient of concentration.

Since they are confined in the cavity of boat with only way out, they reduce the surface tension only at the rear, creating a net force which drives the boat forward.

1.2.4 Surface Tension of a Fluid Containing Particles

Pure fluids are seldom used, especially in biotechnology. Very often micro- and nano-particles are present and transported by the fluid or they are voluntarily added to the fluid. Depending on their concentration and nature, the presence of micro- or nano-particles in the fluid might modify considerably the value of the surface tension (figure 1.11). At the same time, the presence of micro-particles reduces the contact angle. The notion of contact angle will be discussed later on in this chapter.

This decrease in surface tension depends on the concentration, size and nature of the micro-particles; at the molecular scale, it is linked to the interactions between particles and liquid on one hand, and between particles themselves on the other hand [10].

Figure 1.10 Soap boat: a floating body contains a small volume of soap. At first, the soap exits the rear of the boat under Marangoni stress. Hence a low surface tension region is created behind the boat, whereas the unsoaped region in front of the boat has a larger surface tension. This difference of surface tension pulls the boat forward (courtesy MIT: http://web.mit.edu/1.63/www/Lec-notes/Surfacetension/Lecture4.pdf [9]).

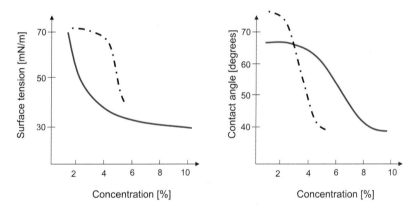

Figure 1.11 Left, surface tension of two different kinds of 10 μm polystyrene particles; right, corresponding equilibrium contact angles.

1.3   Laplace's Law and Applications

Laplace's law is fundamental when dealing with interfaces and micro-drops. It relates the pressure inside a droplet to the curvature of the droplet. This section first describes the mathematical notion of the curvature of a surface, then how it relates to surface tension and pressure, followed by a number of applications.

1.3.1   Curvature and Radius of Curvature

For a planar curve the radius of curvature at a point is the radius R of the osculating circle at that point – the circle which is the closest to the curve at the contact point (figure 1.12). The curvature of the curve at the point is defined by

$$\kappa = \frac{1}{R}. \tag{1.10}$$

Note that the curvature as well as the curvature radius are signed quantities. Curvature radius

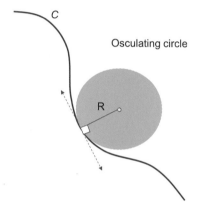

Figure 1.12 Radius of curvature and osculating circle.

can be positive or negative depending on the orientation (convex or concave) of the curve. The curvature may be equivalently defined as the rate of change of the direction angle of the tangent vector with respect to arc length. In the case of a parametric curve $c(t) = (x(t), y(t))$, the curvature is given by the relation [11]

$$\kappa = \frac{\dot{x}\ddot{y} - \ddot{x}\dot{y}}{(\dot{x}^2 + \dot{y}^2)^{\frac{3}{2}}}, \tag{1.11}$$

where the dot denotes a differentiation with respect to t. For a plane curve given implicitly as $f(x, y) = 0$, the curvature is

$$\kappa = \Delta \cdot \left(\frac{\nabla f}{\|\nabla f\|} \right), \tag{1.12}$$

that is, the divergence of the direction of the gradient of f. And for an explicit function $y = f(x)$, the curvature is defined by

$$\kappa = \frac{\frac{d^2 y}{dx^2}}{(1 + (\frac{dy}{dx})^2)^{\frac{3}{2}}}. \tag{1.13}$$

The situation is more complex for a surface. Any plane containing the vector normal to the surface intersects the surface along a curve. Each of these curves has its own curvature, called a sectional curvature, signed with respect to the orientation of the surface. The mean curvature of the surface is defined using the principal (maximum and minimum) curvatures κ_1 and κ_2 (figure 1.13) in the whole set of curvatures:

$$H = \frac{1}{2}(\kappa_1 + \kappa_2). \tag{1.14}$$

It can be shown that the principal curvatures κ_1 and κ_1 are located in two perpendicular planes. In fact, it turns out that the sum of the sectional curvatures in any two perpendicular directions is the same. Introducing the curvature radii in (1.14) leads to

$$H = \frac{1}{2}(\kappa_1 + \kappa_2) = \frac{1}{2}\left(\frac{1}{R_1} + \frac{1}{R_2}\right). \tag{1.15}$$

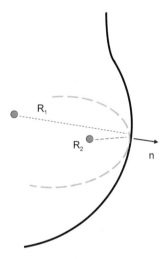

Figure 1.13 Schematic view of the curvature radii of a surface. The dashed and solid curves are the intersections of the surface with two planes perpendicular to the surface and each other.

For a sphere of radius R, the two curvatures are equal to $1/R$ and the mean curvature is $H = 1/R$. For a cylinder of base radius R, the maximum curvature is R and the minimum curvature zero, hence $H = 1/(2R)$. For a plane, the two curvatures are zero and so $H = 0$: a plane has no curvature. At a saddle point of a surface (figure 1.14), one of the curvature radii is positive because it corresponds to a convex arc, whereas the other one is negative, because it corresponds to a concave arc. If $|R_1| = |R_2|$ then the mean curvature H is zero,

$$H = \frac{1}{2}\left(\frac{1}{R_1} + \frac{1}{R_2}\right) = \frac{1}{2}\left(\frac{1}{|R_1|} - \frac{1}{|R_2|}\right) = 0. \tag{1.16}$$

1.3.2 Derivation of Laplace's Law

Suppose a spherical droplet of liquid surrounded by a fluid. Let us calculate the work necessary to increase its volume from the radius R to the radius $R + dR$ (figure 1.15). The part of the work due to the internal volume increase is

$$\delta W_i = -P_0 dV_0, \tag{1.17}$$

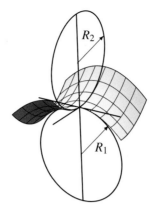

Figure 1.14 The mean curvature at a point is zero if it is a saddle point with equal but opposite sectional curvatures, $|R_1| = |R_2|$.

where dV_0 is the increase of the volume of the droplet,

$$dV_0 = 4\pi R^2 dR. \tag{1.18}$$

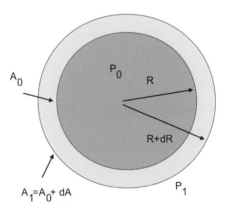

Figure 1.15 Schematic of a liquid drop immersed in a fluid; initially, the droplet radius is R and its surface area A_0. An increase of its radius by a quantity dR corresponds to the new surface area A_1 and the pressure P_1.

The work to pull out the external fluid is

$$\delta W_e = -P_1 dV_1, \tag{1.19}$$

where dV_1 is the decrease of the external volume, equal to $-dV_0$. The work corresponding to the increase of interfacial area is

$$dW_S = \gamma dA, \tag{1.20}$$

where dA is the increase of the surface area. The mechanical equilibrium condition is then

$$\delta W = \delta W_i + \delta W_e + \delta W_s = 0. \tag{1.21}$$

Substituting the values of the work found previously, it follows that

$$\delta P = P_1 - P_0 = 2\frac{\gamma}{R}. \tag{1.22}$$

Equation (1.22) is the Laplace equation for a sphere. The reasoning we have done to obtain equation (1.22) can be generalized,

$$\delta P = \gamma \frac{dA}{dV}. \tag{1.23}$$

For simplicity, we have derived Laplace's equation for the case of a sphere, but we can use (1.23) for an interface locally defined by two (principal) radii of curvature R_1 and R_2; the result would have been then

$$\delta P = \gamma \left(\frac{1}{R_1} + \frac{1}{R_2} \right). \tag{1.24}$$

For a cylindrical interface, as sketched in figure 1.16, one of the two radii of curvature is infinite, and Laplace's equation reduces to

$$\delta P = \frac{\gamma}{R}. \tag{1.25}$$

Equation (1.24) is called Laplace's law. Keep in mind that it is closely linked to the minim-

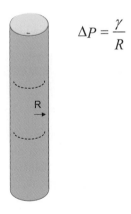

$$\Delta P = \frac{\gamma}{R}$$

Figure 1.16 Laplace's law for a cylindrical interface.

ization of the energy. Laplace's law is fundamental when dealing with interfaces, micro-drops and in digital microfluidics. In the following section, we give some examples of application of Laplace's law.

1.3.3   Examples of Application of Laplace's Law

Amongst other things, Laplace's law explains many phenomena occurring during electrowetting actuation. We will talk about the use of Laplace's law for electrowetting in chapter 10. In this section, we present some applications of Laplace's law outside the electrowetting domain.

1.3.3.1   Pressure in a Bubble

The internal pressure in a bubble can be easily derived from the Laplace law (Fig. 1.17). If we assume that the thickness of the liquid layer is negligible in front of the bubble radius R, Laplace's law yields

$$\delta P = 2\gamma \left(\frac{1}{R_{ext}} + \frac{1}{R_{int}} \right) = \frac{4\gamma}{R}. \tag{1.26}$$

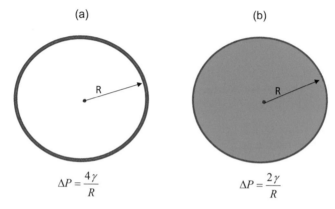

Figure 1.17 Comparison of the internal pressure in a bubble (a) and in a droplet of same radius (b).

The pressure in a bubble can be relatively high. For this reason, in children's kits for blowing bubbles, surfactants are added to the soap solution to facilitate bubble inflation.

1.3.3.2   Liquid Transfer From a Smaller Drop to a Bigger Drop

It has been observed that when two bubbles or droplets are connected together, there is a fluid flow from the small bubble/droplet to the larger one (figure 1.18). This is a direct application of Laplace's law: the pressure inside the small bubble/droplet is larger than that of the larger bubble/droplet, inducing a flow from towards the latter. This flow continues until the smaller bubble/droplet disappears to the profit of the larger one.

In biotechnology, this observation has been used to design microsystems where a microflow in a channel is set up by a difference of size of two droplets placed at both ends [12].

1.3.3.3   Precursor Film and Coarsening

At the beginning of this Chapter, we saw that the concept of an infinitely thin interface and a unique contact angle is a mathematical simplification of reality. When a partially wetting droplet is deposited on a flat solid surface, a very thin film of a few nanometers spreads before the contact line, and the contact between the liquid and the solid resembles the sketch of figure 1.19. The precursor film can be explained by thermodynamic considerations: because a jump between the chemical potential of the gas and of the solid is not physical, liquid molecules intercalate between the gas and the solid [13]. Molecules of the liquid progressively spread under the action of the "disjoining pressure" caused by the van der Waals interactions between

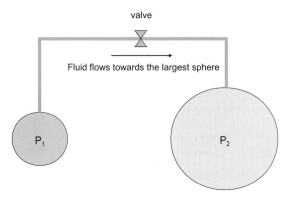

Figure 1.18 Fluid flow from the smaller bubble/droplet to the larger since the smaller bubble has higher curvature and thus higher pressure.

the liquid and solid molecules [14]. Precursor films exist for hydrophilic contact for static droplets as well as for dynamic wetting. Mechanisms of the advancing precursor film are still a subject of investigation [15].

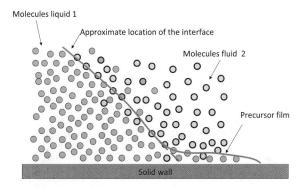

Figure 1.19 Interface with precursor film.

Such precursor films have been observed for different wetting situations, as shown in figure 1.20. Note the extreme thinness of the film in the photographs. When droplets of different size are deposited on a flat solid surface, if the droplets are sufficiently close to one another, it is observed that the smaller droplets disappear to the profit of the large droplets. This phenomenon is called "coarsening". Experimental evidence of coarsening is shown in figure 1.21.

The explanation of the phenomenon requires two steps: first, the existence of a precursor film (an extremely thin film on the solid surface spreading around each droplet) that links the droplets together; second, as in the previous example, the pressure is larger in a small droplet according to Laplace's law, and there is a liquid flow towards the largest droplet (figure 1.22). The precursor film is very thin, thus the flow rate between droplets is very small and mass transfer is extremely slow. Hence experimental conditions require that the droplets do not evaporate.

A numerical simulation of the transfer of liquid from the smaller droplet to the larger droplet can be easily performed using the Evolver as shown in figure 1.23. In the model, the precursor film is not modeled; the two initial droplets are simply united in the same logical volume.

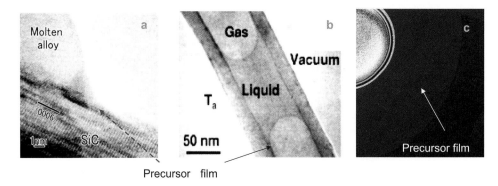

Precursor   film

Figure 1.20 (a) Precursor film of spreading molten alloy (from [16], ©Elsevier, 2002); (b) precursor film of a liquid plug inside a carbon nanotube (from [17], ©AIP, 2005); (c) AFM scan on liquid crystal precursor film (from [18], courtesy Nanolane).

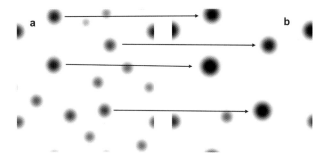

Figure 1.21 Experimental observation of coarsening: the number of droplets diminishes; only the largest droplets remain: (a) beginning of the observation, (b) increase in size of the large droplets and vanishing of the smaller droplets [19] (Courtesy Michael Bestehorn).

1.3.3.4   Pressure in Droplets Constrained Between Two Parallel Plates

When using Laplace's law, one should be careful of the orientation of the curvature. A convex surface has two positive radii of curvature. A "saddle" surface has one positive and one negative curvature radius. Take the example of a water droplet flattened between two horizontal plates (we will see in Chapter 3 that this situation is frequent in EWOD-based microsystems [20]). Suppose that the droplet is placed at the intersection of a hydrophobic band and a hydrophilic band. As a result, the droplet is squished by the hydrophobic band and elongated on the

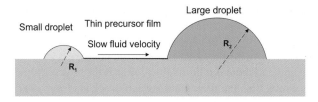

Figure 1.22 When two droplets are linked by a precursor film, a fluid flow is established from the smaller droplet to the larger droplet. The smaller droplet progressively disappears.

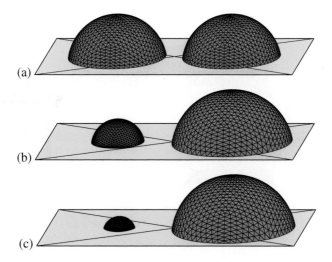

Figure 1.23 Evolver simulation of two droplets of slightly different volumes and supposedly in communication through the precursor film: liquid is transferred from the smaller droplet to the larger one.

hydrophilic band (figure 1.24). The pressure in the droplet is given by the Laplace law

$$P_{drop} - P_0 = \gamma\left(\frac{1}{R_1} + \frac{1}{R_2}\right) = \gamma\left(\frac{1}{R_3} + \frac{1}{R_4}\right), \qquad (1.27)$$

where R_1, R_2, R_3 and R_4 are respectively the horizontal curvature radius in the hydrophilic region, the vertical curvature radius in the hydrophilic region, the horizontal curvature radius in the hydrophobic region, and the vertical curvature radius in the hydrophobic region. Taking into account the sign of the curvatures, we obtain

$$P_{drop} - P_0 = \gamma\left(\frac{1}{|R_1|} - \frac{1}{|R_2|}\right) = \gamma\left(\frac{1}{|R_4|} - \frac{1}{|R_3|}\right), \qquad (1.28)$$

The pressure in the drop being larger than the exterior pressure is equivalent to satisfying either of the relations

$$|R_4| < |R_3| \qquad (1.29)$$

or

$$|R_2| > |R_1|. \qquad (1.30)$$

The vertical curvature radius R_4 in the hydrophobic region is smaller than the concave horizontal radius R_3 and the vertical curvature radius in the hydrophilic region R_2 is larger than the convex horizontal radius R_1. We shall see in Chapter 4 the use of a hydrophobic band to "cut" the droplet into two daughter droplets. For that to happen, the curvature radius R_3 must be sufficiently small, so that the two concave contact lines contact each other. The inequality (1.29) then produces a condition on the level of hydrophobicity required to obtain droplet division.

1.3.3.5   Zero Pressure Surfaces: Example of a Meniscus on a Rod

Laplace's law is often seen as a law determining a pressure difference on the two sides of the interface from the observation of curvature. But it is interesting to look at it the other

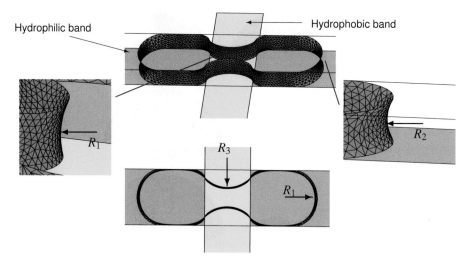

Figure 1.24 Sketch of a water droplet pinched by a hydrophobic surface. Case of a droplet constrained between two parallel planes (the upper plane has been dematerialized for visualization). Simulation performed with the Surface Evolver [21].

way: knowing the pressure difference, what conclusion may be reached on the curvature of the interface?

We consider an example in which the pressure difference is zero: this is the case of a cylindrical rod dipped in a wetting liquid. We suppose that the angle between the surface and the rod (contact angle) is $\theta = 0$. We shall develop the notion of contact angle in section 1.3.5. The liquid rises along the rod, deforming the free surface (figure 1.25). What is the shape of the surface? Laplace's law combined with the hydrostatic pressure yields

$$P_0 + \gamma \left(\frac{1}{R_1} + \frac{1}{R_2} \right) = P_0 - \rho g z, \tag{1.31}$$

where R_1 and R_2 are the two (signed) principal curvature radii. $R_1(z)$ is the (negative) curvature radius of the vertical profile at the elevation z, and $R_2(z)$ is the (positive) radius of the osculating circle perpendicular to the vertical (which is a tilted circle, not the circular horizontal cross-section). Assuming that the system is small enough that the gravity term can be neglected compared to the surface tension term, we are left with

$$\frac{1}{R_1} + \frac{1}{R_2} = 0. \tag{1.32}$$

This is the equation of a zero curvature surface, also called a minimal surface. The equation of the surface can be obtained by writing that the vertical projection of the surface tension force is constant [22]. Using the notations of figure 1.26, we find

$$2\pi r \gamma \cos \theta = 2\pi b \gamma. \tag{1.33}$$

Substituting the relation $\tan \theta = \frac{dr}{dz} = \dot{r}$, we are left with

$$\frac{r}{\sqrt{1 + \dot{r}^2}} = b. \tag{1.34}$$

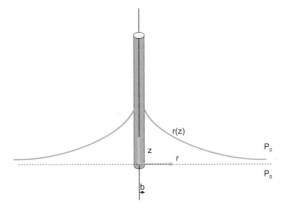

Figure 1.25 Sketch of rod dipped into a liquid (wetting case).

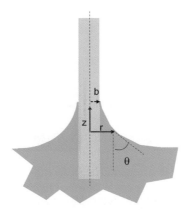

Figure 1.26 Vertical profile of liquid surface at the vicinity of a vertical rod.

Integration of (1.34) yields the equation of the vertical profile

$$r = b^* \cosh\left(\frac{z - h^*}{b^*}\right),$$ (1.35)

where b^* and h^* are constants depending on the contact angle with the vertical cylinder. In the case of a wetting contact ($\theta = 0$) the exact solution is

$$r = b \cosh\left(\frac{z - h_0}{b}\right),$$ (1.36)

where b is the wire radius and h_0 is the height of the interface along the wire. We note that equation (1.36) is the equation of a catenoid. Note that if the boundary of the surface is a circle at $z = 0$ of radius R_0, then z_0 goes to infinity as R_0 goes to infinity. This is not physical if any gravity is present; gravity flattens the surface so that the rise remains bounded for arbitrarily large R_0. We shall see in the following chapter that the expression

$$\kappa^{-1} = \sqrt{\frac{\gamma}{\rho g}}$$ (1.37)

is a characteristic height of capillarity in the presence of gravity, called the capillary length. Assuming that $b \ll h_0$, we can obtain an approximate value for the maximum height h_0 [22] by approximating the surface as a catenoid anchored on a ring of radius $R_0 = \kappa^{-1}$, which at $z = 0$ gives $\kappa^{-1} = b\cosh(-\frac{h_0}{b})$. Using $\cosh x \approx \frac{1}{2}\exp|x|$ for large $|x|$, we get

$$h_0 = b\ln\frac{2\kappa^{-1}}{b}. \tag{1.38}$$

Figure 1.27 shows the deformed surface obtained by a numerical simulation. Relation (1.38) shows that the elevation h_0 along the wire increases with the surface tension. At first sight, this may seem a paradox because the surface is pulled tighter when the surface tension increases. However, we show later in this Chapter that the capillary force exerted by the wire is proportional to the surface tension. The force pulling the surface is thus larger for high surface tension liquids.

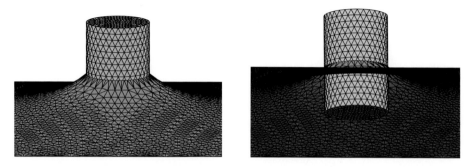

Figure 1.27 Vertical wire dipped into a fluid. The surface is deformed depending on the surface tension of the liquid and on the contact angle (Surface Evolver calculation). Left: view of the surface in the vicinity of the wire. Right: view from below showing the gain in elevation of the surface at the contact of the wire.

1.3.3.6   Self Motion of a Liquid Plug Between Two Non-Parallel Wetting Plates

It was first observed by Hauksbee [23] that a liquid plug between two non-parallel wetting plates moves towards the narrow gap. A sketch of the plug is shown in figure 1.28. Laplace's law furnishes a very clear explanation of this phenomenon. Suppose that figure 1.28 is a wedge (2D situation) and let us write Laplace's law for the left side interface

$$P_0 - P_1 = \frac{\gamma}{R_1}, \tag{1.39}$$

and for the right side interface

$$P_0 - P_2 = \frac{\gamma}{R_2}. \tag{1.40}$$

Subtraction of the two relations leads to

$$P_1 - P_2 = \gamma\left(\frac{1}{R_2} - \frac{1}{R_1}\right). \tag{1.41}$$

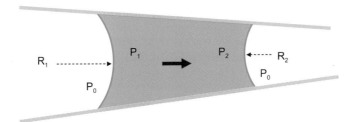

Figure 1.28 Sketch of a liquid plug moving under capillary forces between two plates. The contact angle is $\theta < 90°$.

Next, we show that $R_2 < R_1$. Looking at figure 1.29, we have

$$R_2 \sin \beta = d_2, \tag{1.42}$$

where d_2 is the half-distance between the plates at the narrower contact point. The angle β is

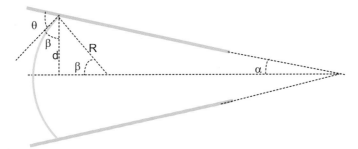

Figure 1.29 Curvature of the interface in a dihedral.

linked to θ and α by the relation

$$\beta = \frac{\pi}{2} + \alpha - \theta. \tag{1.43}$$

Finally we obtain

$$R_2 = \frac{d_2}{\cos(\alpha - \theta)}. \tag{1.44}$$

Using the same reasoning with a meniscus oriented in the opposite direction, we obtain the expression of R_1

$$R_1 = \frac{d_1}{\cos(\alpha + \theta)}. \tag{1.45}$$

Comparing relations (1.44) and (1.45), noting that $d_2 < d_1$ and $\cos(\alpha - \theta) > \cos(\alpha + \theta)$, we deduce that R_2 is then smaller than R_1, and $P_1 > P_2$. The situation is not stable. Liquid moves from the high pressure region to the low pressure region and the plug moves towards the narrow gap region. It has also been observed that the plug accelerates; it is due to the fact that the difference of the curvatures in equation (1.41) is increasing when the plug moves to a narrower region. Bouasse [24] has remarked that the same type of motion applies for a cone, where the plug moves towards the tip of the cone. In reality, Bouasse used a conical frustum (slice of cone) in order to let the gas escape during plug motion.

1.3.3.7   Laplace's Law in Medicine: Normal and Shear Stress in Blood Vessels

1.3.3.7.1   Shear Stress in Vascular Networks   A human body – or any mammalian organism – respects the rules of physics. Take the example of blood vessels. The arrangement of blood vessel networks very often satisfies Murray's law (figure 1.30). In 1926, Murray observed the

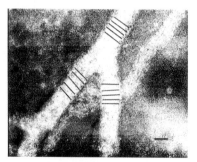

Figure 1.30 Left: Schematic of blood vessel system. Right: photograph of blood vessel division in chickens from [26].

morphology of the blood system and found a very general relation between the dimensions of a "parent" branch and of a "daughter" branch, and he found that the same relation applies at any level of bifurcation. Soon after, he published this discovery [25]. Since that time, this relation is known as Murray's law and can be written as

$$d_0^3 = d_1^3 + d_2^3, \tag{1.46}$$

where d_0, d_1, and d_2 are the "parent" and "daughters" channel diameters. Usually, daughter branches have the same dimension $d_1 = d_2$. A recurrence relation can be deduced from (1.46); it shows that the diameter, flow rates and average velocities at the nth generation, i.e. at a bifurcation of rank n (figure 1.30), are related to the "origin" diameter, flow rate and average velocity by the relations

$$d_n = \frac{d_0}{2^{\frac{n}{3}}}, \tag{1.47}$$

$$Q_n = \frac{Q_0}{2^n}, \tag{1.48}$$

$$V_n = \frac{V_0}{2^{\frac{n}{3}}}. \tag{1.49}$$

These relations can be developed further to show that the wall friction is the same at each level [26,27]. This property simply stems from the expression of the shear stress of a cylindrical duct,

$$\tau_n = \frac{8\eta V_n}{d_n} = \frac{8\eta V_0}{d_0} = \tau_0. \tag{1.50}$$

Murray showed that, on a physiological point of view, such a relation minimizes the work of the blood circulation. The important thing here is that the shear stress is constant in most blood vessel networks. Now what about the normal stress?

1.3.3.7.2   Normal Stress in Vascular Networks   In the particular case of human or mammalian blood systems, the normal stress is simply the internal pressure, because, to a first approximation, the flow is purely axial and there is no radial component of the velocity. It has been observed that the thickness of the walls of blood vessels satisfies Laplace's law (figure 1.31). In this particular case, the surface tension is replaced by the wall tension T, and Laplace's law becomes

$$P = \frac{T}{R}.$$

(1.51)

At a given distance from the heart, the pressure is approximately the same and Laplace's equation (1.51) has the consequence that the wall tension increases together with the radius. As a consequence, arteries have larger wall thickness than veins, and similarly veins compared to capillaries. In medicine, an aneurysm is a localized, blood-filled dilation (balloon-like bulge) of

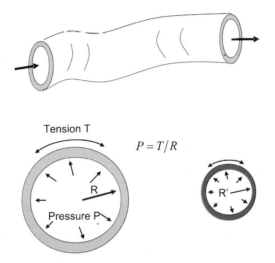

Figure 1.31 Schematic view of a blood vessel: if the internal pressure is P, the tension in the wall is $T = PR$. Small blood vessels have thinner walls than large blood vessels.

a blood vessel caused by disease or weakening of the vessel wall. Aneurysms most commonly occur in arteries at the base of the brain and in the aorta (the main artery coming out of the heart, an aortic aneurysm). It has been observed that, as the size of an aneurysm increases, there is an increased risk of rupture, which can result in severe hemorrhage, other complications or even death. This expansion is a direct consequence of the preceding reasoning: if the thickness of the vascular wall is such as it withstands a given tension T, an increase of the radius will require a higher tension of the wall, and therefore the aneurysm will continue to expand until it ruptures (figure 1.32). A similar logic applies to the formation of diverticuli in the gut [28].

1.3.3.8   Laplace's Law in Medicine: the Example of Lung Alveoli

It is very tempting to refer automatically to Laplace's law because of its simplicity. But one should refrain from doing that uncritically. A striking example is that of lung ventilation. Ventilation of lungs has been widely studied for medical purposes. It was usual to consider the alveoli such as spherical balloons inflated during lung ventilation (fig. 1.33). The problem is

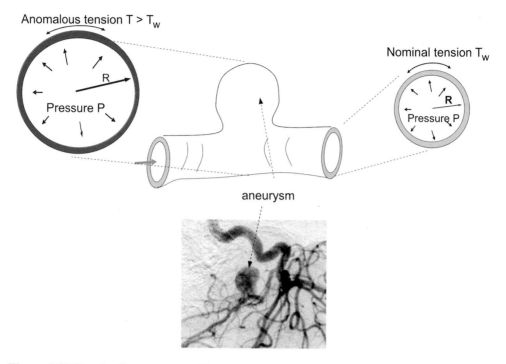

Figure 1.32 Sketch of an aneurysm. The wall tension required at the location of the aneurysm exceeds the wall tension that the thickness of the wall can withstand.

that the alveoli are connected, and when applying Laplace's law, the air in the smaller alveoli should be driven to the larger alveoli and a general collapse of the lungs would occur. Because the collapse of the alveoli does not – luckily – correspond to reality, it has been suggested that the concentration of surfactant in the alveoli is not uniform and compensates for the different pressures. Recently, a different, more realistic analysis has been made [29]: the alveoli are not

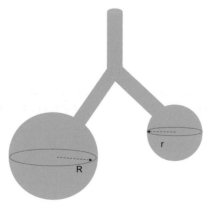

Figure 1.33 Wrong sketch for the alveoli leading to an improper application of Laplace's law.

"free" spheres but are packed together (figure 1.34) and there are pores in the alveoli walls.

Alveoli cannot expand freely, and they are limited in their inflation. As a result, smallest alveoli do not collapse during ventilation, because large alveoli cannot grow indefinitely, and Laplace's law is not the answer in this kind of problem.

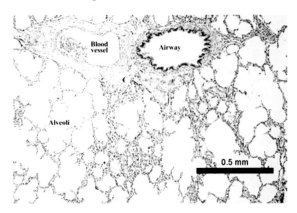

Figure 1.34 Image of lung alveoli; Detail from an original slide generously provided by A. Mescher.

1.3.3.9 Laplace's Law in a Gravity Field

Consider a drop on a flat horizontal surface, a drop large enough that gravity deforms it. For simplicity, let us assume a liquid droplet of density ρ in a free atmosphere. Inside the liquid, this hydrostatic pressure is given by the term $P_1 = const - \rho g z$, where z is the vertical direction oriented upwards, and outside the drop the atmospheric pressure can be considered constant, $P_2 = const$. The Laplace law is then

$$\gamma \left(\frac{1}{R_1} + \frac{1}{R_2} \right) = const - \rho g z. \tag{1.52}$$

The curvature is then smaller at the top of the drop than it is at the bottom of the drop. This effect can well be seen in an Evolver simulation (figure 1.35).

1.3.3.10 Generalization: Laplace's Law in Presence of a Flow Field

The expression of Laplace's law derived in section 1.3 assumed totally static conditions, or at the least that the shear rate of the flow field close to the interface is negligible in comparison with the surface tension. This is indeed often the case since usually capillary numbers are small in microfluidic systems. Recalled that the capillary number is a non-dimensional number characterizing the ratio between inertial forces and capillary forces. However, the flow field effect on the interface cannot be always neglected. For example, systems like flow-focusing devices are currently used in biotechnology to produce emulsions and encapsulates [30-32]. In such systems, the dynamic flow field exerts a considerable force on the interface, as has been shown by Tan et al. [33] (figure 1.36). In such a case the balance of the forces on the surface is more complicated: the complete stress tensor has to be taken into account, and we obtain the generalized Laplace law for liquid 1 and 2:

$$(P_1 - P_2)n_i = \gamma \left(\frac{1}{R_1} + \frac{1}{R_2} \right) n_i + \left(\sigma_{ik}^{'(1)} - \sigma_{ik}^{'(2)} \right) n_k, \tag{1.53}$$

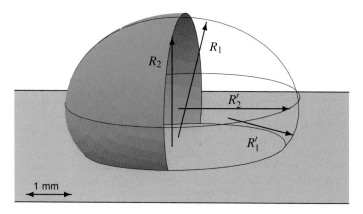

Figure 1.35 Large droplet flattened by the action of gravity: The vertical curvature radius R'_1 is small at the bottom of the drop because the internal hydrostatic pressure is larger and the horizontal curvature radius R'_2 relatively large. At the top, the curvatures R_1 and R_2 are equal and relatively large.

where n_i is the unit normal vector, and σ'_{ik} the viscous part of the stress tensor [34]. Note the implicit summation on the repeated index k on the right side of (1.53).

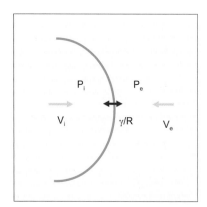
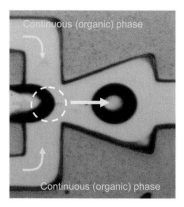

Figure 1.36 Left: sketch of the interface; right: photo of a flow focusing device; the generalized Laplace's law applies for the incoming discontinuous phase (inside the dotted circle).

1.3.4   Wetting - Partial or Total Wetting

So far, we have dealt with interfaces between two fluids. Triple contact lines are the intersections of three interfaces involving three different materials: for example a droplet of water on a solid substrate in an atmosphere has a triple contact line. Liquids spread differently on a horizontal plate according to the nature of the solid surface and that of the liquid. In reality, it depends also on the third constituent, which is the gas or the fluid surrounding the drop. Two different situations are possible: either the liquid forms a droplet, and the wetting is said to be partial, or the liquid forms a thin film wetting the solid surface, the horizontal dimension of the film depending on the initial volume of liquid (figure 1.37). For example, water spreads like a film on a very clean and smooth glass substrate, whereas it forms a droplet on a plastic substrate. In the case of partial wetting, there is a line where all three phases come together. This line is called the contact line or the triple line.

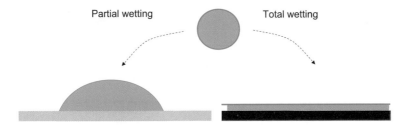

Figure 1.37 Wetting is said to be total when the liquid spreads like a film on the solid surface.

A liquid spreads on a substrate in a film if the energy of the system is lowered by the presence of the liquid film (figure 1.38). The surface energy per unit surface of the dry solid surface is γ_{SG}; the surface energy of the wetted solid is $\gamma_{SL} + \gamma_{LG}$. The spreading parameter S determines the type of spreading (total or partial)

$$S = \gamma_{SG} - (\gamma_{SL} + \gamma_{LG}). \tag{1.54}$$

If $S > 0$, the liquid spreads on the solid surface; if $S < 0$ the liquid forms a droplet.

Figure 1.38 Comparison of the energies between the dry solid and the wetted solid.

When a liquid does not totally wet the solid, it forms a droplet on the surface. Two situations can occur: if the contact angle with the solid is less than 90°, the contact is said to be "hydrophilic" if the liquid has a water base, or more generally "wetting" or "lyophilic". In the opposite case of a contact angle larger than 90°, the contact is said to be "hydrophobic" with reference to water or more generally "not wetting" or "lyophobic" (figure 1.39 and figure 1.40).

1.3.5 Contact Angle - Young's Law

1.3.5.1 Young's Law

We have seen that surface tensions are not exactly forces, their unit is N/m; however they represent a force that is exerted tangentially to the interface. Surface tension can be looked at as a force per unit length. This can be directly seen from its unit. But it may be interesting to give a more physical feeling by making a very simple experiment (figure 1.41) [22]. Take a solid frame and a solid tube that can roll on this frame. If we form a liquid film of soap between the frame and the tube – by plunging one side of the structure in a water-soap solution – the tube starts to move towards the region where there is a liquid film. The surface tension of the liquid

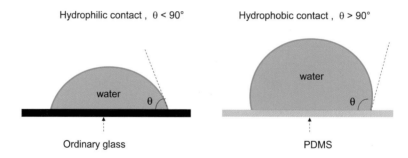

Figure 1.39 Water spreads differently on different substrate.

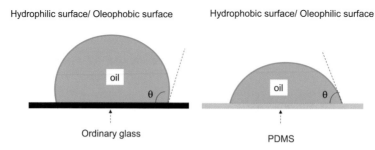

Figure 1.40 Silicone oil has an opposite wetting behavior than water.

film exerts a force on its free boundary. On the other hand, we can increase the film surface by exerting a force on the tube. The work of this force is given by both of the relations

$$\delta W = F dx, \qquad \delta W = \gamma dA = 2\gamma L dx. \qquad (1.55)$$

The coefficient 2 stems from the fact that there are two interfaces between the liquid and the air, on either side of the film. Comparing these relations shows that the surface tension γ is a force per unit length, perpendicular to the tube, in the plane of the liquid and directed towards the liquid.

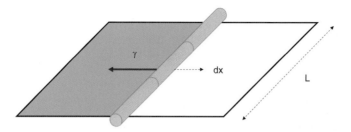

Figure 1.41 A tube placed on a rigid frame whose left part is occupied by a soap film requires a force to be displaced towards the right, this force opposed to the surface tension that tends to bring the tube to the left.

We can then sketch the different forces that are exerted by the presence of a fluid on the triple line (figure 1.42). At equilibrium, the resultant of the forces must be zero. We use a coordinate

Figure 1.42 Schematic of the forces at the triple contact line.

system where the x-axis is the tangent to the solid surface at the contact line (horizontal) and the y-axis is the direction perpendicular (vertical). At equilibrium, the projection of the resultant on the x-axis is zero and we obtain the relation

$$\gamma_{LG} \cos\theta = \gamma_{SG} - \gamma_{SL}. \tag{1.56}$$

This relation is called Young's law and is very useful to understand the behavior of a drop. Especially, it shows that the contact angle is determined by the surface tensions of the three constituents. For a droplet on a solid, the contact angle is given by the relation

$$\theta = \arccos\left(\frac{\gamma_{SG} - \gamma_{SL}}{\gamma_{LG}}\right). \tag{1.57}$$

Young's law can be more rigorously derived from free energy minimization. Consider a sessile droplet large enough for the effect of the triple line to be neglected. The change of free energy due to a change in droplet size can be written as [30]

$$dF = \gamma_{SL}dA_{SL} + \gamma_{SG}dA_{SG} + \gamma_{LG}dA_{LG} = (\gamma_{SL} - \gamma_{SG} + \gamma_{LG}\cos\theta)dA_{SL}, \tag{1.58}$$

where θ is the contact angle. At mechanical equilibrium $dF = 0$ and

$$\gamma_{SL} - \gamma_{SG} + \gamma_{LG}\cos\theta = 0. \tag{1.59}$$

Equation (1.59) is Young's law, identical to (1.56).

    Note that sometimes it happens that, in real experimental situations when we deal with real biological liquids, one observes unexpected changes in the contact angle with time. This is just because biological liquids are inhomogeneous and can deposit a layer of chemical molecules on the solid wall, thus progressively changing the value of the tension γ_{SL}, and consequently the value of θ, as stated by Young's law.

1.3.5.2   Droplet on a Cantilever

Let us come back to the derivation of Young's law. Young's law has been obtained by a projection on the x-axis of the surface tension forces. But the force balance applies also to projection in any direction. We will mention two cases where slanted projection of Young's law is of importance, first that of a cantilever, second that of the contact between three liquids. In the case of a micro-cantilever, the presence of a droplet induces capillary forces along the triple line (figure 1.43). The deformation results from the resultant of the capillary forces perpendicular to the cantilever. At rest, this resultant bends the cantilever. The calculation is lengthy and has been derived by Yu and Zhao [35].

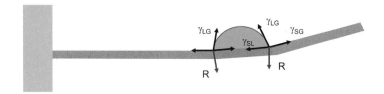

Figure 1.43 Cantilever deformed by the presence of a water droplet, after [35].

1.3.5.3   Contact Between Three Liquids – Neumann's Construction

Take two immiscible liquids, denoted 1 and 2, with the droplet of liquid 2 deposited on the interface between liquid 1 and a gas. Even if the density of liquid 2 is somewhat larger than that of liquid 1, the droplet may "float" on the surface, as shown in figure 1.44. The situation is comparable to that of Young's law with the difference that the situation is now two-dimensional. It is called Neumann's construction, and the following vector equality holds:

$$\vec{\gamma}_{L1L2} + \vec{\gamma}_{L1G} + \vec{\gamma}_{L2G} = 0. \tag{1.60}$$

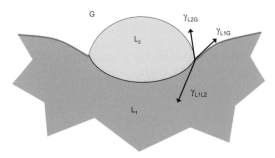

Figure 1.44 Droplet on a liquid surface.

Note that the densities of the two liquids condition the vertical position of the center of mass of the droplet, but at the triple line, it is the y-projection of equation (1.60) that governs the morphology of the contact. In figure 1.45 we show some pictures of floating droplets obtained by numerical simulation (Surface Evolver).

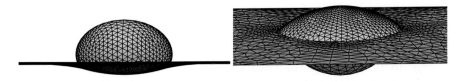

Figure 1.45 Numerical simulations of different positions of a droplet (1 mm) on a liquid surface depending on the three surface tensions. Left: the liquid/liquid surface tension is very large. Right: the surface tension of the droplet with the other liquid has been reduced. Both drops are at equilibrium due to the balance of buoyancy and surface tensions.

1.3.5.4   Nanobubbles on Hydrophobic Walls, Line Tension and the Modified Young Law

It has been observed that bubbles often form along hydrophobic walls, even when the surface is smooth. The size of these bubbles is in the mesoscopic range – between the microscopic and nanoscopic scales: bubble dimension is usually less than 200 nm. A paradox arises when calculating the internal pressure. Using the Laplace law with a curvature radius of the order of the observed contact at the wall, one finds that the internal pressure should be of the order of

$$P \simeq \frac{\gamma}{R} \simeq \frac{70 \times 10^{-3}}{200 \times 10^{-9}} \simeq 3.5 \times 10^5 \, Pa. \tag{1.61}$$

At this level of pressure the gas should dissolve, and the bubble would disappear rapidly in the liquid. So why are these bubbles stable? From a Laplace's law point of view, either their surface tension should be smaller than that of a "macroscopic" bubble, or their curvature radius should be larger. It is easy to see that a reduced surface tension is not sufficient to find a sustainable internal pressure. On the other hand, at this scale, contact angle measurements are very tricky. However, recent measurements [36] have shown that nano-bubbles have very flat profiles – the base radius is 5-20 times larger than the height – because the contact angle of the bubble is much smaller than a macroscopic contact angle of the bubble on the same substrate (figure 1.46).

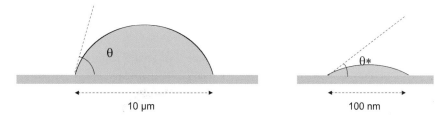

Figure 1.46 Comparative sketches of micro-bubble and nano-bubble (not to scale).

The problem has now shifted from Laplace's law to Young's law. What has changed in Young's law for the contact angle between the nanoscale and the microscopic/macroscopic scale? The answer to this question is not yet clear. A plausible answer is connected to the notion of "line tension" and to the so-called modified Young's law. Young's law has been derived for a triple line without consideration of the interactions near the triple contact line. A sketch of the interactions leading to surface tension and line tension is shown in figure 1.47. The molecules close to the triple line experience a different set of interactions than at the interface. To take into account this effect, a line tension term can be introduced in Young's law [37,38]:

$$\gamma_{SG} = \gamma_{SL} + \gamma_{LG} \cos \theta + \frac{\gamma_{SLG}}{r}, \tag{1.62}$$

where r is the contact radius, γ_{SLG} the line tension (unit N), and θ^* the real contact angle. The contact angle is then changed by the line tension according to

$$\cos \theta^* = \cos \theta - \frac{\gamma_{SLG}}{r \gamma_{LG}}. \tag{1.63}$$

For a droplet contact radius larger than 10 μm, the effect of the line tension is negligible; the value of the correction term is of the order of 10^{-4} [22]. But this is not the case for nano-drops and nano-bubbles, whose contact radii are much smaller.

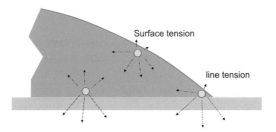

Figure 1.47 Sketch of interactions leading to surface tension and line tension.

1.3.6   Work of Adhesion, Work of Cohesion and the Young-Dupré Equation

In this section, we introduce the notions of work of adhesion and work of cohesion. These notions are valid for solids or immiscible liquids. When applied to a solid and a liquid, the concept of work of adhesion combined with Young's law produces the Young-Dupré equation. Work of adhesion and Young-Dupré's equation have been widely used to determine the surface tension of solids (section 1.4.3).

1.3.6.1   Work of Adhesion

Imagine a body contacting another body on a surface S of area S (figure 1.48). The surface energy of S when there is contact is

$$E_{12} = \gamma_{12}S. \tag{1.64}$$

The work of adhesion is the work required to separate the two bodies. After separation, the surface energies are

$$E = E_1 + E_2 = (\gamma_1 + \gamma_2)S. \tag{1.65}$$

The work of adhesion is then

$$W_a = \gamma_1 + \gamma_2 - \gamma_{12}. \tag{1.66}$$

1.3.6.2   Work of Cohesion

The work of cohesion is obtained similarly, but this time the body being split is homogeneous (figure 1.49). The same reasoning yields

$$W_c = 2\gamma_1. \tag{1.67}$$

In other words, the surface energy is half the work of cohesion.

1.3.6.3   The Young-Dupré Equation

Let us express the work of adhesion for a liquid and a solid (figure 1.50). Using (1.66) with the surface tensions $\gamma_1 = \gamma_{LG} = \gamma$, $\gamma_2 = \gamma_{SG}$, and $\gamma_{12} = \gamma_{SL}$, we obtain

$$W_a = \gamma + \gamma_{SG} - \gamma_{SL}. \tag{1.68}$$

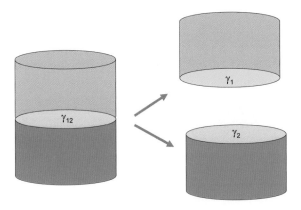

Figure 1.48 Work of adhesion is the work done to separate two surfaces of incompatible substance.

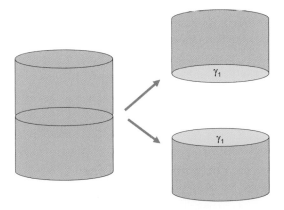

Figure 1.49 Work of cohesion is the work done to separate two surfaces of the same substance.

Upon substitution of Young's law, we derive the Young-Dupré equation

$$W_a = \gamma(1 + \cos\theta). \tag{1.69}$$

For a super-hydrophobic contact, $\theta = \pi$ and $\cos\theta = -1$; we deduce that $W_a = 0$: there is no work needed to separate a super-hydrophobic liquid from a solid. Concretely, a droplet of water rolls freely over a super-hydrophobic surface. The Young-Dupré equation indicates that the more hydrophobic (non-wetting) is the contact between a liquid and a solid, the smaller is the work of adhesion.

On the other hand, adhesion maintains a droplet suspended below a solid, as shown in figure 1.51.

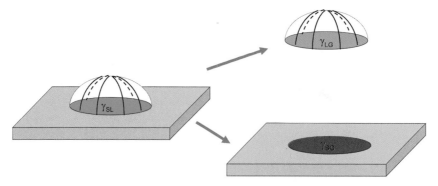

Figure 1.50 Sketch of the Young-Dupré equation for the work of adhesion of a liquid and a solid: $W_a = \gamma(1 + \cos\theta)$.

Figure 1.51 Sketch of a suspended droplet.

1.3.7   Capillary Force, Force on a Triple Line

1.3.7.1   Introduction

We have seen in section 1.3.5 the correspondence between surface tension and capillary forces. One example is that of a coin floating on water. Figure 1.52 shows a small coin floating on water, even if the buoyancy forces are not sufficient to maintain it at the water surface. The capillary forces all along the edge of the coin add a supplementary vertical force which counterbalances the apparent weight of the coin.

Capillary forces are still more important at a micro-scale. We have all seen insects "walking" on the surface of a water pond (figure 1.53). Their hydrophobic legs do not penetrate the water surface and their weight is balanced by the surface tension force. More than that, it is observed that some insects can walk up a meniscus, i.e. can walk on an inclined water surface. The explanation of this phenomenon was recently given by Hu *et al.* [39] and refers to complex interface deformation under capillary forces.

In the domain of microfluidics, capillary forces are predominant; some examples of the action of capillary forces are given in the following sections.

1.3.7.2   Capillary Force Between Two Parallel Plates

We all have remarked that two parallel plates squishing a liquid film make the plates very adhesive. For instance, when using a microscope to observe objects in a small volume of liquid deposited on a plate and covered by a secondary glass plate, it is very difficult to separate the plates. We sketch this problem in figure 1.54. We will assume the liquid does not reach the

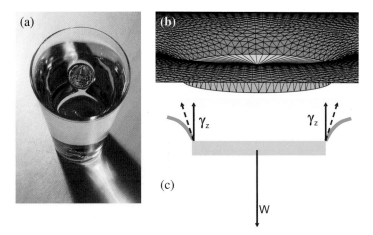

Figure 1.52 A small coin floats on water, even though the buoyancy forces are not sufficient to maintain it at the water surface (a). Surface tension forces act on the coin edge to counterbalance gravity: (b) Surface Evolver simulation and (c) sketch of the forces.

Figure 1.53 Left: Capillary forces make the water surface resist the weight of an insect. Right: an insect walking up a meniscus. Reprinted with permission from [33], ©Nature Journal, 2005.

plate edges, for simplicity. First we remark that the meniscus has a circular shape horizontally, radius R, in order to minimize the free energy.

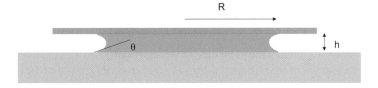

Figure 1.54 Film of water between two glass plates.

We use Laplace's law at the free interface. The first (horizontal) radius of curvature is approximately R. The second (vertical) radius of curvature, shown in figure 1.55, is calculated by

$$R_2 \sin\left(\frac{\pi}{2} - \theta\right) = \frac{h}{2},$$
(1.70)

or

$$R_2 = \frac{h}{2\cos\theta}.$$
(1.71)

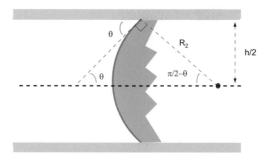

Figure 1.55 Calculation of the vertical curvature.

Laplace's law states that

$$\Delta P = \gamma\left(\frac{1}{R} - \frac{2\cos\theta}{h}\right).$$
(1.72)

In (1.72), the minus sign derives from the concavity of the interface. Because the vertical gap h is much less than the horizontal dimension R, we have the approximation

$$\Delta P \simeq -\frac{2\gamma\cos\theta}{h}.$$
(1.73)

And the capillary force that links the plates together is

$$F \simeq \frac{2\gamma\cos\theta}{h}\pi R^2.$$
(1.74)

This capillary force can be quite important; for $h = 10~\mu m$ and $R = 1$ cm, the force F is of the order of 2.5 N.

1.3.7.3 Capillary Rise in a Tube

When a capillary tube is plunged into a volume of wetting liquid, the liquid rises inside the tube under the effect of capillary forces (figures 1.56 and 1.57). It is observed that the height reached by the liquid is inversely proportional to the radius of the tube.

Historically, many scientists have investigated this phenomenon, from Leonardo da Vinci, to Hauksbee, to Jurin. This property is now referred to as Jurin's law [40]. Using the principle of minimum energy, one can conclude that the liquid goes up in the tube if the surface energy of the dry wall is larger than that of the wetted wall. We define the impregnation criterion I by

$$I = \gamma_{SG} - \gamma_{SL}.$$
(1.75)

Figure 1.56 Capillary rise in tubes of different internal cross section.

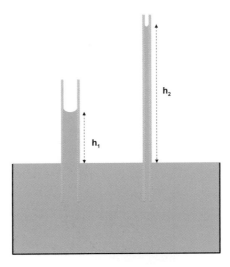

Figure 1.57 Capillary rise is inversely proportional to the capillary diameter.

The liquid rises in the tube if $I > 0$, else the liquid descends in the tube. Using Young's law, the impregnation criterion can be written in the form

$$I = \gamma \cos \theta. \tag{1.76}$$

When the liquid rises in the tube, the system gains gravitational potential energy – because of the elevation of a volume of liquid – and loses capillary energy due to the reduction of the surface energy. The balance is [22]

$$E = \frac{1}{2}\rho g h V_{liquid} - S_{contact} I = \frac{1}{2}\rho g h (\pi R^2 h) - 2\pi R h I = \frac{1}{2}\rho g \pi R^2 h^2 - 2\pi R h \gamma \cos \theta. \tag{1.77}$$

Note that we have not taken into account in (1.77) the detailed shape of the meniscus for the height h. The equilibrium elevation of the liquid is given by

$$\frac{\delta E}{\delta h} = 0, \tag{1.78}$$

which results in

$$h = \frac{2\gamma\cos\theta}{\rho g R} = 0, \tag{1.79}$$

Equation (1.79) is called Jurin's law. The capillary rise is inversely proportional to the tube radius. Jurin's law can also be applied to the case where the liquid level in the tube decreases below the outer liquid surface; this situation happens when $\theta > 90$ degrees. The maximum possible height that a liquid can reach corresponds to $\theta = 0$ and is $h = \frac{2\gamma}{\rho g R}$. In microfluidics, capillary tubes of 100 μm diameter are currently used; if the liquid is water ($\gamma = 72$ mN/m), and using the approximate value $\cos\theta \approx 1/2$, the capillary rise is of the order of 14 cm, which is quite important at the scale of a microcomponent. We have just given an expression for the capillary rise and we have seen that the capillary rise is important. What is the corresponding capillary force? The capillary force balances the weight of the liquid in the tube. This weight is given by

$$F = \rho g \pi R^2 h. \tag{1.80}$$

Replacing h by its value from equation (1.79), we find the capillary force

$$F = 2\pi R \gamma \cos\theta. \tag{1.81}$$

The capillary force is the product of the length of the contact line $2\pi R$ times the line force $f = \gamma\cos\theta$. This line force is sketched in figure 1.58.

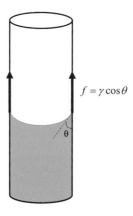

$$f = \gamma\cos\theta$$

Figure 1.58 Sketch of the capillary force of a liquid inside a tube.

Note that the capillary force per unit length f is identical to the impregnation criterion

$$f = \gamma\cos\theta = I. \tag{1.82}$$

For $f > 0$ the liquid goes up in the tube and for $f < 0$ the liquid goes down. Note that figure 1.58 is not quite exact: the liquid surface outside the tubes is not totally horizontal. There are also capillary forces on the outside of the tube, as shown in figure 1.59. To derive the expression of the capillary rise inside the tube, a control volume corresponding to the liquid volume inside the tube was first considered. Let us consider now a control volume defined by the pipette (figure 1.60). The force to maintain the pipette is

$$F = P - P_A + P_{c,e} + P_{c,i}, \tag{1.83}$$

where P is the weight of the tube, P_A the floatation force and $P_{c,i}$ and $P_{c,e}$ are respectively the interior and exterior capillary forces exerted on the solid.

$$F = P - P_A + 2\pi R_{int}\gamma\cos\theta + 2\pi R_{ext}\gamma\cos\theta. \tag{1.84}$$

This force is a function of the surface tension γ. In section 1.4, we will see that the measure of such a force constitutes a way to determine the surface tension.

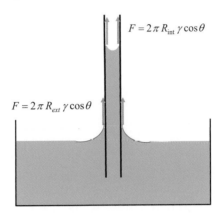

Figure 1.59 Capillary forces act also on the tube exterior, raising the level of the liquid around the tube (if the liquid wets the solid; it would be the opposite if the liquid were not wetting).

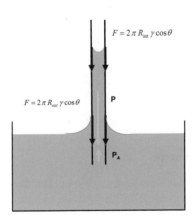

Figure 1.60 Forces acting on the pipette.

1.3.7.4   Capillary Rise Between Two Parallel Vertical Plates

The same reasoning can be done for a meniscus rising between two vertical parallel plates (figure 1.61) separated by a distance $d = 2R$. It is easy to show that in this case

$$h = \frac{\gamma\cos\theta}{\rho g R} = 0. \tag{1.85}$$

If we introduce the capillary length κ defined by

$$\kappa^{-1} = \sqrt{\frac{\gamma}{\rho g}}, \tag{1.86}$$

we can rewrite (1.86) in the form

$$h = \kappa^{-2} \frac{\cos\theta}{R} = 0, \tag{1.87}$$

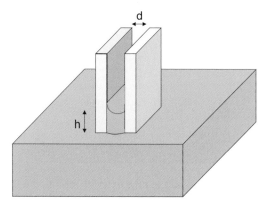

Figure 1.61 Capillary rise between two parallel vertical plates.

Remark: The expressions for the two geometries (cylinder and two parallel plates) are similar. If we use the coefficient c, with $c = 2$ for a cylinder and $c = 1$ for parallel plates [41], we have

$$h = c\kappa^{-2} \frac{\cos\theta}{R} = 0, \tag{1.88}$$

where R is either the radius of the cylinder or the half-distance between the plates.

1.3.7.5 Capillary Rise in a Pipette

The analysis of the capillary rise – or descent – in cylindrical tubes or between two parallel plates has been recently extended by Tsori [41] to the case where the walls are not parallel, as for instance a conical pipette (figure 1.62).

The mechanical equilibrium states that the Laplace pressure is balanced by the hydrostatic pressure

$$P_0 + \frac{c\gamma}{r} = P_0 - \rho g h, \tag{1.89}$$

where c is the index defined previously, $c = 2$ for cones and $c = 1$ for wedges, and r is the curvature radius of the meniscus. Note that the depth h is counted negatively from the surface. Using the same approach as that of section 1.3.3.6, the curvature radius is expressed by

$$r(h) = -\frac{R(h)}{\cos(\theta + \alpha)}, \tag{1.90}$$

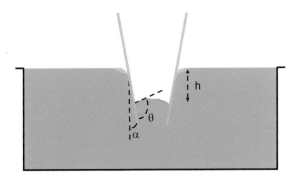

Figure 1.62 Capillary descent inside a hydrophobic conical pipette.

with

$$R(h) = R_0 + h\tan\alpha, \tag{1.91}$$

where R_0 is the internal radius of the pipette at $h = 0$. Substituting (1.91) in (1.90) and then in (1.89), we obtain

$$\cos(\theta + \alpha) = -\frac{1}{c}h(R_0 + h\tan\alpha)\frac{\rho g}{\gamma}. \tag{1.92}$$

Using the capillary length κ^{-1} to scale the variables, we obtain the non-dimensional variables $\bar{R} = \kappa R$ and $\bar{h} = \kappa h$. Equation (1.92) then becomes

$$(\tan\alpha)\bar{h}^2 + \bar{R}_0\bar{h}c\cos(\theta + \alpha) = 0. \tag{1.93}$$

This is a quadratic equation in $\bar{h}$. The discussion of this equation is somewhat complex. Depending on the values of α and θ, the meniscus may be stable or not stable; in the latter case, the meniscus jumps to the top or the bottom of the pipette, where it gets stabilized by pinning (anchoring to an angle). The diagram of figure 1.63 summarizes the meniscus behavior.

The important information here is that there are two domains where the meniscus "jumps" inside the pipette until it finds a pinning edge. The first case is that of a cone/wedge angle α larger than a critical value α^* (and the contact angle θ sufficiently large); the meniscus stays pinned at the bottom of the pipette, and no liquid penetrates the pipette, unless a negative pressure is established. On the other hand, when the angles α and θ are sufficiently small (α smaller than a negative critical value $-\alpha^*$), the liquid jumps to the top of the cone/wedge. The critical values depend on the internal radius R_0 of the pipette at $h = 0$. In conclusion, a cone-shaped micro-pipette dipped into a liquid does not have always the expected behavior, i.e. there might not be the expected capillary rise.

1.3.7.6 Force on a Triple Line

The analysis of the capillary rise in tubes has shown the expression of the capillary force on the triple contact line [42]. This expression can be generalized to any triple contact line [43]. For a triple contact line Ω – as sketched in figures 1.64 and 1.65 – the capillary force is

$$F_x = \int_\Omega \vec{f}\cdot\vec{i}\,dl = \int_\Omega \gamma\cos\theta\,\vec{n}\cdot\vec{i}\,dl. \tag{1.94}$$

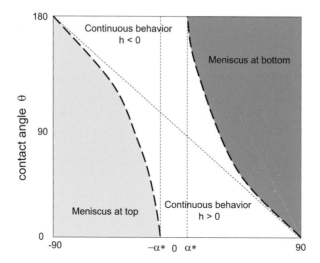

Figure 1.63 Diagram of meniscus behavior in the (θ, α) plane. The angle α^* is defined by $\sin \alpha^* = \frac{\bar{R}_0^2}{4c}$. Depending on the contact angle, if the angle of a pipette is sufficiently large, whether it is hydrophobic or hydrophilic, the meniscus will stay at the bottom. No liquid will penetrate the pipette unless a negative pressure is established in the pipette.

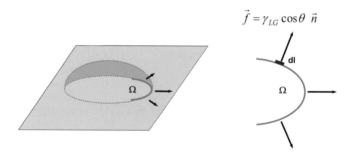

Figure 1.64 Schematic of the capillary force on a triple line.

Suppose that we want to find the value of the resultant of the capillary forces in a particular direction, say the x-direction. The projection along the x-direction of equation (1.90) can be written as

$$F_x = \int_\Omega \vec{f} \cdot \vec{i} \, dl = \int_\Omega \gamma \cos\theta \, \vec{n} \cdot \vec{i} \, dl. \tag{1.95}$$

Equation (1.95) can be simplified and cast in the form

$$F_x = \int_\Omega \gamma \cos\theta \, \vec{n} \cdot \vec{i} \, dl = \gamma \cos\theta \int_\Omega \vec{n} \cdot \vec{i} \, dl = \gamma \cos\theta \int_\Omega \cos\alpha \, dl = \gamma \cos\theta \int_0^e dl'. \tag{1.96}$$

Finally we obtain the expression

$$F_x = e \gamma \cos\theta. \tag{1.97}$$

Equation (1.97) shows that the resulting force on a triple contact line in any direction does not depend on the shape of the interface [43]; it depends only on the distance between the two ends of the triple line normal to the selected direction.

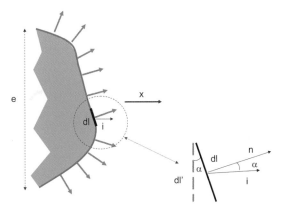

Figure 1.65 Capillary force on a triple contact line in the *x*-direction.

1.3.7.7   Examples of Capillary Forces in Microsystems

It is very common in biotechnology to use plates comprising thousands of micro-holes or cusps. The position of the free surface of the liquid in the cusps is of utmost importance. In particular, the liquid must not exit the holes under the action of capillary forces. As an example, figure 1.66 shows a free liquid interface in a square hole, calculated with the Surface Evolver. In the following Chapter, we will study in more detail the position of an interface in a hole, as a function of the wetting characteristics of the solid surfaces and the shape of the hole.

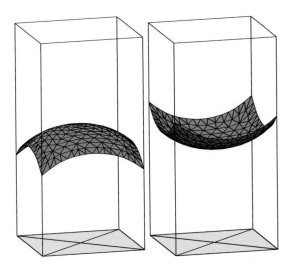

Figure 1.66 The surface of a liquid in a micro-well is not flat due to capillary forces. The figure is a simulation with the Surface Evolver. Left: case of water in a hydrophobic well (contact angles of 130°); right: case of water in a hydrophilic well (contact angles 50°). The "free" surface is tilted downwards or upwards depending on the contact angle. The walls have been dematerialized for clarity.

1.4   Measuring the Surface Tension of Liquids

We have remarked that surface tension can be seen as a force. Surface tension can then be measured by comparison with another known and calibrated force. This other force can be pressure as in the "bubble pressure method", also known as the Schrödinger method, or it can be a capillary force as in the Wilhelmy plate method, or it can be gravity as in the pendant drop method, and more recently it can be fluid stress as in the drop deformation method. In the following, we analyze these methods. Devices that are used to measure the surface tension are called tensiometers.

1.4.1   Using Pressure (Bubble Pressure Method)

The idea of using pressure to balance surface tension was first proposed by Erwin Schrödinger. The principle is based on the maximum pressure that an interface can support. A sketch of the experimental set up is shown in the figure 1.67. A tube filled with liquid 1 is plunged into a

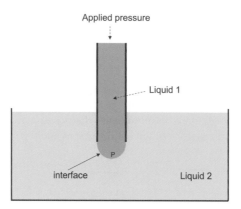

Figure 1.67 Principle of surface tension measurement using pressure: the maximum pressure which a liquid interface can support is when the interface has the shape of a half-sphere. This pressure is related to the surface tension by the Laplace law.

beaker containing liquid 2. As pressure in liquid 1 is increased, the interface at the tube outlet deforms till it reaches the form of a half-sphere. Above this maximum pressure, the interface blows out and liquid 1 breaks down into droplets flowing through liquid 2. The maximum pressure is related to the surface tension between liquid 1 and liquid 2 by the Laplace relation

$$P = 2\frac{\gamma_{L1L2}}{R_{tube}}. \tag{1.98}$$

This value can be found numerically by using the Surface Evolver (figure 1.68). We have used a capillary tube of $100 \ \mu m$ radius so that the gravity force does not introduce a bias in the result. Using a surface tension value of 72 mN/m (water in air), we find a maximum critical pressure of 1440 Pa, which is the value expected from (1.98). Remark: Theoretically, when the pressure is increased above the value defined by equation (1.93), the interface is no longer stable. It expands suddenly (figure 1.69) and breaks down. Physically, it corresponds to the instability of the interface and the formation of droplets. Experimentally the maximum pressure before the interface breaks is difficult to determine very precisely. This is the reason why other methods to measure surface tension have been developed.

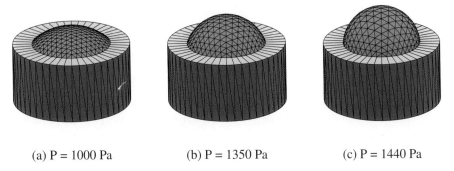

(a) P = 1000 Pa   (b) P = 1350 Pa   (c) P = 1440 Pa

Figure 1.68 Result of a calculation with Surface Evolver: the surface stably inflates with increasing pressures, until it reaches a half-sphere.

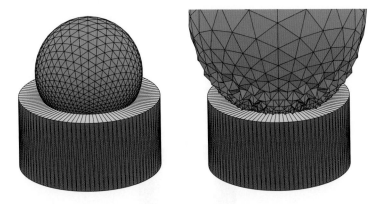

Figure 1.69 Above the threshold value the interface is no longer stable at constant pressure; it starts expanding and the numerical evolution reaches a point where the surface starts to explode.

1.4.1.1   Using the Capillary Rise on a Plate - Wilhelmy Plate

Surface tension of a liquid can be determined by measuring the capillary force exerted by the liquid on a solid plate [44]. The principle is that already described in section 1.3.7.3: the surface of the liquid rises along a vertical plate and the height reached by the liquid on the plate is proportional to the surface tension as shown by equation (1.79). In the standard method, a thin plate (perimeter about 40 mm) is lowered to the surface of a liquid and the downward force directed to the plate is measured. Surface tension is directly the force divided by the perimeter of the plate. A couple of very important points with this method must be noted. First, the plate must be completely wetted before the measurement to ensure that the contact angle between the plate and the liquid is zero. If this is not true the Wilhelmy method is not valid. Secondly, one must be sure that the position of the plate is correct, meaning that the lower end of the plate is exactly on the same level as the surface of the liquid. Otherwise the buoyancy effect must be calculated separately. Figure 1.70 shows a numerical simulation of the Wilhelmy method.

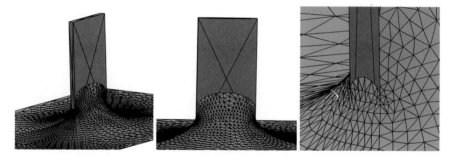

Figure 1.70 Simulation of the capillary rise along a Wilhelmy plate using the Surface Evolver: left, perspective view; middle, front view; right close-up on the contact line at a corner. Note how the contact line curves tangent to the corner in order to make a well-defined tangent plane at the corner.

1.4.2   Using Gravity: the Pendant Drop Method

By definition, a pendant drop is a drop suspended from a fixed solid, as shown in figure 1.71. The two forces acting on the drop are gravitation and surface tension. The shape of such a drop is then a function of the surface tension (figure 1.72). The pendant drop method consists of extracting the surface tension from an image of the drop shape.

Figure 1.71 Image of a pendant drop.

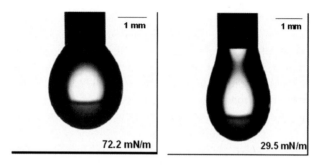

Figure 1.72 Experimental view of two drops: left: water droplet in air; right: oil droplet in air.

1.4.2.1 Bond Number

For a pendant drop, the ratio of the gravitational and surface tension forces is scaled by a non-dimensional number, the Bond number, defined by

$$Bo = \frac{\Delta \rho g R^2}{\gamma},\tag{1.99}$$

where $\Delta \rho$ is the difference of the density of the liquid and the surrounding fluid, g the gravitational constant, γ the surface tension and R a typical dimension of the droplet. Here we choose R to be the maximum horizontal radius of the pendant drop. The shape of the pendant drop is shown in figure 1.73 for different Bond numbers. Figure 1.73 shows that, for a well chosen

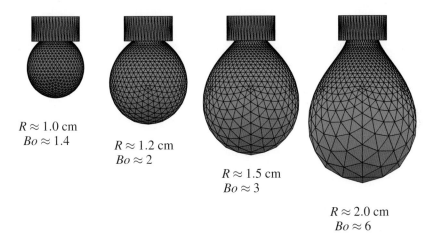

$R \approx 1.0$ cm
$Bo \approx 1.4$

$R \approx 1.2$ cm
$Bo \approx 2$

$R \approx 1.5$ cm
$Bo \approx 3$

$R \approx 2.0$ cm
$Bo \approx 6$

Figure 1.73 Shape of a pendant drop vs. Bond number: the drop shape departs from that of a sphere as its volume increases, until it detaches from the solid support. The liquid of the drop is water and is immersed in silicone oil; the surface tension is $\gamma = 33$ mN/m. Calculation performed with Surface Evolver.

Bond number, the competition between the gravitational force and the surface tension determines the shape of the droplet. On one hand, the gravitational force, i.e. the weight of the drop, tends to elongate the droplet vertically; on the other hand, the surface tension tends to minimize the interface by making it spherical. From an image analysis, the volume – and consequently the weight – of the drop can be determined, and also a vertical profile, which is a function of the surface tension and the weight of the drop (figure 1.74). In the following section we indicate the numerical approach that is used to determine the surface tension.

1.4.2.2 Method

There exist different numerical schemes to extract the surface tension from the pendant drop shape. A well known software is that pioneered by del Rio *et al.* called ADSA [45,46]. We do not give here the details of the method – which would be long - but just the main lines. The drop pressure is determined by writing the Laplace equation at the bottom of the droplet. Due

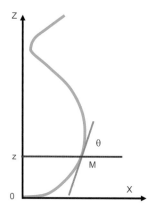

Figure 1.74 Typical drop contour in the pendant drop method.

to axisymmetry

$$\Delta P = \gamma\left(\frac{1}{R} + \frac{1}{R'}\right) = 2\frac{\gamma}{R}. \tag{1.100}$$

The curvature radius being derived from an image analysis, equation (1.100) produces the internal pressure of the drop. Second, we write the equilibrium equation of the drop in any horizontal section. Because of axisymmetry, this equilibrium imposes a zero vertical resultant:

$$2\pi R \gamma \sin\theta = V(\rho_h - \rho_l)g + \pi R^2 P. \tag{1.101}$$

In (1.101), R is the horizontal radius in the considered section, θ is the angle of the tangent at M to the contour of the image of the drop, V is the volume of the fluid under the plane of altitude z, ρ_h and ρ_l are the densities of the two fluids, and g the gravitational acceleration. The left term in equation (1.101) corresponds to the surface tension force; the first term on the right hand side, to the weight of the liquid below the considered section; and the second term on the right hand side to the pressure force. For each section, (1.101) produces a value for the surface tension γ. An averaging of all the values of γ determines precisely the real value of the surface tension.

Because the pendant drop method reacts very quickly, it is even possible to determine the surface tension as a function of the surface concentration in surfactants [47]. When CMC is reached, the shape of the drop does not evolve anymore. In conclusion, the advantages of the pendant drop method are:

- Small volume of liquid.

- Easy to spread a known amount of surfactant at the surface.

- Rapid rates of surface area change compared to Langmuir-Wilhelmy balance.

However, one must be very cautious that the capillary is very smooth and cylindrical to preserve axisymmetry.

1.4.2.3 Using Shear Stress in a Microflow

Hydrodynamic forces are usually difficult to control and to monitor accurately. However, the situation has changed drastically with the use of microflows. Such flows are completely lam-

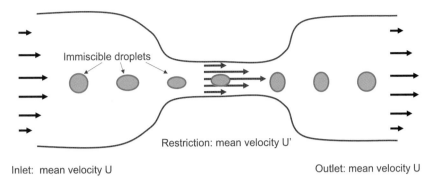

Figure 1.75 Schematic of droplet stretching in a contracted section for surface tension measurement.

inar and their velocity is largely predictable [18]. With this in mind, and observing that a droplet/bubble changes its shape according to the change in the flow mean velocity, a method for measuring surface tensions between two immiscible liquids or between a liquid and a gas has been proposed [49,50]. The principle is sketched in figure 1.75.

An experimental view of the phenomenon is shown in figure 1.76. A droplet traveling in the middle of the tube experiences a very small shear stress. However, when it reaches a constricted section, the flow accelerates and the shear stress increases. As a result the droplet takes an ellipsoidal shape oriented along the axis of the tube. At the entrance of the divergent section, the shear stress acts oppositely and the droplet takes an ellipsoidal shape oriented perpendicularly to the tube axis. After a while, the spherical shape is regained.

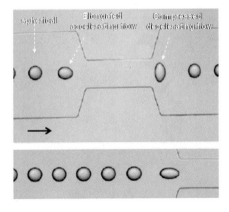

Figure 1.76 Experimental image of droplets in accelerated and decelerated fluid flows, reprinted with permission from [50], ©RSC, 2006, http://dx.doi.org/10.1039/B511976F.

Let us define the drop deformation by

$$D = \frac{a-b}{a+b},$$

(1.102)

where a and b are the two vertical dimensions of the droplet. When the droplet is spherical,

$D = 0$. Let $\dot{\varepsilon}$ be the extension rate given by the change of axial velocity u,

$$\dot{\varepsilon} = \frac{du}{dx}. \tag{1.103}$$

Then it can be shown that D is given by the solution of the differential equation

$$\frac{dD}{dt^*} = \frac{5}{2\hat{\eta}+3}\tau\dot{\varepsilon} - D, \tag{1.104}$$

where t* is a nondimensional time defined by $\tau^* = t/\tau$, and $\dot{\eta} = \eta_{drop}/\eta_{carrier}$ where η denotes the dynamic viscosity. Finally τ is defined by

$$\tau = \frac{\alpha\eta_{carrier}a_0}{\gamma}, \tag{1.105}$$

where a_0 is the undistorted radius and α a rational function of $\hat{\eta}$. Instead of solving (1.105), it is convenient to re-write it in the form

$$\alpha\eta_{carrier}\left(\frac{5}{2\hat{\eta}+3}\dot{\varepsilon} - u\frac{\delta D}{\delta x}\right) = \frac{\gamma D}{a_0}. \tag{1.106}$$

D and its axial evolution $\frac{\delta D}{\delta x}$ are estimated by image analysis. Then, a plot of the quantity $\alpha\eta_{carrier}(\frac{5}{2\hat{\eta}+3}\dot{\varepsilon} - u\frac{\delta D}{\delta x})$ as a function of $\frac{D}{a_0}$ is a linear curve with slope γ. Figure 1.77 shows experimental results of the method. This microfluidic tensiometer is particularly interesting because it produces the instantaneous surface tension on line.

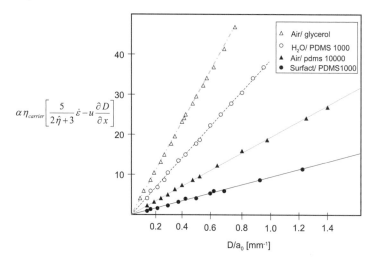

Figure 1.77 Plot of the flow extension rate versus the droplet deformation. The slope is the surface tension.

1.4.3   Surface Free Energy

1.4.3.1   Introduction

In the preceding sections, we have used the concept of surface tension of solids. However, surface tension of a solid is a complex notion; on one hand, it is very difficult to measure intrinsically and on the other hand different analytical relation have been proposed, but their results are still not completely satisfactory [51]. In a microfluidic approach, we avoid as much as possible having to use surface tension of solids. We use Young's law as soon as it is possible to eliminate the solid surface tension, and make our approaches depend on only two values: γ_{LG} and θ. Still, we present the notion of surface free energy and in particular the notion of critical surface tension, which is important for wetting analysis.

1.4.3.2   Method of Good-Girifalco

The approach of Good and Girifalco [52] consists of expressing the work of adhesion in terms of γ_{LG} and γ_S,

$$W_a = 2\Phi(\gamma_{LG}\gamma_S)^{\frac{1}{2}}, \tag{1.107}$$

where Ψ is a function of the molar volumes of the liquid and the solid. Inserting equation (1.107) in the Young-Dupré relation (1.56) yields

$$W_a = 2\Phi(\gamma_{LG}\gamma_S)^{\frac{1}{2}} = \gamma_{LG}(1 + \cos\theta), \tag{1.108}$$

leading to

$$\gamma_S = \gamma_{LG}\frac{(1 + \cos\theta)^2}{4\Phi^2}. \tag{1.109}$$

The remaining problem is to estimate Φ. Kwok and Neumann [53] made a systematic study of the contact between organic fluids and low energy surfaces. They based their approach on a dependency of the type

$$\Phi = \exp(-\beta(\gamma_{LG} - \gamma_S)^2) \simeq 1 - \beta(\gamma_{LG} - \gamma_S)^2, \tag{1.110}$$

where β is a constant slightly depending on the substrate. Substituting of (1.110) in (1.108) and using Young's law yields

$$\cos\theta = -1 + 2\sqrt{\frac{\gamma_S}{\gamma_{LG}}}\exp(-\beta(\gamma_{LG} - \gamma_S)^2). \tag{1.111}$$

For any given substrate γ_S is constant, and by using many different liquids and measuring γ_{LG} and θ, a least-square analysis leads to the values of β and γ_S. The parameter β was found to be of the order of 0.000100 to 0.000130 $(m^2/mJ)^2$ for low energy substrate such as polystyrene and polymethacrylate.

1.4.3.3   Fowkes Method

The preceding approach does not take into consideration the polar or apolar nature of the solid. The different components of the surface energy can be taken into account by writing [54]

$$\gamma_S = \gamma_S^D + \gamma_S^P, \tag{1.112}$$

where the superscripts D and P respectively stand for "diffusive" (non-polar) and polar. Neumann's approach [55] consists in subdividing the polar component into a Lewis acid and a Lewis base, so that the work of adhesion can be cast in the form

$$W_a = \gamma_{LG}(1 + \cos\theta) = 2\sqrt{\gamma_{LG}^D\gamma_S^D} + 2\sqrt{\gamma_{LG}^-\gamma_S^-} + 2\sqrt{\gamma_{LG}^+\gamma_S^+}. \tag{1.113}$$

Equation (1.113) contains 8 parameters. The contact angle θ can be measured, and the properties of the liquid are also known (they can be measured). Then we are left with 3 unknowns describing the solid. If we first use a non-polar liquid ($\gamma_{LG}^{-} = \gamma_{LG}^{+} = \gamma_{LG}^{P} = 0$) then (1.113) simplifies to

$$W_a = \gamma_{LG}(1 + \cos\theta) = 2\sqrt{\gamma_{LG}^{D}\gamma_{S}^{D}}, \qquad (1.114)$$

and we deduce γ_{S}^{D}. More generally if we use three liquids – including a non-polar one – we find a system of three equations with the three unknowns ($\gamma_{S}^{D}, \gamma_{S}^{+}, \gamma_{S}^{-}$). The total solid surface tension is then given by the relation

$$\gamma_S = \gamma_S^D + \gamma_S^P = \gamma_S^D + 2\sqrt{\gamma_S^+\gamma_S^-}. \qquad (1.115)$$

A very instructive example has been given by Combe *et al.* for measuring the surface tension of the enamel of human teeth [56].

1.4.3.4   Critical Surface Tension and Surface Free Energy

A very general question in capillary studies is: would a given liquid completely wet a given solid or not? In other words, what is the condition for spreading? Zisman and coworkers pioneered this problem in the 1950s, with the introduction of the notion of "critical surface tension", denoted here CST for simplicity, and defined by the proposition: a solid surface cannot be completely wetted by a liquid if the value of the liquid surface tension is above the critical surface tension.

1.4.3.4.1   Zisman Plot   Zisman and coworkers established an empirical connection between the cosine of the contact angle cos θ and the liquid/air interfacial tension γ [57]. For a given low energy solid surface, they measured the contact angle θ for many different liquids. The plot of cos θ as a function of γ approximates a straight line; it has the shape shown in figure 1.78.

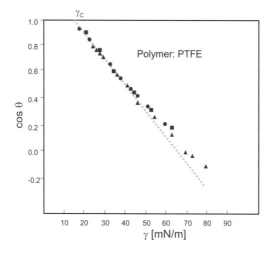

Figure 1.78 Zisman plot of the contact angle of different liquids on a PTFE surface.

When the contact angle is such that $\cos\theta = 1$, i.e. $\theta = 0$, the liquid wets the solid surface. If we denote γ_C the point where $\cos\theta = 1$, obtained by a linear extrapolation of Zisman's curves, then a liquid with a surface tension smaller than γ_C wets the surface. As a matter of fact, this nearly linear behavior is valid not only for pure liquids, but also for aqueous solutions. Zisman's approach was later refined by Bargeman et al. [58]. Observing that Zisman's curves were not exactly linear, especially for aqueous solutions, they have shown that linearity could be achieved by considering the relation between $\gamma\cos\theta$ and γ instead of the relation between $\cos\theta$ and γ (figure 1.79). In summary, there is a linear relation between adhesion tension and surface tension of aqueous solutions,

$$\gamma_{LG}\cos\theta = a\gamma_{LG} + b, \tag{1.116}$$

where a and b are constants depending on the solid surface.

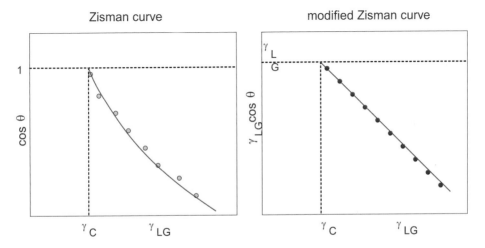

Figure 1.79 Modified Zisman's plot for PMMA (polymethyl methacrylate) and PTFE (polytetrafluoroethylene).

Zisman's approach is very practical; it easily produces the value of the CST. It has been observed that the value of the CST is often close to that of the surface tension of the solid, i.e. the Surface Free Energy (SFE). For a liquid whose surface tension approaches the critical surface tension, Young's law can be written

$$\gamma_{LG} = \gamma_C = \gamma_{SG} - \gamma_{SL}. \tag{1.117}$$

We have seen that in partial wetting conditions, there is a precursor film on the solid substrate; this is evidently the case at the onset of total wetting. This precursor film has a film pressure P_e (we discuss the film pressure in the next section) so that the surface tension γ_{SG} is given by

$$\gamma_S - P_e = \gamma_{SG}, \tag{1.118}$$

and (1.112) can be cast in the form

$$\gamma_S = \gamma_C + (\gamma_{SL} + P_e). \tag{1.119}$$

For a liquid having a surface tension γ_C, the term inside the parenthesis on the right hand side of (1.119) is often small. This is in particular the case when the solid surface is apolar, i.e. the polar component of the surface free energy is negligible. For example PMMA has a CST of 33.1 mN/m and a SFE of 39 mN/m; PTFE has a CST of 20 mN/m and a SFE of 23.4 mN/m. In such cases the CST is a good approximation of the SFE. The advantage is that the critical surface tension can easily be obtained by Zisman plots.

1.4.3.4.2 Disjoining Pressure   When a liquid film on a solid is very thin (like the precursor film), all the liquid molecules have an interaction with the solid wall, as sketched in figure 1.80. We are then at a very small scale (less than 1 nm). The energy of the film takes a special form. For a thick film, the energy is $\gamma + \gamma_{SL}$. For a thin film of thickness e the film energy is [22]

$$E = \gamma + \gamma_{SL} + P_e(e), \tag{1.120}$$

where $P_e(e)$ is a function of e such that $P_\infty = 0$ and $P_0 = \gamma_S - (\gamma + \gamma_{SL})$.

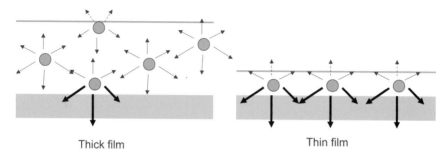

Thick film   Thin film

Figure 1.80 In a thin film all the molecules in the liquid interact with molecules of the solid wall.

From a thermodynamical point of view, Derjaguin has introduced the notion of "disjoining pressure" in the chemical potential [59],

$$\mu = \mu_0 + \frac{dP_e}{de} V_0 = \mu_0 - \Pi_e V_0, \tag{1.121}$$

where V_0 is the molar volume in the liquid phase. The quantity Π_e in (1.122), defined by

$$\Pi_e = -\frac{dP_e}{de}, \tag{1.122}$$

has the dimension of a pressure. Π_e is called the "disjoining pressure". With this definition, and using (1.119), de Gennes [22] shows that the total surface tension of a thin film is given by

$$\gamma_{film} = \gamma_{LG} + \gamma_{SL} + P_e + e\Pi_e. \tag{1.123}$$

We have seen that bubbles and droplets at equilibrium tend to take a spherical shape. If this were not the case, the internal pressure would not be uniform since the Laplace law applied in regions of different curvatures would produce different internal pressures. In consequence, there is liquid motion inside the bubble/droplet in order to equilibrate the internal pressure (figure 1.81).

But what about adjacent droplets or bubbles? The sketch of figure 1.81 shows that a droplet is deformed by the presence of the neighboring droplet and its curvature is not constant. A

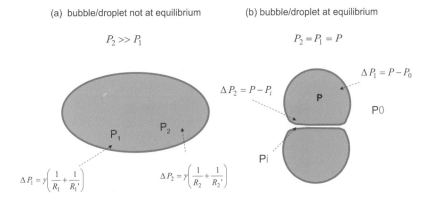

Figure 1.81 Schematic of a droplet out of equilibrium (left) and an assembly of two droplets/bubbles at equilibrium (right).

first analysis using Laplace's law would conclude that the internal pressure is not uniform. In reality, the pressure in the thin film between the drops/bubbles is not the external pressure P_0 but a different pressure related to the "disjoining pressure". According to Laplace equation, the film pressure equilibrates the pressure jump across the interface, and we have the relations

$$\Delta_{P_1} = \gamma \left(\frac{1}{R_1} + \frac{1}{R_1'} \right) = P - P_0 \Delta_{P_2} = \gamma \left(\frac{1}{R_2} + \frac{1}{R_2'} \right) = P - P_i, \qquad (1.124)$$

where indices 1 and 2 refer to two locations, a first one far from the deformed interface and the other close to the deformed interface. In (1.112), P_i is the film pressure.

1.5 Minimization of the Surface Energy and Minimal Surfaces

"Minimal surfaces" are a particular type of surface. From a mathematical standpoint, a minimal surface is officially just a surface with zero mean curvature. This definition is very restrictive: there are not many minimal surfaces with known equations. One can find a description of these surfaces in reference [60]. These surfaces locally minimize their area. In the domain of physics of liquid interfaces, we have already seen an example of a minimal surface in section 1.3.3.5 with the surface of a liquid rising along a cylindrical vertical tube (figure 1.82). Another example of a minimal surface is that of the catenoid (figure 1.83). Physically, a catenoid is the surface formed by a liquid film spanning two solid rings. It is a general thermodynamic principle in physics that systems evolve to their minimal energy level. In particular, this is the case of interfaces. Surfaces of droplets in static equilibrium have a minimal free energy. Thus, in the domain of liquid interfaces, we are tempted to give a less restrictive definition for minimal surfaces. A minimal surface is a surface that minimizes its area under some constraints. The goal is to minimize the energy of the system (surface, gravitational, etc.) under some constraints imposed by external conditions, like walls, wires, fixed volume or fixed pressure. With this definition, minimal surfaces comprise a much larger set of surfaces and describe liquid films and interfaces [43,61]; when gravity is negligible, these surfaces have a constant mean curvature [62].

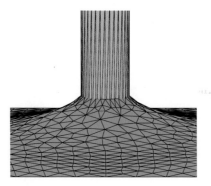

Figure 1.82 Minimal surface formed by the interface of a liquid with a vertical rod.

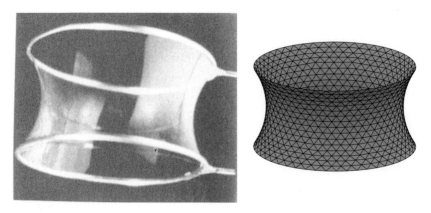

Figure 1.83 Left: Image of a soap film between two solid circles; right: result of this calculation with the Surface Evolver.

1.5.1   Minimization of the Surface Energy

According to the extended definition of a minimal surface, the droplet interface can be calculated by minimization of its surface energy. In the absence of a droplet on the surface of the solid substrate, the surface energy is

$$E_{SG,0} = \gamma_{SG} S_{SG,0}. \tag{1.125}$$

After deposition of the droplet, the surface energy is the sum of the three surface energies

$$E = E_{LG} + E_{SL} + E_{SG,1}, \tag{1.126}$$

where $E_{SG,1}$ is the surface energy of the solid surface in contact with the gas. Then, we have

$$E = E_{LG} + \int\int_{S_{SL}} (\gamma_{SL} - \gamma_{LG}) \, dA + E_{SG,0}. \tag{1.127}$$

The last term on the right hand side of (1.123) does not depend on the drop shape. Thus, we have to minimize

$$E = \gamma_{LG} S_{LG} + \int\int_{S_{SL}} (\gamma_{SL} - \gamma_{LG}) \, dA. \tag{1.128}$$

Taking into account Young's law, the energy to be minimized is

$$E = \gamma_{LG} S_{LG} + \gamma_{LG} \int \int_{S_{SL}} \cos \theta \, dA. \qquad (1.129)$$

As we have mentioned earlier, the parameters intervening in equation (1.129) are θ and γ_{LG}. Thanks to Young's equation, we don't need the surface tension of the solid with the liquid or the gas. This is a real simplification since we have shown that θ and γ_{LG} can be measured relatively easily.

Equation (1.129) constitutes the basis for the calculation of droplet shapes that we develop in the next Chapter.

1.5.2 Conclusion

This chapter was devoted to the study of surface tensions and capillary forces. The main notions presented in this chapter are the relation between curvature and pressure expressed by Laplace's law, and the relation between the different surface tensions and contact angle at the triple line expressed by Young's law. From these two relations, an expression for the capillary force on a triple line has been deduced. Such an expression has a key role in determining the behavior of droplets on different substrates and geometry in microsystems that we shall consider in the next Chapter.

This Chapter has shown the essential role of surface tension and capillarity at the microscale. These forces often dominate forces such as gravity or inertia, which are predominant at the macroscopic scale. Although we have taken the stance of presenting capillarity and surface tension from an engineering point of view by considering global effects, one has to keep in mind that interactions at the nanoscopic scale are the real underlying causes of these global effects.

1.6   References

[1] J. Berthier, *Microdrops and digital microfluidics*. William Andrew Publishing, 2008.

[2] J. Israelachvili, *Intermolecular and surface forces*. Academic Press, 1992.

[3] Table of surface tension for chemical fluids: http://www.surface-tension.de/.

[4] G. Navascues, "Liquid surfaces: theory of surface tension," *Rep. Prog. Phys.* **42**, pp. 1133–1183, 1979.

[5] R. Eötvös, *Wied. Ann. Phys.*, **27**, pp. 445-459, 1886.

[6] E.A. Guggenheim, "The principle of corresponding states," *J. Chem. Phys.* **13**, pp. 253–261, 1945.

[7] A. Alexeev, T. Gambaryan-Roisman, P. Stephan, "Marangoni convection and heat transfer in thin liquid films on heated walls with topography: Experiments and numerical study," *Phys. Fluids* **17**(6), p. 062106, 2005.

[8] K. Szymczyk, A. Zdiennicka, B. Janczuk, W. Wocik, "The wettability of polytetrafluoroethylene and polymethyl methacrylate by aqueous solution of two cationic surfactants mixture," *J. Colloid. Interface Science* **293**, pp. 172–180, 2006.

[9] MIT, Lecture 4, Marangoni flows, http://web.mit.edu/1.63/www/Lec-notes/Surfacetension/Lecture4.pdf.

[10] F. Ravera, E. Santini, G. Loglio, M. Ferrari, L. Liggieri, "Effect of Nanoparticles on the Interfacial Properties of Liquid/Liquid and Liquid/Air Surface Layers," *J. Phys. Chem. B* **110 (39)**, pp. 19543–19551, 2006.

[11] E. Weisstein, http://mathworld.wolfram.com/Curvature.html.

[12] E. Berthier, D.J. Beebe. "Flow rate analysis of a surface tension driven passive micropump," *Lab On a Chip* **7** (11), pp.1475–1478, 2007.

[13] L.M. Pismen, B.Y. Rubinstein, I. Bazhlekov, "Spreading of a wetting film under the action of van der Waals forces," *Physics of Fluids* **12 (3)**, p. 480, 2000.

[14] P.G. deGennes, "Wetting: statistics and dynamics," *Rev. Mod. Phys.* **57**, p. 827, 1985.

[15] Quanzi Yuan, Ya-Pu Zhao, "Precursor film in dynamic wetting, Electrowetting and Electro-Elasto-Capillarity," *Phys. Rev. Let.* **104**, p.246101, 2010.

[16] C. Iwamoto, S. Tanaka, "Atomic morphology and chemical reactions of the reactive wetting front," *Acta Materiala* **50**, pp. 749, 2002.

[17] A.L. Yarin, A.G. Yazicioglu, C. M. Megaridis, "Thermal stimulation of aqueous volumes contained in carbon nanotubes: experiment and modelling," *Appl. Phys. Lett.* **86**, p.013109, 2005.

[18] Nanolane: http://www.nano-lane.com/wetting-afm.php.

[19] L. Pismen, "Dewetting patterns, moving contact lines, and dynamic diffuse interfaces," *CISM lecture notes*, Udine, Italy, 16-20 October 2006.

[20] J. Berthier, Ph. Clementz, O. Raccurt, P. Claustre, C. Peponnet, Y. Fouillet, "Computer aided design of an EWOD microdevice," *Sensors and Actuators A* **127**, pp. 283-294, 2006.

[21] K. Brakke, "The Surface Evolver," *Exp. Math.* **1**, pp. 141–165, 1992.

[22] de Gennes, P-G., F. Brochard-Wyart, D. Quèrè, *Drops, bubbles, pearls, waves.* Springer, New York, 2004.

[23] A.A. Darhuber, S.M. Troian,"Principles of microfluidic actuation by modulation of surface stresses," *Annu. Rev. Fluid Mech.* **37**, pp. 425–455, 2005.

[24] H. Bouasse, *Capillarité, phénomènes superficiels.* Delagrave, Paris, 1924.

[25] C.D. Murray,"The physiological principle of minimum work. 1. The vascular system and the cost of blood volume," *Proc. Natl. Acad. Sci. USA* **12**, pp. 207–214, 1926.

[26] L.A. Taber, S. Ng, A.M. Quesnel, J. Whatman, C.J. Carmen, "Investigating Murray's law in the chick embryo," *J. Biomech.* **34**(1), pp. 121–124, 2001.

[27] K.A. Mc Culloh, J.S. Sperry, F.R. Adler, "Water transport in plants obeys Murray's law," *Nature* **421**, pp. 939–942, 2003.

[28] E. Goljan, *Pathology*, 2nd ed., Mosby Elsevier, Rapid Review Series, 1998.

[29] H.D. Prange, "Laplace's law and the alveolus: a misconception of anatomy and a misapplication of physics," *Adv. Physiol. Educ.* **27**, pp. 34–40, 2003.

[30] J. Atencia, D.J. Beebe, "Controlled microfluidic interfaces," *Nature* **437**, pp. 648–655, 2005.

[31] T. Thorsen, R.W. Roberts, F.H. Arnold, S.R. Quake, "Dynamic pattern formation in a vesicle-generating microfluidic device," *Phys. Rev. Lett.* **86**(18), pp. 4163–4166, 2001.

[32] J. Berthier, S. Le Vot, P. Tiquet, N. David, D. Lauro, P.Y. Benhamou, F. Rivera, "Highly viscous fluids in pressure actuated Flow Focusing Devices," *Sensors and Actuators A* **158**, pp. 140–148, 2010.

[33] Y.-C. Tan, K. Hettiarachchi, M. Siu, Y.-R. Pan, A.P. Lee, "Controlled microfluidic encapsulation of cells, proteins and microbreads in lipid vesicles," *JACS* **128**(17), pp. 5656-5658, 2006.

[34] L. Landau, E. Lifchitz, *Fluid mechanics*, 2nd edition: Volume 6, Reed Educational and Professional Publishing, Oxford, 2000.

[35] Y.S. Yu, Y.P. Zhao, "Deformation of PDMS membrane and microcantilever by a water droplet: Comparison between Mooney-Rivlin and linear elastic constitutive models,"

Journal of Colloid and Interface Science **332**, pp. 467–476, 2009.

[36] X.H. Zhang, X.D. Zhang, S.T.Lou, Z.X. Zhang, J.L.Sun, J. Hu, "Degassing and temperature effects on the formation of nanobubbles at the mica/water interface," *Langmuir* **20**(9), pp. 3813–3815, 2004.

[37] J.Y. Wang, S. Betelu, B.M. Law, "Line tension approaching a first-order wetting transition: Experimental results from contact angle measurements," *Physical Review E* **63**, pp. 031601-1, 031601-10, 2001.

[38] J. Drehlich, "The significance and magnitude of the line tension in three-phase (solid-liquid-fluid) systems," *Colloids and Surfaces A: Physicochemical and Engineering Aspects* **116**, pp. 43–54, 1996.

[39] D.L. Hu, J. W. M. Bush, "Meniscus-climbing insects," *Nature* **437**, pp. 733–736, 2005.

[40] J. Jurin, "An account of some experiments shown before the Royal Society; with an enquiry into the cause of the ascent and suspension of water in capillary tubes," *Philosophical Transactions of the Royal Society* **30**, pp.739–747, 1719.

[41] Y. Tsori, "Discontinuous liquid rise in capillaries with varying cross-sections," *Langmuir* **22**, pp. 8860–8863, 2006.

[42] Jun Zeng, T. Korsmeyer, "Principles of droplet electrohydrodynamics for lab-on-a-chip," *Lab Chip* **4**, pp. 265–277, 2004.

[43] J. Berthier, Ph. Dubois, Ph. Clementz, P. Claustre, C. Peponnet, Y. Fouillet, "Actuation potentials and capillary forces in electrowetting based microsystems," *Sensors and Actuators, A: Physical* **134**(2), pp. 471-479,2007.

[44] K. Holmberg, *Handbook of Applied Surface and Colloid Chemistry*. New York, Wiley and Sons, 2002.

[45] O.I. del Rio, A.W. Neumann,"Axisymmetric drop shape analysis: computational methods for the measurement of interfacial properties from the shape and dimensions of pendant and sessile drops," *J. Coll. Int. Sci.* **196**, pp. 134–147, 1997.

[46] M. Hoofar, A.W. Neumann, "Axisymmetric drop shape analysis (ADSA) for the determination of the surface tension and contact angle," *J. Adhesion* **80**(8), pp. 727–743, 2004.

[47] H.A. Wege, J.A. Holgado-Terriza, M.J. Galvez-Ruiz, M.A. Cabrerizo-Vilchez, "Development of a new Langmuir-type pendant drop film balance," *Colloids and Surfaces B: Biointerfaces* **12**, pp 339–349, 1999.

[48] J. Berthier, P. Silberzan, *Microfluidics for Biotechnology*. Artech House, 2005.

[49] S.D. Hudson, J.T. Cabral, W.J. Goodrum, Jr., K.L. Beers, E.J. Amis, "Microfluidic interfacial tensiometry," *Applied Physics Letters* **87**, p. 081905, 2005.

[50] J.T. Cabral, S.D. Hudson, "Microfluidic approach for rapid multicomponent interfacial tensiometry," *Lab Chip* **6**, pp. 427–436, 2006.

[51] L. Makkonen,"On the methods to determine surface energies," *Langmuir* **16**, pp. 7669–7672, 2000.

[52] R.J. Good, L.A. Girifalco,"A theory for estimation of surface and interfacial energies. 3. Estimation of surface energies of solids from contact angle data," *J. Phys. Chem.* **64**, pp. 561–565, 1960.

[53] D.Y. Kwok, A.W. Neumann,"Contact angle interpretation in terms of solid surface tension," *Colloids and Surfaces A: Physicochemical and Engineering Aspects* **161**(15), pp. 31–48, 2000.

[54] F.M. Fowkes, "Additivity of intermolecular forces at interfaces. 1. Determination of contribution to surface and interfacial tensions of dispersion forces in various liquids," *J. Phys. Chem.* **67**(12), p. 2538, 1963.

[55] A.W. Neumann, R.J. Good, C.J. Hope, M. Seipal, "Equation of state to determine surface

tensions of low energy solids from contact angles," *J. Colloid Interf. Sci.* **49**, pp. 291–304, 1974.

[56] E.C. Combe, B. A. Owen, J.S. Hodges,"A protocol for determining the surface energy of dental materials," *Dental Materials* **20**, pp. 262–268, 2004.

[57] W.A. Zisman, "Contact angle, wettability and adhesion," *Advances in Chemistry Series* **43**, p. 1, 1964.

[58] D. Bargeman, F. van Voorst Vader, "Tensile strength of water," *J. Coll. Sci.* **42**, pp. 467–472, 1973.

[59] A. Adamson, A. Gast, *Physical Chemistry of Surfaces*, 6th edition. John Wiley and Sons Inc., 1997.

[60] Brandeis University: http://rsp.math.brandeis.edu/3D-XplorMath/Surface/gallery_m.html.

[61] K. Brakke, "Minimal Surfaces, Corners, and Wires," *J. Geom. Anal.* **2**(1), pp. 11–36, 1992.

[62] D.E. Hewgill, "Computing surfaces of constant mean curvature with singularities," *Computing* **32**, pp. 81–92, 1984.

2

Minimal Energy and Stability Rubrics

2.1 Abstract

This chapter describes basic themes of stability for liquids in contact with various geometries of walls. A few simple rules can lead to a well-developed intuition of the stability of liquid shapes in general circumstances. There are some configurations for which it is possible to prove mathematically what the absolute minimum energy shapes are. Obviously, minimum energy shapes are stable.

A brief summary of the rubrics (very loosely stated):

- Spherical drops are usually minimal energy.

- Rotational symmetry means lower energy.

- Pressure increasing with volume means stability.

- Pressure decreasing with volume can give rise to a "double-bubble" instability.

The first section focuses on spheres, and works through a gallery of examples where a spherical section is the minimum energy surface. Included are droplets on planes, droplets in wedges, bridges between parallel planes, and droplets outside convex bodies. It is also possible to include configurations with mobile walls. The second section describes how "symmetrization" to rotational symmetry lowers the energy. This leads to the classic rotationally symmetric surfaces of constant mean curvature: the sphere, the cylinder, the catenoid, the unduloid, and the nodoid. Sections of these surfaces often turn up as bridges between parallel substrates. The third section describes a general method of showing stability in cases where the pressure increases as volume increases. This applies, for example, to concave liquid shapes in corners and grooves. The fourth section discusses the "double-bubble" instability, which arises when pressure decreases with volume, and there are separate parts of the liquid that can somewhat independently grow and shrink. The famous Plateau-Rayleigh instability of a liquid cylinder may be seen as an example. The last section contains a more precise summary of the rubrics.

The basic formulations of the rubrics omit gravity, but some of them do apply when gravity is present; that will be noted in the sections below. Otherwise, for microdrops gravity can be imagined as making small corrections.

The common geometric properties of the surfaces in this chapter are first, constant mean curvature (due to constant pressure, since gravity is not being included), and second, constant contact angle at any boundary wall (although the angle may vary on different walls). In two dimensions, that means that all interface curves are arcs of circles, which makes it possible to prove minimal energy configurations by enumerating all possible geometries of circular arcs. But we cannot take that approach in three dimensions, since constant mean curvature surfaces need not be sections of spheres.

This chapter contains a few mathematical proofs at the level of multivariable calculus. They are included because they are so short and wondrous. For those who worry that calculus only applies to smooth surfaces and we are not assumming that here, rest assured that the mathematical field known as Geometric Measure Theory [1][2] makes it possible to apply all the standard concepts of multivariable calculus to arbitrary geometrical shapes, no matter how rough or nondifferentiable.

2.2   Spherical Shapes as Energy Minimizers

2.2.1   The Sphere Theorems

It is well-known that a sphere is the least-area surface that can enclose a given volume. Spherical shapes turn up a lot in contact-angle problems because a sphere has the nice property that its intersection with a given plane has a constant contact angle, which depends only on the distance of the plane from the center of the sphere and not on the orientation of the plane.

A proof of area-minimality due to Gromov [3] can be generalized to cover the case where there are walls with given contact angles. Gravity is assumed to be absent, so the total energy of the system is the free surface area plus the wall contact energy. Basically, the theorem says that if a spherical drop happens to hit the walls of a convex region with exactly the right contact angles, then the spherical drop minimizes the total energy.

Theorem 2.1. *Suppose Ω is a polyhedral convex region (possible infinite, possibly even all of space) defined by boundary planes T_i, with each plane T_i having a prescribed contact surface tension γ_i. For any region U of Ω, define the energy $E(U)$ of U to be the sum of the area of the free surface of U, $\partial U_0 = \partial U \cap interior(\Omega)$, and the sum of the areas of U intersecting the boundary planes T_i, each multiplied by the appropriate tension γ_i:*

$$E(U) = area(\partial U_0) + \sum_i \gamma_i \, area(\partial U \cap T_i). \tag{2.1}$$

Suppose S is a sphere such that the angle of intersection of S with each boundary plane T_i satisfies Young's Law, assuming the free surface of the sphere has surface tension 1, and if S does not intersect T_i then $\gamma_i \geq 1$. Suppose W is the intersection of a sphere S with Ω. Then the energy of W is least among all regions U of Ω with equal volume, and any region of equal energy must be a translation of W.

Proof. The theorem is true in any ambient dimension, but we will state the proof in three dimensions for simplicity. We may assume the center of the sphere S is at the origin. We may further assume the radius of S is one. Let the boundary ∂W of W be made up of a free surface

∂W_0 and a set of plane pieces $\partial W_i = \partial W \cap T_i$. Then the energy of of W is equal to 3 times its volume:

$$E(W) = area(\partial W_0) + \sum_i \gamma_i\, area(\partial W_i) \tag{2.2}$$

$$= \int_{\partial W_0} 1\, dA + \sum_i \int_{\partial W_i} \gamma_i\, dA \tag{2.3}$$

$$= \int_{\partial W_0} \vec{x} \cdot \vec{N}\, dA + \sum_i \int_{\partial W_i} \vec{x} \cdot \vec{N}_i\, dA \tag{2.4}$$

$$= \int_{\partial W} \vec{x} \cdot \vec{N}\, dA \;=\; \int_W div\,\vec{x}\, dV \tag{2.5}$$

$$= \int_W 3\, dV \tag{2.6}$$

$$= 3\, volume(W). \tag{2.7}$$

Here $\vec{N}$ is the unit normal vector, $\vec{x}$ is simply the position vector $\vec{x} = (x,y,z)$, and $\gamma_i = \vec{x} \cdot \vec{N}_i$ is exactly the sphere intersection assumption.

Now we prove any competitor region U of the same volume has equal or greater energy. Let a volume-preserving map $\vec{f} : U \to W$ be defined by $\vec{f}(x,y,z) = (f_1(x), f_2(x,y), f_3(x,y,z))$, with each component map monotone increasing in its last variable. This map can be constructed dimension by dimension, with $f_1(x)$ defined first so the volume in W of $x \le x_0$ is the same as the volume of $x \le f_1(x_0)$, then defining $f_2(x,y)$ with area on each x slice, etc. This makes the Jacobian matrix of f triangular, so the Jacobian determinant is the product of the main diagonal. Thus by volume conservation,

$$\frac{\partial f_1}{\partial x} \frac{\partial f_2}{\partial y} \frac{\partial f_3}{\partial z} = 1, \tag{2.8}$$

and hence by a standard mathematical inequality

$$\frac{\partial f_1}{\partial x} + \frac{\partial f_2}{\partial y} + \frac{\partial f_3}{\partial z} \ge 3, \qquad \text{i.e. } div\,\vec{f} \ge 3, \tag{2.9}$$

with equality if and only if all the partials concerned are 1. Then

$$E(U) = area(\partial U_0) + \sum_i \gamma_i\, area(\partial U_i)$$

$$= \int_{\partial U_0} 1\, dA + \sum_i \int_{\partial U_i} \gamma_i\, dA$$

$$\ge \int_{\partial U_0} \vec{f} \cdot \vec{N}\, dA + \sum_{\partial U_i} \vec{f} \cdot \vec{N}_i\, dA$$

$$\ge \int_{\partial U} \vec{f} \cdot \vec{N}\, dA \tag{2.10}$$

$$\ge \int_U div\,\vec{f}\, dV$$

$$\ge \int_U 3\, dV$$

$$\ge 3\, volume(U)$$

$$\ge 3\, volume(W)$$

$$\ge E(W).$$

Uniqueness up to translation follows from tracing the conditions of equality. Equality in equation (2.9) requires the diagonal elements of the Jacobian to be 1, and equality in equation (2.10) then requires the off-diagonal elements of the Jacobian to be zero on the boundary. Hence the Jacobian is the identity matrix around the boundary (and inside, it can be shown), so the regions are identical except possibly for translation.   □

The theorem as stated here skips over a lot of technical niceties, such as the definitions of "area" and "boundary" and divergence of a nonsmooth vectorfield, but these can be handled with geometric measure theory as mentioned earlier, and it turns out that everything works just like it does for nice smooth surfaces and vectorfields.

Theorem 2.1 can be generalized to compare energies of regions in two different convex domains, if the second domain Φ has contact angles depending on wall orientation the same as domain Ω. That is, the walls of Φ are parallel to the walls of Ω, with the same contact angles.

Theorem 2.2. *Suppose Ω and W are as in Theorem 2.1. Suppose Φ is a convex polyhedral domain such that the contact surface tension on each boundary plane of Φ is the same as on a parallel boundary plane of Ω. Then the energy of W is less than or equal to any region of Φ with equal volume, with equality only if U is a translation of W.*

Proof. The proof is the same as for Theorem 2.1, which really didn't use the fact that the competitor region U was in the same domain.   □

One application of Theorem 2.2 is that if a boundary plane is mobile (for example, a chip floating on a solder drop), then the solder drop will push or pull on the chip to make a spherical drop.

2.2.2   Sphere

The simplest application of Theorem 2.1 is the case where the convex region Ω is the whole space and W is a complete sphere. There is no boundary of Ω, so no contact angles, and the theorem simply says the sphere is the least area surface enclosing a given volume.

2.2.3   Spherical Cap on a Plane

The next simplest case is when the convex region Ω is a half-space bounded by a single plane with a constant contact angle. Then Theorem 2.1 says the minimum energy drop is a spherical cap with the center of the sphere located to make the intersection of the sphere and plane happen at the given contact angle, as shown in figure 2.1.

2.2.4   Drop Between Parallel Plates

A droplet of liquid between two parallel plates has least energy when it has a spherical shape, according to Theorem 2.2. For a given volume and given contact angles, there is only one separation distance of the plates that permits a spherical drop with the given contact angles. If the distance between the plates is changed, then by Theorem 2.2, the energy will increase. Hence only for the spherical case will the net force on the plates be zero; plates closer together produce a nodoid outer region and will be pushed apart, while plates farther apart will produce an unduloid region and be pulled together. The spherical drop is possible only if the sum of the contact angles is greater than 180°; otherwise, the liqiud wants to spead infinitely between infinitesimally close plates.

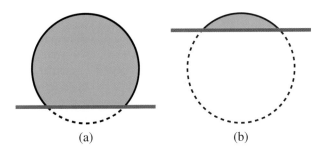

Figure 2.1 (a) Hydrophobic drop as section of sphere. (b) Hydrophilic drop as section of sphere.

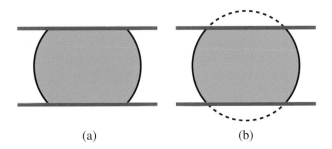

Figure 2.2 (a) Bridge between parallel planes. (b) Bridge as section of a sphere.

2.2.5 Droplet in a Wedge

Closely related to the case of parallel planes is the case of two planes making a wedge. The minimum energy region is a sphere, regardless of volume, since the sphere can adjust to arbitrary volume by moving in or out. There are no unduloid-like or nodoid-like shapes in a wedge.

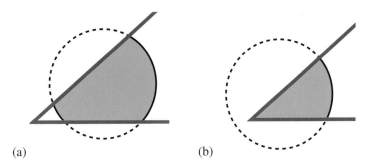

Figure 2.3 (a) Droplet in a wedge where the sum of the contact angles exceeds $180°$ plus the wedge angle. (b) Droplet in a wedge where the sum of the contact angles is less than $180°$ plus the wedge angle, but exceed $180°$ minus the wedge angle.

Note that in case (a) of figure 2.3 the radius of the sphere and the energy of the configuration are independent of the wedge angle. The spherical caps cut off depend only on the contact angles on the planes, and the orientations of the planes make no difference. Thus there is no force trying to open or close a hinged wedge; whatever the wedge angle, the drop can move in or out until it finds the place where the proper size caps are cut off. Likewise, if a drop is between two planes, there is no torque trying to make the planes parallel or non-parallel.

A related configuration is a drop inside a circular cone. Theorem 2.1, applied in its most general version with a continuum of boundary planes, says the minimal energy region is spherical, if the contact angle is greater than 90° minus the cone angle (from axis). If the contact angle is greater than 90° plus the cone angle, then the tip of the cone is dry, as in figure 2.3(a). Otherwise, the liquid fills the tip, as in figure 2.3b. For smaller contact angles, the liquid still wets the cone tip, but the surface is a concave spherical cap; see below.

2.2.6   Droplet on an Exterior Corner

A spherical case where Theorem 2.1 does not apply and a spherical droplet is not the absolute minimum is a drop on the exterior of a corner, as shown in figure 2.4(a). The domain involved is the exterior of the corner, which is not convex, thus making Theorem 2.1 inapplicable. The corner droplet has higher energy than a spherical droplet fully on one plane side of the corner. The corner droplet is an equilibrium shape, but an unstable equilibrium; it is unstable to creeping to one side or the other.

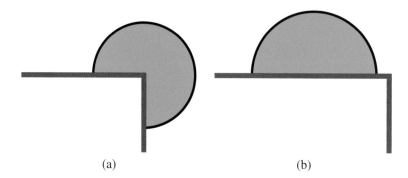

(a)   (b)

Figure 2.4 Two competing configurations for minimum energy exterior to a corner. (a) Symmetric round bubble at the corner. (b) Round bubble entirely on one side. The latter has lower pressure, and lower energy.

2.3   Symmetrization and the Rouloids

2.3.1   Steiner Symmetrization

Unfortunately, spherical shapes do not solve all liquid problems, particularly the common case of a liquid bridge between fixed parallel planes. But is is possible to show that the minimum energy bridge must be rotationally symmetric. Jakob Steiner in 1842 published [4] an argument (made rigorous by others later) that if one took a non-rotationally-symmetric liquid shape, sliced it into infinitesimally thin parallel layers, and re-formed each layer into a circular disk of equal area about a common axis (fig. 2.5). then the result will have less surface area than the original. It works regardless of holes or interior bubbles or disconnected pieces. This process is known as Steiner symmetrization. It does not itself produce the minimum energy configuration, but applying it to any proposed minimum shows that the minimum must be rotationally symmetric.

If the slicing is horizontal, then the presence of gravity does not affect the argument, since pieces of liquid only move horizontally. Note that even though the symmetrization process does not necessarily preserve contact angles, it does preserve contact areas on the planes, and hence symmetrization does decrease total energy.

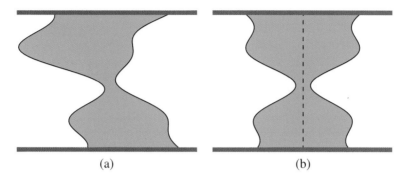

(a)   (b)

Figure 2.5 2D example of Steiner symmetrization. (a) Initial shape between parallel lines. (b) Symmetrized shape, with less surface length.

2.3.2   Rouloids

Delaunay [5] found all the surfaces of constant mean curvature (i.e. zero gravity) that are surfaces of revolution about an axis. They are spheres, cylinders, catenoids, unduloids, and nodoids. Planes perpendicular to the axis are also technically part of this group. These surfaces collectively are known as "rouloids", since their generating curves, known as roulettes (fig. 2.6), may be formed by tracing the path of the focus of a conic section as it rolls along the axis. A java applet illustrating the rolling conics may be found at http://www.susqu.edu/brakke/roulades/roulades.html.

Refer to the papers of Vogel [6]-[10] for an extensive discussion of the stability of rouloid liquid bridges.

2.3.2.1   The Sphere

The sphere is well-known to be the surface of least area that encloses a given volume. There are numerous proofs of this, but my favorite is given in Theorem 2.3 below.

As a roulette, the generating curve is a semi-circle, which may be generated by rotating a radial segment about the center of the circle. The segment may be viewed as a degenerate ellipse of zero minor axis, with its foci at its endpoints. Technically, a segment that rotates indefinitely (as in the other roulettes) generates a sequence of tangent spheres.

2.3.2.2   The Cylinder

The roulette for a cylinder of radius R is the straight line parallel to the axis generated by the center of a circle of radius R rolling on the axis. The sectional curvatures of the cylinder are zero parallel to the axis and $1/R$ perpendicular to the axis.

If a cylinder of liquid gets too long, it develops an instability known as the Rayleigh-Plateau instability (more commonly, just Rayleigh instability). If the ends are fixed, then the critical

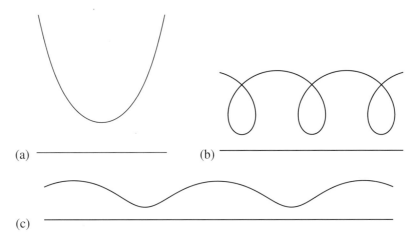

Figure 2.6 Roulette curves: (a) The catenary. (b) A nodary. (c) An undulary. The straight horizontal lines are the axes the conic rolls on.

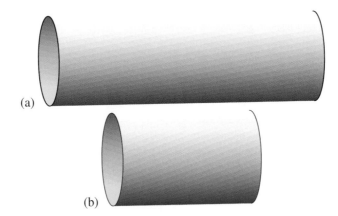

Figure 2.7 Cylinder of radius R. (a) Critical length $2\pi R$ for fixed ends. (b) Critical length πR for free ends on planes with 90 degree contact angles.

length is equal to the circumference, and the mode of instability is a neck forming near one end and a bulge forming near the other, as shown in fig. 2.8(a). Once the instability starts, the neck will continue to shrink until it pinches off. It does not reach a stable unduloid shape.

If the cylinder ends are free to move on parallel planes with contact angle 90°, then the critical length is half the circumference, with the unstable mode having a neck at one end and a bulge at the other, as in figure 2.8b.

The other rouloid surfaces described here have the same general stability pattern, complicated a bit by the non-uniformity along the axis.

2.3.2.3   The Catenoid

The catenary curve (fig. 2.6a) is famous as the curve made by a hanging chain, and it has the simple formula $y = \cosh x$. It is the roulette formed by the focus of a parabola rolling on the axis. The corresponding surface of revolution, the catenoid, is notable for having zero mean

(a)   (b)

Figure 2.8 Rayleigh instability modes. (a) Fixed ends. (b) Free ends.

curvature. It is the shape made by a soap film between two parallel rings.

Figure 2.9 Catenoid.

2.3.2.4   Unduloids

The undulary curve (fig. 2.6c) is formed by the focus of an ellipse rolling on the axis. An undulary can be expressed in parametric form using elliptic integrals [11]. The corresponding surface of revolution is called an unduloid. For fixed ends, the critical length for stability is one period, as shown in figure 2.10b. For free ends, the critical length is about half that; Vogel [6] has shown the critical length corresponds to the points of inflection in the undulary curve.

2.3.2.5   Nodoids

A nodary curve (fig. 2.6b) is generated by a focus of a hyperbola rolling on the axis. The rolling actually involves both branches of the hyperbola alternately being tangent to the axis, with the point of contact moving from one infinity to the other as the focus moves a finite distance. A nodary intersects itself. A nodary curve can be expressed in parametric form using elliptic integrals [11].

The corresponding surface of revolution is the nodoid, as shown in figure 2.11(a). Due to the self-intersections, arbitrary length nodiods do not make sense as liquid surfaces. Two situations arise: the surface is an "outer" piece of nodoid (fig. 2.11b) or it is an "inner" piece of nodoid (fig. 2.11c). Note the interior of the inner piece corresponds to the exterior of the outer piece, so the inner piece has negative mean curvature. Note that the inner nodoid is the only rouloid that has negative mean curvature, hence contains a liquid with negative pressure.

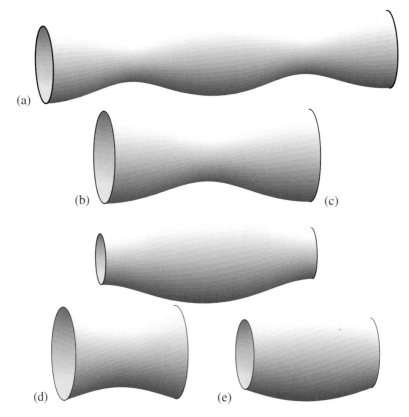

Figure 2.10 Unduloids. (a) Two periods of an unduluoid. (b),(c) Critically stable unduloids for fixed ends. (d),(e) Critically stable unduloids for free ends on planes with appropriate contact angles.

Nodiods are stable up to where the surface tangent goes perpendicular to the axis, both fixed and free ends. For an unstable outer nodoid with fixed ends, bulging back beyond the fixed ends, the mode of instability is to form an asymmetric bulge to one side, which becomes stable. An inner nodoid that goes beyond stability does a symmetric neck collapse.

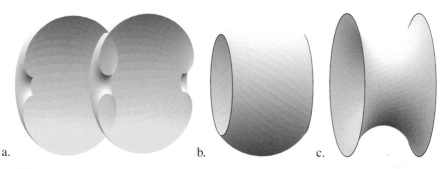

Figure 2.11 Nodoids. (a) Two periods of a nodoid, cross-section. (b) A segment of "outer" nodoid. (c) A segment of "inner" nodoid, enlarged.

2.3.3 Rouloid Summary

A rouloid may be conveniently specified by giving its radius at a point where the generator is parallel to the axis and the pressure. The following table lists the rouloids in pressure order for said radius being 1.

Roulade	Conic	Pressure	Major axis	Minor axis	Comments
Nodoid	hyperbola	$P > 2$	$a = 1/P$	$b = \sqrt{1 - 2/P}$	outer arc
Sphere	line segment	$P = 2$	$a = 1/2$	$b = 0$	
Unduloid	ellipse	$1 < P < 2$	$a = 1/P$	$b = \sqrt{2/P - 1}$	
Cylinder	circle	$P = 1$	$a = 1$	$b = 1$	
Unduloid	ellipse	$0 < P < 1$	$a = 1/P$	$b = \sqrt{2/P - 1}$	
Catenoid	parabola	$P = 0$	$a = \infty$	$b = \infty$	parabola is $y = x^2/4$
Nodoid	hyperbola	$P < 0$	$a = -1/P$	$b = \sqrt{1 - 2/P}$	inner arc

Table 2.1 Properties of rouloids, in pressure order, for radius 1 where the roulette curve tangent is parallel to the axis. The major and minor axis refer to the conic section that does the rolling.

2.4 Increasing Pressure and Stability

There is another technique of proving regions are absolutely energy minimizing that applies to a distinct class of surfaces from the Gromov proof. In the Gromov proof, the regions were spherical, and have the property that the pressure decreases as the volume increases. If one has a family of surfaces that behave in the opposite manner, having pressure increase as volume increases, then one can conclude they are absolutely energy minimizing surfaces. Also, in contrast to Gromov's proof, this technique can incorporate gravitational energy.

The theorem involves the notion of "foliation" of a region by surfaces. A family of surfaces foliates a region if every point of the region is on exactly one of the surfaces; think of the leaves of a book as foliating the volume of the book, or elevation contours foliating a topographic map. Note that gravity is easily incorporated in the theorem (the $G(X)$ can be taken to be the gravitational potential energy).

Theorem 2.3. *Let Ω be a domain of space with its boundary $\partial\Omega$ having designated contact surface tension $\gamma(\vec{x})$. Suppose $G(\vec{x})$ is a scalar potential energy density on Ω. Suppose that W_t is a nested family of regions of Ω such that Ω is foliated with surfaces ∂W_t such that each ∂W_t meets the contact angle condition at the triple line where it meets $\partial\Omega$. Further suppose there is a monotone increasing scalar function $h(t)$ such that the mean curvature of ∂W_t is $h(t) - G(\vec{x})$ on the free part of ∂W_t (i.e. $h(t)$ is the pressure at height zero). Then each W_t is the minimum energy region in Ω among all regions of the same volume.*

Proof. Let $\vec{w}$ be the exterior unit normal vectorfield at each point of each ∂W_t. Then we can deduce the energy of a region W_t is the integral of h over the interior:

$$E(W_t) = \int_{\partial W_t \cap \partial \Omega} \gamma dA + \int_{\partial W_t \backslash \partial \Omega} 1 \, dA + \int_{W_t} G \, dV \qquad (2.11)$$

$$= \int_{\partial W_t \cap \partial \Omega} \vec{w} \cdot \vec{N} \, dA + \int_{\partial W_t \backslash \partial \Omega} \vec{w} \cdot \vec{N} \, dA + \int_{W_t} G \, dV \qquad (2.12)$$

$$= \int_{\partial W_t} \vec{w} \cdot \vec{N} \, dA + \int_{W_t} G \, dV \qquad (2.13)$$

$$= \int_{W_t} div \, \vec{w} \, dV + \int_{W_t} G \, dV \qquad (2.14)$$

$$= \int_{W_t} h(t) \, dV. \qquad (2.15)$$

Since each point of Ω is on exactly one leaf ∂W_t, one may also view h as a function of $\vec{x} \in \Omega$, so one can refer to $h(\vec{x})$. Let it be noted that W_t occupies the exactly the region of $h(\vec{x}) \le h(t)$ and thus has the lowest integral of $h(\vec{x})$ over all regions of equal volume.

Now let V be a region of Ω with the same volume as some W_t. Then

$$E(V) = \int_{\partial V \cap \partial \Omega} \gamma dA + \int_{\partial V \backslash \partial \Omega} 1 \, dA \qquad (2.16)$$

$$\ge \int_{\partial V \cap \partial \Omega} \vec{w} \cdot \vec{N} \, dA + \int_{\partial V \backslash A} \vec{w} \cdot \vec{N} \, dA \qquad (2.17)$$

$$\ge \int_{\partial V} \vec{w} \cdot \vec{N} \, dA \qquad (2.18)$$

$$\ge \int_{V} div \, \vec{w} \, dV \qquad (2.19)$$

$$\ge \int_{V} h \, dV \qquad (2.20)$$

$$\ge \int_{W_t} h \, dV \qquad (2.21)$$

$$\ge E(W_t). \qquad (2.22)$$

$\square$

2.4.1   Wedge

As an application of Theorem 2.3, consider liquid in a wedge such that the sum of the wedge angle and the two contact angles is less than $180°$. Then there is a concave circular equilibrium (this could be a purely 2D example, or the cross-section of a 3D wedge with $90°$ contact-angle side planes). By geometric similarity, as the amount of liquid increases, the circle radius gets larger and the pressure increases (note the liquid is on the concave side of the circular arc, hence the pressure is proportional to the negative inverse of the radius). Thus Theorem 2.3 shows that this configuration is minimal energy, hence stable. In particular, in 3D such a liquid shape is stable for any length wedge.

2.4.2   In a Square

Consider a liquid in a square 2D box, where the contact angle is less than $45°$. Then the liquid can make concave regions in the corners. A foliation can be done using circular arcs

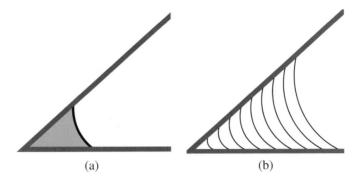

Figure 2.12 (a) Negative pressure liquid in a wedge. (b) Foliation of wedge by increasing pressure surfaces (i.e. less negative pressure).

of increasing radius in the corners, at least until the arcs meet in the middles of the edges. Thus Theorem 2.3 shows that having the liquid spread equally in the corners is the minimum energy configuration, as opposed to, say, unequal liquid in the corners. For larger fluid volumes, corners can join up. If the contact angle is 0°, then the foliation can be continued to where the corners join up and further into circles in the interior, showing there is a continuum of stable shapes. But if the contact angle is not 0°, then the foliation cannot be continued, and there is a jump in the stable shape when the liquid meets at the centers of edges. But as long as the fluid stays within the corner foliation, it is stable, even though there is a lower energy shape extending outside the foliation. The same kind of analysis can be done with a small amount of liquid in the corners and edges of a 3D box.

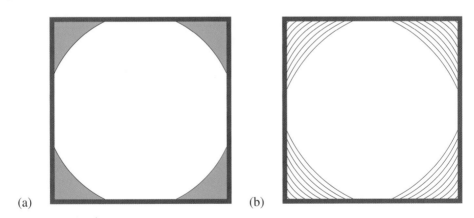

Figure 2.13 Liquid in corners of a square. (a) Since smaller corner drops have more negative pressure, the minimum energy is equalized corners. (b) A foliation of the neighborhood of the fluid.

2.4.3 Round Well

Consider liquid in the bottom of a round well round well (fig. 2.14). If the contact angles on the floor and wall add up to more than 90°, then it is possible for the liquid surface to intersect the floor/wall junction and make an isolated blob (fig. 2.14a). For a larger volume, the minimal energy configuration will be a ring around the bottom (much like fig. 2.14b), but note that if the volume of the ring gets too small, it will be subject to the double-bubble instability discussed below. If the sum of the floor and wall contact angles add to less than 90°, then the isolated blob of figure 2.14a is impossible, and a foliation argument shows that the uniform ring as in figure 2.14b is stable no matter how little liquid there is. As in the box example, if the contact angle is 0°, a single foliation can handle the transition to a fully covered bottom, otherwise there is a jump to the fully covered bottom, as in 2.14c. Note that once the bottom is covered, the foliation consists of congruent parallel surfaces until the contact line reaches the top of the well, but such equal-pressure surfaces are fine for Theorem 2.3. After the contact line reaches the top, there can be further spherical cap foliation surfaces pinned at the top of the well, with increasing pressure until the foliations reach a hemisphere. For volume beyond that, the foliation argument does not work.

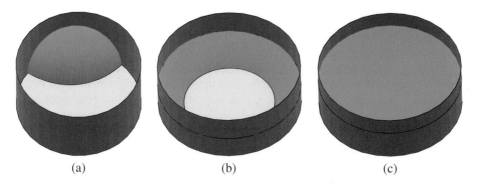

(a)   (b)   (c)

Figure 2.14 Flat-bottomed round well. (a) Drop on side of bottom, needs sum of contact angles greater than 90 degrees. (b) Ring of liquid around bottom of well, sum of contact angles less than 90 degrees, or larger volume. (c) Full bottom of well.

2.4.4 Square Well

Consider liquid in the bottom of a square well (fig. 2.15). If the sum of the floor and wall contact angles is greater than 90°, then the contact line can cross the wall/floor junction, and the liquid might cover 1, 2, 3, or 4 corners (or even be in separated pieces in various combinations). But if the sum of the contact angles is less than 90°, then only a concave version of figure 2.15.d is possible. As with the round well, a foliation argument can be made to show figure 2.15.d is minimal energy, along with various levels of symmetric filling.

2.5 The Double-Bubble Instability

The double-bubble instability is epitomized by two equal-diameter spherical soap bubbles joined by a tube (fig. 2.16). The configuration is in equilibrium, since the bubble radii, and hence pressures, are equal. But if there is the slightest disturbance so that one bubble becomes larger, then the pressure in the smaller bubble is higher, and air will be pumped from the smaller bubble to the larger, amplifying the difference. The transfer of air will stop only when the

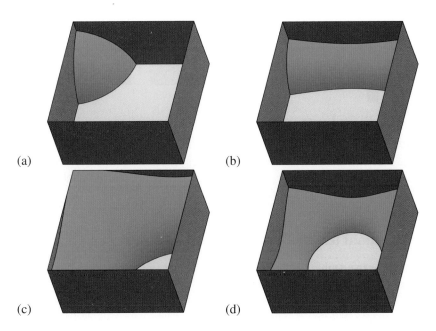

Figure 2.15 Liquid in the bottom of a square well (a) Occupying one corner. (b) Occupying two corners. (c) Occupying three corners. (d) Occupying four corners.

smaller bubble has collapsed to a small spherical cap on the tube end of equal radius to the large bubble.

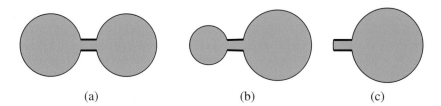

Figure 2.16 The double bubble instability. (a) Two equal bubbles connected by a pipe. (b) Collapse of the left bubble in progress, as it pumps air to the right bubble. (c) Final stable state, with both bubbles having the same curvature.

A "double-bubble pipe" (fig. 2.17) makes a nice demonstration of both the principles of increasing-pressure stability and the double-bubble instability. When the pipe is first drawn out of a soap solution, there are two flat, zero-pressure films across the pipe openings. As one blows air into the pipe, the two bubbles start inflating equally. This is because initially the volume increase leads to increasing pressure, so if one bubble is larger than the other, the larger bubble pumps air into the smaller bubble. But when the bubbles reach hemispheres, larger volume means lower pressure, so the double-bubble instability takes over, and as one blows more air in, one bubble shrinks and the other gets larger. For a double-bubble pipe, this transition happens smoothly, but for a triple-bubble pipe there is a sudden jump as the bubbles reach hemispheres, with two bubbles collapsing and one bubble enlarging.

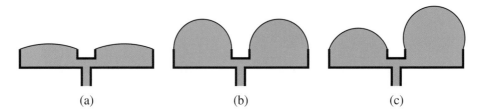

(a)   (b)   (c)

Figure 2.17 The double bubble pipe for blowing two soap bubbles simultaneously. (a) For low volume, there are two identical spherical cap bubbles. (b) Hemispherical bubbles are the maximum volume for symmetric bubbles. (c) Above the critical volume, the stable configuration is one small bubble and one large bubble.

2.5.1   Conditions for the Double-bubble Instability

The condition for the double-bubble instability to happen is for there to be two connected regions of liquid at the same pressure, but for each of which the pressure decreases as the volume increases. The regions do not have to be spherical bubbles, and they don't have to be the same shapes. Any shapes will do, as long as pressure decreases with volume. The argument still applies that any perturbation increasing the volume of one region at the expense of the other will lead to higher pressure in the smaller region, making it pump liquid to the larger region until pressures are equalized. The regions need not be separated by a rigid pipe; as examples below show, two parts of one liquid shape may function as separate regions and give rise to a double-bubble instability.

2.5.2   Plateau-Rayleigh Instability

The Plateau-Rayleigh instability (often called just the Rayleigh instability) says that a liquid cylinder with fixed ends becomes unstable to necking when its length reaches π times its diameter. This may be seen as an instance of the double-bubble instability. Consider two separate cylinders of liquid of the same radius. If they were connected by a small tube, obviously they would be subject to the double bubble instability. If one considered two distant segments of one cylinder, one could almost consider them separate regions, since the segments could change size with only a very gradual change in the radius between them, and hence only a small additional area due to a sloped surface. For a short cylinder, the extra area due to the slope between the regions overcomes the area savings due to the double-bubble instability, so short cylinders are stable. It so happens that the critical point where the double-bubble instability overcomes the slope area happens when the cylinder length is π times the radius, for fixed cylinder ends, say on perfectly wetting circular pads.

2.5.3   Plateau-Rayleigh Instability in Corners and Grooves

Liquid in a corner or groove between two planes can form an equilibrium convex cylindrical surface, if the sum of the corner angle and the contact angles on the walls exceed 180° (the Concus-Finn condition). Here it is assumed the contact lines on the walls are free to move. Just as for a complete cylinder, if the surface is long enough, a necking instability will develop. Evolver calculations show that the ratio of critical length to radius of curvature grows as the angle sum approaches 180°.

2.5.4   Instability of a Cylinder on a Hydrophilic Strip

Consider a perfectly wetting horizontal strip with liquid forming a cylindrical section above it, with contact lines pinned at the sides of the strip. As with an isolated cylinder, different parts of the liquid on the strip may be considered different regions. The analogy is to the double-bubble pipe, rather than the original double bubble. For small volumes of liquid, the top of the liquid is nearly flat, with small pressure. As volume increases, the pressure increases, and hence the cylindrical shape is stable for an arbitrary length strip. But when the liquid goes beyond a semicircular shape, the double-bubble instability becomes active and necking takes place. For a long enough strip, necking may start at several places, but the analogy is more like the triple-bubble pipe, and one large blob grows suddenly at the expense of other potential blobs.

2.5.5   Double-bubble Instability in Bulging Liquid Bridge

Consider a liquid bridge between two circular parallel pads, such that the pads are close enough that the liquid squeezes outside the ideal cylinder between the pads. As long as the bulge is small enough that it does not protrude below the level of the lower pad or above the level of the upper pad, then the Steiner symmetrization argument shows that the circularly symmetric shape (a section of unduloid, sphere, or nodiod) will be stable. But if the bulge does penetrate the pad levels, then it becomes susceptible to a double-bubble instability, with one side bulging out and the opposite side moving inward.

2.5.6   Lower-pressure Comparison Theorem

A version of the double-bubble argument can often be used to decide between two competitors for the lowest energy configuration without an explicit calculation of energies. In the corner-drop example above (§2.2.6), a simple way to see the one-plane bubble is lower energy than the corner bubble is to note the one-plane bubble clearly has a larger radius and thus a lower pressure. Imagining a thin pipeline between the two, the higher pressure in the corner would drive liquid to the one-plane drop, ultimately leaving the one-plane drop with all the liquid. This argument is not quite air-tight, but can be made so if the two shapes behave well as their size changes. In particular, if one can grow each shape from zero volume and zero energy such that for each volume the first shape has lower pressure than the second, then the first shape will have lower energy. This follows directly from the fact that pressure is the rate of change of energy with respect to volume.

2.6   Conclusion

This chapter has discussed various methods of detecting the stability of liquid shapes. To summarize:

- A spherical drop in a convex region minimizes energy when the sphere is located so it intersects all the walls at the proper contact angle.

- If the walls of a convex region are mobile, the liquid will move them to a shape where the drop is spherical.

- If the configuration is such that Steiner symmetrization can be done, then the minimum energy shape will be rotationally symmetric.

- If a region can be foliated with constant-pressure surfaces whose pressure increases with volume, then each foliation is the surface of a region with minimum energy.

- If a liquid has separate regions where the pressure decreases as volume increases, then the liquid may be subject to the double-bubble instability, in which one region increases at the expense of the other.

- If there are two competing shapes for minimum energy, the lower-pressure shape is likely to be best.

2.7   References

[1] H. Federer. *Geometric Measure Theory*. Springer-Verlag, New York, 1969.

[2] F. Morgan. *Geometric Measure Theory: A Beginner's Guide*. 4th ed. Academic Press, San Diego, 2008.

[3] M. Gromov, "Isoperimetric inequalities in Riemannian manifolds," Appendix I to *Asymptotic Theory of Finite Dimensional Normed Spaces* by V. D. Milman and G. Schechtman, Lecture Notes in Mathematics No. 1200, Springer-Verlag, New York, 1986.

[4] Jakob Steiner, presented to Berlin Academy of Science, 1836.

[5] C. Delaunay, "Sur la surface de revolution dont la courbure moyenne est constante," *Jour. de Mathématiques* **6**, pp. 309–320, 1841.

[6] T. Vogel, "Stability of a drop trapped between two parallel planes", *SIAM J. Appl. Math.* **47**(3), pp. 516–525, 1987.

[7] T. Vogel, "Stability of a drop trapped between parallel planes II: general contact angles," *SIAM J. Appl. Math.* **49**(4), pp. 1009-1028, 1989.

[8] T. Vogel and R. Finn, "On the volume infimum for liquid bridges," *Zeitschrift fuer Analysis und Ihre Anwendungen* **11**, pp. 3-23 (1992).

[9] T. Vogel, "Comments on radially symmetric liquid bridges with inflected profiles", *Discrete and Continuous Dynamical Systems*, Supplemental Volume, pp. 862-867, 2005.

[10] T. Vogel, "Convex, rotationally symmetric bridges between spheres", *Pac. J. Math.* **224**(2), pp. 367–377, 2006.

[11] M. Ćirić, "Notes on constant mean curvature surfaces and their graphical presentation," *Filomat* **23**(2), pp. 97–107, 2009.

3

Droplets: Shape, Surface and Volume

3.1 Abstract

Microfluidics is a science dedicated not only to the study of continuous microflows in microchannels but also to the study of micro-drops. As a matter of fact, many applications of microfluidics in biotechnology use droplets. This is the case in planar microfluidics or digital microfluidics, where discrete droplets are moved individually on a locally planar surface [1,2], and in droplet microfluidics, where droplets are produced in microdevices such as T-junctions or flow focusing devices (FFDs) to form capsules or emulsions [2-6]. On the other hand, individual droplets might be dispensed from a robot tip by electric means [7,8]. For biological purposes, droplets provide a small, bounded environment for the study of cells (fig. 3.1). The behavior of micro-droplets depends on their surface energy and volume. This is the reason why this chapter is devoted to the determination of the shape, surface area and volume of static droplets.

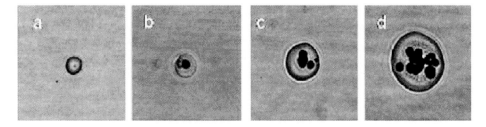

Figure 3.1 Cells contained in droplets (deposited on a substrate by means of ECC), reprinted with permission from [8], ©NSTI, 2003).

3.2   The Shape of Micro-drops

3.2.1   Sessile Droplets – the Bond Number

Large droplets deposited on horizontal surfaces have a flattened shape, whereas small droplets have a spherical shape (fig. 3.2). This observation stems from a balance between gravity and surface tension. A microscopic droplet is governed solely by surface tension, whereas the shape of a larger droplet results from a balance between gravity and surface tension. The scale length of this transition is the "capillary length". This length is defined by the ratio of the Laplace pressure – related to the size of a microscopic droplet – to the hydrostatic pressure [9]. If we compare the two pressures for a drop, we obtain

$$\frac{\Delta P_{Laplace}}{\Delta P_{Hydrostatic}} = \frac{\gamma/l}{\Delta\rho g l}, \tag{3.1}$$

where γ is the surface tension, l a characteristic length, $\Delta\rho$ the relative density of the liquid, i.e. the difference between the density of the droplet and that of the surrounding fluid, and g the gravitational constant. The two pressures are of the same order when

$$l = \sqrt{\frac{\gamma}{\Delta\rho g}}, \tag{3.2}$$

which is called the capillary length. A droplet of dimension smaller than the capillary length has a shape resembling that of a spherical cap. A droplet larger than the capillary length is flattened by gravity (fig. 3.2).

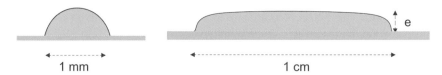

1 mm   1 cm

Figure 3.2 Comparison of the shape between micro-drops and macro-drops (not to scale): micro-drops have the shape of spherical caps whereas larger drops are flattened by the action of gravity and their height is related to the capillary length.

A dimensionless number – the Bond number – can be derived from (3.2) bearing a similar meaning. The Bond number is expressed by

$$Bo = \frac{\Delta\rho g R^2}{\gamma} \tag{3.3}$$

where R is a characteristic dimension (usually the droplet radius, or the droplet height). If $Bo \ll 1$, the droplet is spherical, else the gravitational force flattens the droplet on the solid surface. A numerical simulation of the two shapes of droplets obtained with the Surface Evolver is shown in figure 3.3.

The capillary length is of the order of 2 mm for most liquids, even for mercury. In the following sections we analyze successively the characteristics of drops having respectively large and small Bond number.

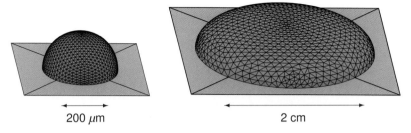

$\overleftrightarrow{\qquad}$ 200 μm $\overleftrightarrow{\qquad}$ 2 cm

Figure 3.3 Numerical simulations of a micro-drop ($Bo \ll 1$) and a larger drop ($Bo \gg 1$) obtained with Surface Evolver (not to scale).

3.3 Electric Bond Number

Electric methods are often used to manipulate droplets and their load. Methods like electro-hydrodynamic generation (EHD) [5] and electric charge concentration (ECC) [6] are used to deposit droplets on a solid substrate. When electric fields are present, there is a deformation of the droplet under the action of the electric forces, which is opposed by the surface tension force. An electric Bond number characterizes the importance of the deformation due to the electric field:

$$Bo_e = \frac{\varepsilon |E|^2 R}{\gamma}, \tag{3.4}$$

where R is a characteristic length scale of the droplet (approximately the radius, when the droplet is close to spherical), γ the surface tension, ε the electric permittivity (of the surrounding fluid), and E the applied electric field. It is recalled that the permittivity ε is the product of the relative permittivity ε_r by the vacuum permittivity ε_0,

$$\varepsilon = \varepsilon_r \varepsilon_0, \tag{3.5}$$

where $\varepsilon_0 = 8.854187817 \times 10^{-12}$ F/m. For a 500 μm water droplet, the gravity Bond number is of the order of 0.034; on the other hand, an electric Bond number of 0.034 is obtained with an electric field of 81.7 V/mm. A "floating" droplet is elongated in the direction of the electric field as soon as the electric field strength is high enough to overcome the interfacial surface tension, as shown in figure 3.4 [10,11].

Using the slender-body approximation, Sherwood [11] has found that the shape of the droplet was nearly spheroidal with sharper ends and related to the Bond number by

$$Bo_e = \left[\frac{2^{2/3} 15^{1/3}}{\pi^2} \right] \left[\frac{\varepsilon_{ext}}{\varepsilon_{int} - \varepsilon_{ext}} \right]^{1/2} \left[\frac{l/b}{\log(l/b)} \right], \tag{3.6}$$

where ε_{ext} and ε_{int} are respectively the surrounding fluid and droplet fluid permittivities. For very high electric fields, the droplet may break up into micro droplets. This observation constitutes the basis for electrohydrodynamic generation of droplets (EHD).

3.4 Shape, Surface Area and Volume of Sessile Droplets

In the following we focus on sessile droplets only subject to the Earth's gravity field (no electric field).

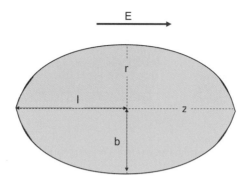

Figure 3.4 Shape of a water droplet in an electric field for a $Bo_e = 2.56$ [8].

3.4.1   Height of a "Large" Droplet: $Bo \gg 1$

According to the observation of the preceding section, a "large" droplet has a flat upper surface and its shape is shown in figure 3.5. In order to account for the density of the two liquids, we denote by $\Delta\rho$ the density difference between the droplet and the surrounding fluid.

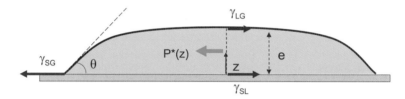

Figure 3.5 Equilibrium of the forces (per unit length) on a control volume of the drop.

Let us calculate the height of such a droplet as a function of contact angle and surface tension. Take the control volume shown in figure 3.5 and write the balance of the forces that act on this volume [12]. The surface tension contribution is

$$S = \gamma_{LG} - (\gamma_{SL} + \gamma_{LG}),\tag{3.7}$$

and the hydrostatic pressure contribution is

$$P^* = \int_0^e \Delta\rho(e-z)dz = \frac{1}{2}\Delta\rho g e^2.\tag{3.8}$$

The equilibrium condition yields

$$P^* + S = 0,\tag{3.9}$$

which results in the relation

$$\frac{1}{2}\Delta\rho g e^2 + \gamma_{LG} - (\gamma_{SL} + \gamma_{LG}) = 0.\tag{3.10}$$

Recall that Young's law imposes a relation between the surface tensions

$$\gamma_{SG} - \gamma_{SL} = \gamma_{LG}\cos\theta.\tag{3.11}$$

Upon substitution of (3.11) into (3.10), we obtain

$$\gamma_{LG}(1 - \cos\theta) = \frac{1}{2}\Delta\rho g e^2. \tag{3.12}$$

Using the trigonometric expression

$$(1 - \cos\theta) = 2\sin^2\frac{\theta}{2}, \tag{3.13}$$

we find

$$e = 2\sqrt{\frac{\gamma_{LG}}{\Delta\rho g}}\sin\frac{\theta}{2}. \tag{3.14}$$

Introducing the capillary length given by (3.2),

$$e = 2l\sin\frac{\theta}{2}. \tag{3.15}$$

Relation (3.15) shows that the height of a "large" droplet is proportional to the capillary length. The capillary length being of the order of 2 mm, the height of large droplets is less than 4 mm.

3.4.2   Microscopic Drops: $Bo \ll 1$

3.4.2.1   Shape of the Droplet

As was mentioned earlier, a micro-drop has the form of a spherical cap. A spherical cap is a surface of minimum energy if only surface tension is taken into account. This can be checked by using the Surface Evolver. The result is shown in figure 3.6 for a contact angle of 110°.

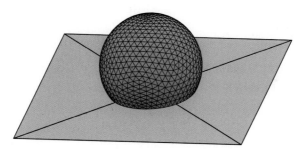

Figure 3.6 Shape of a micro-drop calculated with the Surface Evolver.

3.4.2.2   Volume of a Sessile Droplet

Figure 3.7 shows a cross section of the droplet, on non-wetting and wetting surfaces. Four parameters characterize the droplet: the contact radius a, i.e. the radius of the circular base, the sphere radius R, i.e. the curvature radius of the surface, the Young contact angle θ and the height of the droplet h.

The volume V of such a droplet is a function of two parameters in the set θ, a, R, h. First, we observe that the height of the droplet is expressed by the same relation in the wetting and non-wetting case. In the non-wetting case, the height h is given by

$$h = R + R\cos(\pi - \theta) = R(1 - \cos\theta), \tag{3.16}$$

and in the wetting case

$$h = R + R\cos\theta = R(1 - \cos\theta). \tag{3.17}$$

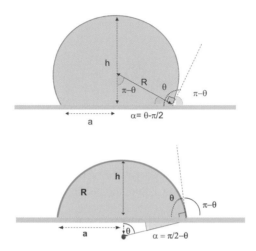

Figure 3.7 Cross section of a micro-drop sufficiently small to have a spherical shape; Top: Non-wetting droplet; bottom: wetting droplet.

The contact radius a – radius of the circular base – is in both cases

$$a = R\sin(\pi - \theta) = R\sin\theta. \tag{3.18}$$

Thus in the set $[\theta, a, R, h]$, h and a are functions of R and θ. Another useful relation can be obtained by combining (3.16) and (3.17):

$$h = a\frac{(1 - \cos\theta)}{\sin\theta} = a\tan\frac{\theta}{2}. \tag{3.19}$$

The volume of the spherical cap is calculated by the formula applied to the sketch of figure 3.8 [3,4]

$$V = \int_{R-h}^{R} \pi r^2 dz. \tag{3.20}$$

Integration of (3.20) leads to

$$V = \int_{R-h}^{R} \pi(R^2 - z^2)dz = \frac{\pi h^2}{3}(3R - h). \tag{3.21}$$

The droplet height h can be eliminated by substitution of (3.14) in (3.18); the expression for the volume as a function of R and θ is then

$$V(R, \theta) = \frac{\pi R^3}{3}(2 - 3\cos\theta + \cos^3\theta). \tag{3.22}$$

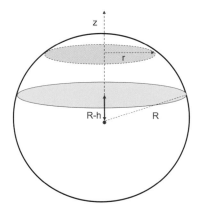

Figure 3.8 Schematic of the volume for the integration of relation (3.18).

Besides expression (3.18) and (3.19), other expressions can be derived for the volume of the droplet. The Pythagorean theorem yields

$$|h - R| = R^2 - a^2. \qquad (3.23)$$

Relation (3.24) produces the expression of R as a function of a and h:

$$R = \frac{a^2 + h^2}{2h}. \qquad (3.24)$$

Substitution of (3.25) in (3.22) yields

$$V(a,h) = \frac{\pi}{6} h(h^2 + 3a^2). \qquad (3.25)$$

If we make use of the Pythagorean theorem (3.20) and substitute in (3.23), we find the expression

$$V(a,R) = \frac{\pi}{6} \left(R \pm \sqrt{R^2 - a^2} \right) \left[3a^2 + \left(R \pm \sqrt{R^2 - a^2} \right)^2 \right]. \qquad (3.26)$$

Note that in (3.26) the $+$ sign corresponds to a non-wetting case (lyophobic) and the $-$ sign corresponds to the wetting case (lyophilic). For $\theta = 90°$ and $a = R$, (3.26) reduces to the half sphere volume $V = 2\pi R^3 / 3$. Using (3.14) and (3.15) the other expressions of the volume are

$$V(a,\theta) = \frac{\pi}{3} a^3 \frac{(2 - 3\cos\theta + \cos^3\theta)}{\sin^3\theta} \qquad (3.27)$$

and

$$V(h,\theta) = \frac{\pi}{3} h^3 \frac{(2 - 3\cos\theta + \cos^3\theta)}{1 - \cos^3\theta}. \qquad (3.28)$$

3.4.2.3   Surface Area of a Spherical Cap

Surface energy is proportional to the surface area. Hence formulas giving the surface area of spherical caps are useful. The spherical cap is a surface of revolution obtained by rotating a segment of a circle. More generally, any axisymmetric surface obtained by rotating the curve $y = f(x)$ about the x-axis has the following expression for area [13]

$$S = 2\pi \int f(x)\sqrt{1 + f'(x)^2}\, dx. \tag{3.29}$$

In our case, with the same notations as figure 3.8, the curve is defined by $r = f(z)$ and is rotated about the z axis, so that the preceding formula becomes

$$S = 2\pi \int r\sqrt{1 + r'^2}\, dz. \tag{3.30}$$

Upon integration between the two limits $R - h$ and R one obtains

$$S = 2\pi R h. \tag{3.31}$$

Other forms of the expression of the surface of the spherical cap are useful. Using (3.14), we find

$$S(h, \theta) = \frac{2\pi h^2}{1 - \cos\theta}. \tag{3.32}$$

Using (3.22) and the relations (3.14) and (3.15), we obtain

$$S(a, h) = \pi(a^2 + h^2). \tag{3.33}$$

After eliminating h in the preceding relation,

$$S(a, \theta) = \frac{2\pi a^2}{1 + \cos\theta}. \tag{3.34}$$

Using (3.20), the surface area can also be cast in the form

$$S(a, R) = 2\pi R \left(R \pm \sqrt{R^2 - a^2} \right). \tag{3.35}$$

where the $+$ sign corresponds to the lyophobic case and conversely the $-$ sign to the lyophilic case.

Finally, the surface energy of the spherical cap with surface tension γ is

$$E_{surf} = \gamma S, \tag{3.36}$$

where S is given by any of the preceding relations.

3.4.2.4   Example: Measurement of the Volume of a Droplet from a Single Image (Hydrophobic Case)

During the observation of a water droplet on a electrowetting substrate with the electrodes switched off, the droplet is non-wetting and two concentric circles appear in its microscope image from the bottom (fig. 3.9). A simulation with Evolver leads us to think that the large circle corresponds to the horizontal cross section of the droplet at its equator, and the smaller circle to the contact circle (fig. 3.10).

The radius of the spherical cap is

$$R = \frac{d_1}{2}, \tag{3.37}$$

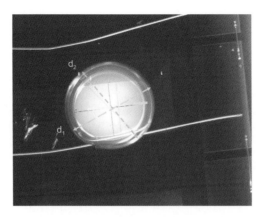

Figure 3.9 Microscope image of a sessile droplet on a hydrophobic surface: the larger diameter d_1 is the diameter of the sphere and d_2 is the diameter of the contact circle of the drop with the substrate (photo Ph. Clementz, CEA/LETI).

(a)   (b)

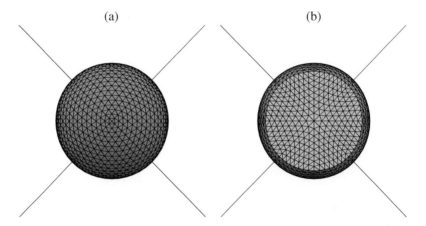

Figure 3.10 Surface Evolver numerical simulation of a sessile droplet: the two diameters are clearly seen on an image taken from below (right).

and the contact radius

$$a = \frac{d_2}{2}.$$
(3.38)

The information contained in these two relations is enough to determine completely the characteristics of the drop. Using the relation

$$h = R + \sqrt{R^2 - a^2},$$
(3.39)

we find the height of the drop and consequently the drop volume is given by

$$V = \frac{\pi}{48} \left(d_1 + \sqrt{d_1^2 - d_2^2} \right) \left[3d_2^2 + \left(d_1 + \sqrt{d_1^2 - d_2^2} \right)^2 \right].$$
(3.40)

Thus the contact angle is

$$\theta = \arcsin \frac{a}{R} = \arcsin \frac{d_2}{d_1}. \tag{3.41}$$

In conclusion, a single image (from below) of the drop produces all the characteristics of the droplet. The method can be quite accurate depending on the quality of the image. For example, an estimate of the diameters in the image of figure 3.9 is $d_1 = 1332 \ \mu m$, and $d_2 = 1146 \ \mu m$, leading to a drop volume of $V = 1.05 \ \mu l$ and a contact angle $\theta = 120.6°$. The expected values were $V = 1.0 \ \mu l$ and $\theta = 120°$.

Finally relation (3.31) produces the surface area

$$S = \frac{\pi}{4} \left[d_1 \pm \sqrt{d_1^2 - d_2^2} \right]. \tag{3.42}$$

3.4.3   Droplets Between Two Parallel Plates

It happens very often in biotechnology that digital microfluidic devices have a planar cover, and the droplets are constrained between two horizontal solid surfaces. Such droplets have a relatively smaller free energy than sessile droplets and are easier to handle. This is particularly the case for electrowetting. We consider here only the case of microsystems where the vertical gap δ is small (usually 50 to 500 μm); their Bond number is given by

$$Bo = \frac{\Delta \rho g \delta^2}{\gamma}. \tag{3.43}$$

Usually the value of the surface tension γ is in the range 5 mN/m to 72 mN/m, and $\Delta \rho$ is less or equal to 1000 kg/m^3. Hence, the Bond number is less than 0.5, and the free interfaces have approximately circular cross sections.

3.4.4   Shape of a Droplet Flattened Between Two Horizontal Planes

A droplet between two parallel plates has two contact surfaces with the solid surfaces, the first one with the bottom plate and another with the top plate; these contacts can be either wetting or non-wetting, depending on the material of the substrates. Figure 3.11 shows different possibilities and the corresponding shapes calculated using the Surface Evolver.

In all cases – due to a small Bond number and the flat upper and lower surfaces – the profile of the droplet in a vertical cross section is nearly a circular arc. Taking advantage of this observation, we can derive a formula for the calculation of the volume of the droplet.

3.4.5   Curvature Radius of the Free Interface

Here, "free interface" denotes the liquid/gas interface as opposed to the liquid/solid interface. Referring to the Laplace law, the pressure in a constrained droplet is deduced from the two curvature radii. The expression of the curvature radius in the vertical plane is derived in this section. This curvature radius is shown in figure 3.12. Let us assume a convex surface (case (a) of figure 3.12). Remark that in this case $\theta_1 + \theta_2 > \pi$. Using the geometric property that angles with perpendicular sides are equal, we can write

$$R \cos \theta_1 = H - \delta, \tag{3.44}$$

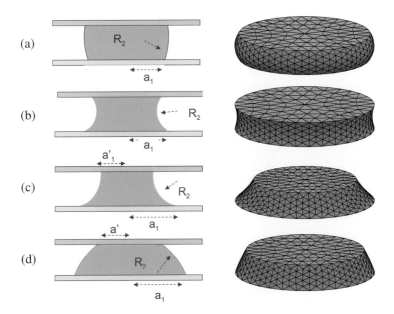

Figure 3.11 Sketch of the shape of a drop between two horizontal plates. (a) Non-wetting droplet – slightly hydrophobic wetting for a water drop. (b) Partial wetting – hydrophilic wetting for a water drop. (c) Drop not wetting the top plate and wetting the bottom plate, showing a concave interface (negative curvature radius), corresponding to the relation. (d) Same situation as (c), but the interface is convex because of the exact degrees of wetting.

where H is the distance between the curvature center and the upper plate and δ the gap between the plates; similarly

$$R\cos(\pi - \theta_2) = -R\cos\theta_2 = H. \tag{3.45}$$

We deduce the value of the curvature radius R,

$$R = -\frac{1}{\cos\theta_1 + \cos\theta_2}. \tag{3.46}$$

The curvature radius R is positive since $\cos\theta_1 + \cos\theta_2 < 0$. In the case of a concave interface corresponding to $\theta_1 < \pi/2$, $\theta_2 > \pi/2$ and $\theta_1 + \theta_2 < \pi$ (case (b) in figure 3.12), similar reasoning leads to the curvature radius

$$R = -\frac{1}{\cos\theta_1 + \cos\theta_2}, \tag{3.47}$$

which is a negative number. Using the notation $c = +1$ when the surface is convex, and $c = -1$ when it is concave, we obtain the general formula

$$R = c\left|\frac{1}{\cos\theta_1 + \cos\theta_2}\right|. \tag{3.48}$$

In the particular case where $\theta_1 + \theta_2 = \pi$ (case (c) in figure 3.12), the vertical profile of the interface is flat (the interface has a conical shape) and the vertical curvature radius is infinite.

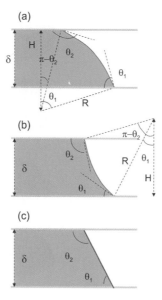

Figure 3.12 Schematic of the geometry of a droplet constrained between two parallel planes; (a) case of a convex interface; (b) case of a concave interface; (c) case of a flat interface.

3.4.6   Convex Droplet Shape

In this section, we consider a convex droplet.

3.4.6.1   Volume of a Droplet Flattened Between Two Parallel Planes and Having the Same Contact Angle with Both Planes (Convex Case)

The volume of a droplet between two parallel surfaces is useful to know, especially for EWOD. The calculation is complicated except in the case where the two contact angles are equal ($\theta_1 = \theta_2 = \theta$) [2]. We develop here a method to calculate the volume of the droplet sketched in figure 3.13. We will see that we obtain a relation of the type

$$V = f(a, \theta, \delta) = g(R, \theta, \delta), \tag{3.49}$$

where a is the contact radius, θ the contact angle, R the maximum horizontal radius, i.e., the horizontal curvature radius, and δ the vertical gap between the plates.

In general, the vertical gap δ is known and so is the contact angle θ. It is interesting to remark that the knowledge of the droplet volume V produces the value of the internal pressure: for a given volume V, the horizontal curvature radius Rh is obtained by inverting (3.38). On the other hand, the vertical curvature radius is given by (3.36). Then the internal pressure is given by

$$\Delta P = \gamma \left(\frac{1}{R_{horizontal}} + \frac{1}{R_{vertical}} \right). \tag{3.50}$$

Note that, in this case where the contact angle θ is larger than $\pi/2$, the Laplace pressure is positive because the two curvatures are positive. If the top surface is a solid plate, it "floats" on

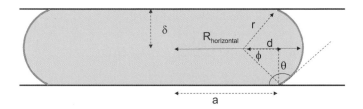

Figure 3.13 Cross section of a droplet constrained by two parallel plates and having the same contact angle with both planes.

top of the droplet [14]. The calculation of the droplet volume is somewhat lengthy. First, let us introduce the angle ϕ defined by $\phi = \theta - \pi/2$.

The curvature radius r of the interface is given by

$$\cos \theta = -\sin \phi = -\frac{\delta}{R}. \tag{3.51}$$

On the other hand, we have

$$\sin \theta = \cos \phi = \frac{d}{r}, \tag{3.52}$$

and we obtain a relation between d and δ:

$$d = -\delta \tan \theta. \tag{3.53}$$

The contact radius a is then

$$a = R - r + d = R + \delta \frac{1 - \cos \theta}{\sin \theta}. \tag{3.54}$$

It can be verified that if $\theta = 90°$ the radius r is infinite and $a = R$. At first glance, the relation (3.54) is indeterminate, but using a Taylor development, we find the limit

$$\lim_{\theta \to \frac{\pi}{2}} \frac{1 - \cos \theta}{\sin \theta} = \frac{-(\frac{\pi}{2} - \theta)^2}{2!(\frac{\pi}{2} - \theta)} = 0, \tag{3.55}$$

and the couple $(\theta, a) = (90°, R)$ verify relation (3.54).

The droplet volume is obtained through the following integral:

$$V = 2 \int_0^\delta \pi \tilde{R}^2 \, dz, \tag{3.56}$$

where $\tilde{R}$ is the horizontal droplet radius at the vertical coordinate z. Using the equation for a circular segment, one finds that $\tilde{R}$ is given by

$$\tilde{R} = (R - r) + \sqrt{r^2 - z^2}, \tag{3.57}$$

and the volume is then

$$V = 2 \int_0^{\frac{\delta}{2}} \pi \left[(R - r) + \sqrt{r^2 - z^2} \right]^2 dz. \tag{3.58}$$

Integration of (3.58) is somewhat lengthy. We finally find

$$V = 2\pi \left\{ (R^2 - 2rR + 2r^2) - \frac{\delta^3}{3} + (R-r)r^2 \left[\theta - \frac{\pi}{2} + \frac{2\theta - \pi}{2} \right] \right\}. \tag{3.59}$$

We notice that, for a droplet constrained by two parallel planes, the droplet volume depends on three independent parameters R, δ and θ.

A similar relation can be found by replacing the equatorial radius R by the contact radius a

$$V = 2\pi a^2 \delta + 2\pi \left(\frac{\delta}{\cos \theta} \right)^2 [af_1(\theta) + \delta f_2(\theta)], \tag{3.60}$$

where

$$f_1(\theta) = \theta - \frac{\pi}{2} + \frac{\sin(2\theta - \pi)}{2} + 2\sin\theta\cos\theta, \tag{3.61}$$

and

$$f_2(\theta) = \tan\theta \left[\theta - \frac{\pi}{2} + \frac{\sin(2\theta - \pi)}{2} + 2\cos\theta \right] - \frac{1}{3}\cos^2\theta + (1 - \sin\theta)^2. \tag{3.62}$$

Remark: relation (3.62) requires the knowledge of the contact angle θ. When the angle θ is not well known – for example in the case of a thin film of another liquid between the droplet and the wall – a simplified formula has been proposed by Nie *et al.* for a hydrophobic (lyophobic) contact angle [15]:

$$V = \frac{\pi}{6}[D^3 - (D - 2\delta)^2(D + \delta)], \tag{3.63}$$

where $D = 2R$. This formula assumes a perfectly spherical shape as shown in figure 3.14. It is straightforwardly deduced from (3.18), noticing that in present case $h = R - d$. In the spherical formulation, two parameters only are sufficient to determine the volume V, but the Young contact angle is not respected. The discrepancy between the spherical and the exact formulas depends on the Young contact angle and on the vertical gap between the plates. The difference between the exact and approximate expressions is shown in figure 3.15 for $\theta = 150°$ and $\delta = 150\mu m$.

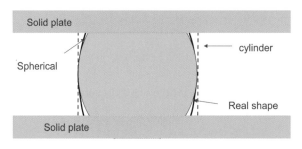

Figure 3.14 The different approaches to the calculation of the volume of a droplet between two parallel horizontal plates.

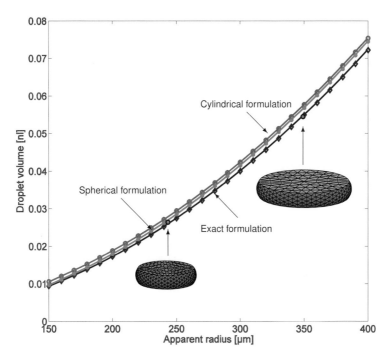

Figure 3.15 Comparison between cylindrical, spherical (Nie *et al.* [15]) and exact formulations for droplet volume; the two dots are the results of Evolver calculations. Contact angle $\theta = 130°$, half-height $\delta = 150\,\mu$m.

3.4.7   Volume of a Droplet Flattened by Two Horizontal Planes with Different Contact Angles with the Two Planes (Convex Case)

Let us consider the case of a convex surface: $\theta_1 < \pi/2$, $\theta_2 > \pi/2$ and $\theta_1 + \theta_2 < \pi/2$, as shown in figure 3.16. The volume V of the droplet now depends on four parameters a_1 (or a_2), δ, θ_1 and θ_2:

$$V = V(a_1, \delta, \theta_1, \theta_2). \tag{3.64}$$

In the case of a convex surface, the figure can be symmetries in order to obtain a symmetrical droplet; the droplet volume is then given by

$$V(a_1, \delta, \theta_1, \theta_2) = \frac{V_{sym}(R, 2\delta_2, \theta_2) - V_{sym}(R, 2\delta_1, \pi - \theta_1)}{2}, \tag{3.65}$$

where V_{sym} denotes the volume of a symmetrical droplet and is given by relation (3.59), or by the simplified relation (3.63). Note that if the surface is concave, an analogous symmetrization can be done in the opposite direction. We now have to express the parameters R, δ_1, δ_2 as functions of a, θ_1, θ_2 and δ. Using equation (3.37), we can write

$$\delta_1 = -\frac{\delta \cos \theta_1}{\cos \theta_1 + \cos \theta_2}, \tag{3.66}$$

$$\delta_2 = \frac{\delta \cos \theta_2}{\cos \theta_1 + \cos \theta_2}. \tag{3.67}$$

Using (3.41), and substituting (3.65), we find

$$R = a_2 - \delta_2 \frac{1 - \sin\theta_2}{\cos\theta_2} = a_2 - \delta\frac{1 - \sin\theta_2}{\cos\theta_1 + \cos\theta_2}. \tag{3.68}$$

Similarly,

$$R = a_1 - \delta_1 \frac{1 - \sin(\pi - \theta_1)}{\cos(\pi - \theta_1)} = a_1 - \delta\frac{1 - \sin\theta_1}{\cos\theta_1 + \cos\theta_2}. \tag{3.69}$$

So, starting from the knowledge of the vector $[a_1, \delta, \theta_1, \theta_2]$, the volume of the droplet is obtained by solving (3.49) and (3.50) or (3.51) to obtain $[R, \delta_1, \delta_2]$, then by substituting these values in (3.44) and finally by using (3.48).

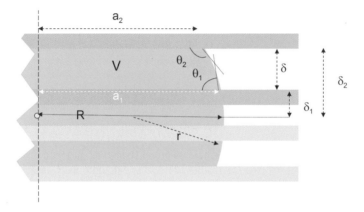

Figure 3.16 Sketch of the droplet: the volume V of the droplet is obtained by considering the volume of symmetrical droplet.

It can be tempting to use the Nie formulation in (3.63) and that would lead to the following relation for the volume

$$V = \frac{\pi}{3}R^3 \left[\left(1 - \frac{\delta_1}{R}\right)^2 \left(2 + \frac{\delta_2}{R}\right) - \left(1 - \frac{\delta_2}{R}\right)^2 \left(2 + \frac{\delta_2}{R}\right) \right]. \tag{3.70}$$

Because Nie's formula is approximated, the difference of two of such expressions might have a noticeable error as shown in figure 3.17. Hence, the knowledge of the contact angles is essential to determine the volume of an asymmetrical droplet and it is recommended to use the exact formula (3.44) or (3.45).

3.4.8   Surface Area (Convex Case)

Let us consider the case of a convex surface: $\theta_1 < \pi/2$, $\theta_2 > \pi/2$ and $\theta_1 + \theta_2 > \pi$, as shown in figure 3.18. Knowing the values of θ_1, θ_2, δ and a_1 (or a_2), δ_1, δ_2, and R are determined by (3.49) and (3.50), and the curvature radius r is given by

$$r = \frac{\delta_1}{\cos\theta_1} = -\frac{\delta_2}{\cos\theta_2}. \tag{3.71}$$

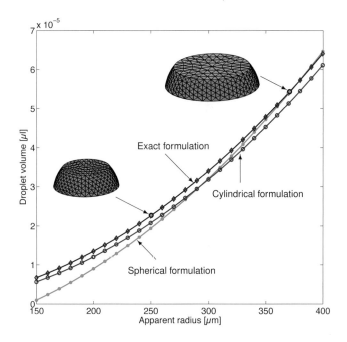

Figure 3.17 Volume of an asymmetric droplet between two parallel planes: comparison between the different expressions; the two dots are the results of the Evolver calculation. $\theta_1 = 80°$, $\theta_2 = 130°$, $\delta = 150\,\mu m$.

Using (3.28) and (3.41), the surface area is given by

$$S = 2\pi \int \tilde{R}\sqrt{1+\tilde{R}^2}\,dz, \qquad (3.72)$$

where $\tilde{R} = R - r + r\cos\alpha$. Using the relation $z = r\sin\alpha$, the differential term dz becomes $dz = r\cos\alpha\,d\alpha$. Substitution in 3.54 leads to

$$S = 2\pi \int_{\frac{\pi}{2}-\theta_1}^{\theta_2 - \frac{\pi}{2}} r(R-r)d\alpha + 2\pi \int_{\frac{\pi}{2}-\theta_1}^{\theta_2 - \frac{\pi}{2}} r^2 \cos\alpha\,d\alpha, \qquad (3.73)$$

whose solution is

$$S = 2\pi r[(R-r)(\theta_1 + \theta_2 - \pi) - r(\cos\theta_1 + \cos\theta_2)]. \qquad (3.74)$$

It is easily verified that expression (3.54) reduces to the surface of the half-sphere $S = 2\pi R^2$ by setting $R = r$, $\theta_1 = 0$, and $\theta_2 = -\pi$. If the vertical gap δ is sufficiently small, the surface area must be close to that of a cone [16]:

$$S_{cone} = \pi(a_1 + a_2)\sqrt{(a_1 - a_2)^2 + \delta^2}. \qquad (3.75)$$

A comparison between the two formulas (3.74) and (3.75) is shown in figure 3.19.

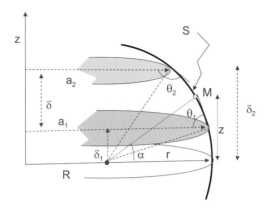

Figure 3.18 Sketch of the surface of a droplet between two parallel planes with arbitrary contact angles.

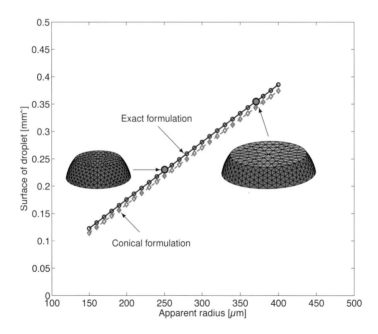

Figure 3.19 Comparison between exact expression and conical approximation for the surface area of the droplet. $\theta_1 = 80°$, $\theta_2 = 130°$, $\delta = 150\,\mu m$.

3.4.9 Concave Droplet Shape

In this section a "concave" droplet shape is considered. The same arguments as in the preceding section can be used to derive the expression of the volume and surface. We just indicate the results.

3.4.9.1   Volume of a Droplet Flattened Between Two Parallel Planes and Having the Same Contact Angle with Both Planes (Concave Case)

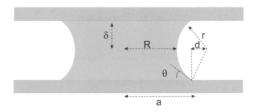

Figure 3.20 Sketch of a symmetrical concave droplet.

In such a case, the droplet is "symmetrical" (fig. 3.20). The vertical curvature radius is negative and its magnitude is

$$r = \frac{\delta}{\cos\theta}. \tag{3.76}$$

The Laplace pressure is then

$$\Delta P = \gamma\left(\frac{1}{R} - \frac{1}{r}\right). \tag{3.77}$$

Usually the horizontal dimension is large and the Laplace pressure is negative: there is a force that attracts the two plates. The contact radius is then given by the same expression as (3.41)

$$a = R + r - d = R + \delta\frac{1 - \sin\theta}{\cos\theta}. \tag{3.78}$$

Integration of expression (3.43) yields

$$V = 2\pi(R^2 + 2rR + 2r^2)\delta - \frac{\delta^3}{3} - (r+R)r^2\left[\theta - \frac{\pi}{2} + \frac{\sin(\pi - 2\theta)}{2}\right]. \tag{3.79}$$

3.4.9.2   Volume of a Droplet Flattened by Two Horizontal Planes with Different Contact Angles with the Two Planes (Concave Case)

We follow an approach similar to that of section 3.4.7 using the geometry shown in figure 3.21. In this case we have $\theta_1 > \pi/2$, $\theta_2 < \pi/2$ and $\theta_1 + \theta_2 < \pi$. The volume of the (concave) droplet is still given by the same expression as in the previous case:

$$V(a_1, \delta, \theta_1, \theta_2) = \frac{V_{sym}(R, 2\delta_2, \theta_2) - V_{sym}(R, 2\delta_1, \pi - \theta_1)}{2}, \tag{3.80}$$

with

$$\delta_1 = -r\cos\theta_1 = -\frac{\delta\cos\theta_1}{\cos\theta_1 + \cos\theta_2}, \tag{3.81}$$

$$\delta_2 = r\cos\theta_2 = \frac{\delta\cos\theta_2}{\cos\theta_1 + \cos\theta_2}, \tag{3.82}$$

and

$$R = a_1 - \delta\frac{1 - \sin\theta_1}{\cos\theta_1 + \cos\theta_2} = a_2 - \delta\frac{1 - \sin\theta_2}{\cos\theta_1 + \cos\theta_2}. \tag{3.83}$$

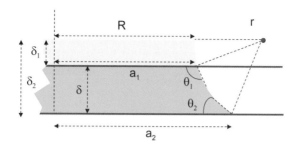

Figure 3.21 Sketch of an asymmetric droplet with a concave interface.

3.4.9.3   Surface Area (Concave Case)

The calculation is based on the relation

$$S = 2\pi \int \tilde{R}\sqrt{1+\tilde{R}^2}\, dz, \tag{3.84}$$

where $\tilde{R} = R + r - r\cos\alpha$, $\alpha \in [\theta_1 - \pi/2, \pi/2 - \theta_2]$. Hence

$$S = 2\pi[(R+r)(\pi - \theta_1 - \theta_2) - r(\cos\theta_1 + \cos\theta_2)]. \tag{3.85}$$

A comparison with the cone formula (3.75) is shown in figure 3.22.

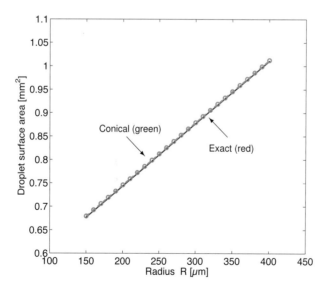

Figure 3.22 Comparison between exact formulation and conical expression for a concave droplet surface. $\theta_1 = 130°$, $\theta_2 = 40°$, $\delta = 150\,\mu m$.

3.5   Conclusion

Relations for curvature radius, volume and surface of sessile and flattened droplets have been derived in this chapter. The volume and surface expressions for convex and concave flattened droplets have been derived for the two following cases:

- Convex shape for $\theta_1 < \pi/2$, $\theta_2 > \pi/2$ and $\theta_1 + \theta_2 > \pi$.

- Concave shape for $\theta_1 > \pi/2$, $\theta_2 < \pi/2$ and $\theta_1 + \theta_2 < \pi$.

All other cases of figure 3.11 can easily be derived from the two cases treated here.

3.6   References

[1] H. Moon, S. K. Cho, R.L. Garrell, C-J Kim, "Low voltage electrowetting-on-dielectric," *Journal of Applied Physics* **92**(7), pp. 4080–4087, 2002.

[2] J. Berthier, *Microdrops and digital microfluidics.* William Andrew Publishing, 2008.

[3] F. Mugele, J-C Baret, "Electrowetting: from basics to applications," *J. Phys.: Condens. Matter* **17**, pp. R705–R774, 2005.

[4] M. Joanicot, A. Ajdari, "Droplet Control for Microfluidics," *Science* **5**, pp. 887-888, 2005.

[5] S.L. Anna, N. Bontoux, H.A. Stone, "Formation dispersions using "flow focusing" in microchannels," *Appl. Phys. Lett.* **82**(3), pp. 364–366, 2003.

[6] P. Garstecki, M.J. Fuerstman, H.A. Ston, G.M. Whitesides, "Formation of droplets and bubbles in a microfluidic T-junction: scaling and mechanism of break-up," *Lab Chip* **6**, pp. 437–446, 2006.

[7] Sung Jae Kim, Yong-Ak Song, P. L. Skipper, Jongyoon Han, "Electrohydrodynamic Generation and Delivery of Monodisperse Pico-Liter Droplets Using PDMS Microchip," *Anal. Chem.* **78**(23), pp. 8011–8019, 2006.

[8] B.S. Lee, J.-G. Lee, H.-J. Cho, N. Huh, C. Ko, "Electrohydrodynamic Micro-droplet Generation on Both Conducting and Non-conducting Surfaces by Electric Induction," *Proceedings of the NSTI-Nanotech 2009 Conference, Houston*, 3-7 May 2009, pp. 489-493.

[9] R. Clift, J.R. Grace, M.E. Weber. *Bubbles Drops and Particles*, Academic Press: New York, 1979, p. 26.

[10] N.Dodgson, C. Sozou, "The deformation of a liquid drop by an electric field," *Journal of Applied Mathematics and Physics* **38**, p. 425, May 1987.

[11] J.D. Sherwood, "The deformation of a fluid drop in an electric field: a slender-body analysis," *J. Phys. A: Math. Gen.* **24** 4047-4053, 1991.

[12] de Gennes, P-G., F. Brochard-Wyart, D. Quèrè, *Drops, bubbles, pearls, waves,* Springer, New-York, 2004.

[13] J.G. Hocking, G.S. Young, *Topology,* New York: Dover, 1988.

[14] K. Suzuki, "Flow resistance of a liquid droplet confined between two hydrophobic surfaces," *Microsystem Technology* **11**, pp. 1107–1114, 2005,

[15] Z.H. Nie, M.S. Seo, S.Q. Xu, P.C. Lewis, M. Mok, E. Kumacheva, G.M. Whitesides, P. Garstecki, H.A. Stone, "Emulsification in a microfluidic flow-focusing device: effect of the viscosities of the liquids," *Microfluid. Nanofluid.,* **5**, pp. 585–594, 2008.

[16] http://www.vom.be/FR/Oppervlaktetools-18.php.

4

Sessile Droplets

4.1 Abstract

This chapter is dedicated to the analysis of the behavior of sessile droplets. A sessile droplet is a droplet sitting on a solid substrate (fig. 4.1).

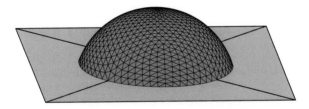

Figure 4.1 Simulation of a sessile droplet on a flat substrate.

The volume and surface area of a sessile droplet on a horizontal, uniform and smooth substrate have been derived in the preceding chapter. In this chapter we analyze more complicated situations where the sessile droplet is placed on an inclined, or an inhomogeneous, or rough substrate. Many different situations may occur: a droplet placed on an inclined plane, or at the boundary between a hydrophilic and hydrophobic substrate, or on a vertical step of the substrate. We derive the notions of pinning and canthotaxis, which may contribute to immobilize a droplet that would otherwise move. Finally, the case of a droplet on a chemically or morphologically inhomogeneous substrate is investigated, and the fundamental Wenzel and Cassie-Baxter laws are presented. These considerations lead to the notions of super-hydrophobicity and super-hydrophilicity.

4.2 Droplet Self-motion Under the Effect of a Contrast or Gradient of Wettability

Microsystem substrates are often composed of different materials which have different wettabilities. Capillary forces are then locally different. Droplets deposited on such inhomogeneous substrates experience unequal capillary forces on their triple line, leading to droplet motion towards the most wettable materials. In this section we focus on different spatial wettability contrasts, from a sharp transition on a flat surface, to a transition on a tilted plate and to a change over a step. Finally we investigate the case of a wettability gradient.

4.2.1 Drop Moving Over a Sharp Transition of Wettability

Consider a water droplet placed on a perfectly smooth horizontal plane at the boundary between two chemically different regions: hydrophilic on one side and hydrophobic on the other. This configuration is sketched in figure 4.2. Note that the precise shape of the triple line is not needed for the reasoning, since only the crossing length e plays a role, as was demonstrated in chapter 1.

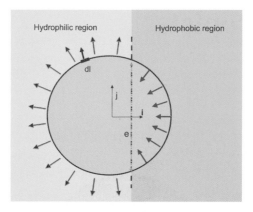

Figure 4.2 Schematic view of a water droplet standing above a hydrophilic/hydrophobic contact: the resulting force is directed towards the hydrophilic region.

From this scheme, it follows that the droplet moves towards the hydrophilic surface because the resultant of the capillary forces is directed towards the hydrophilic region. If L_1 and L_2 are the contact lines respectively in the hydrophilic and hydrophobic planes, and θ_1 and θ_2 the contact angles, the force acting on the drop is

$$F_x = \int_{L_1} \gamma_{LG} \cos\theta_1 (\vec{i} \cdot \vec{dl}) - \int_{L_2} \gamma_{LG} \cos\theta_2 (\vec{i} \cdot \vec{dl}) = -\gamma_{LG} e(\cos\theta_1 - \cos\theta_2), \qquad (4.1)$$

where $\vec{i}$ is the unit vector perpendicular to the triple line in the plane. The resultant F_x is directed towards the hydrophilic region because $\cos\theta_1$ is positive and $\cos\theta_2$ is negative (in fact a mere wettability contrast would also lead to the same conclusion). So the resulting force is directed towards the left in the scheme of figure 4.2, and the drop moves to the left, if the plate is sufficiently smooth to be considered frictionless. The motion stops when the resultant of the contact forces is zero, i.e., when the drop is entirely on the hydrophilic region, as shown in figure 4.3.

Experimental evidence confirms the preceding analysis. In figure 4.4, a water droplet is deposited with a micropipette on a flat horizontal surface, at the boundary of two regions with different contact angles. The droplet moves to the hydrophilic (or the more hydrophilic) region.

The same result is obtained by a calculation with the Evolver (figure 4.5). The starting point is a volume of water (of arbitrary shape) placed over a flat plane, with a 120° contact angle. After a few iterations, the spherical shape of the drop is obtained, and then the contact angle in a half-plane is switched to 70°, while the contact angle in the other half-plane remains

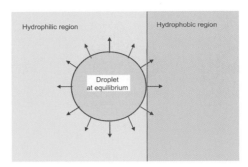

Figure 4.3 The water droplet is at equilibrium when it is entirely located on the hydrophilic region.

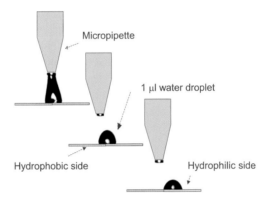

Figure 4.4 Experimental view of the relocation of a micro-drop (1 μl) deposited on a hydrophilic/hydrophobic boundary.

120°. The droplet is not at equilibrium because the resultant of the forces is directed towards the hydrophilic region. The drop evolves to find its equilibrium position, which is located just at the boundary of the transition line, on the hydrophilic side.

Note that the direction of the motion of a water drop is towards the hydrophilic region, whereas the motion of an oil (or organic liquid) droplet would be towards the hydrophobic region in the same geometrical conditions. Note also that the drop behavior would be similar if the transition between the two regions were not abrupt, but smooth, and if there were a wettability gradient between the two regions [1].

4.2.2   Drop Moving Uphill

In this section we show that capillary forces may be sufficient to make a droplet go uphill.

Chaudhury and colleagues [2] have shown that a droplet can go up a slightly inclined plate along a wettability gradient (fig. 4.6). If the substrate surface has a spatial gradient in wettability – i.e. in surface free energy – drops could move uphill. This motion is the result of an imbalance in the forces due to surface tension acting on the liquid-solid contact line on the

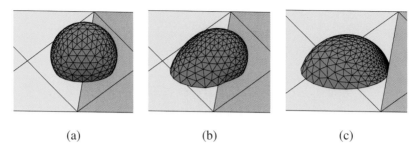

(a)   (b)   (c)

Figure 4.5 Motion of a drop towards the hydrophilic region (Surface Evolver): (a) the droplet is initially deposited over the hydrophilic/hydrophobic transition line (maintaining artificially a 120° contact angle on both regions); (b) after switching the contact angle to 70° on the hydrophilic region (left side), the drop is not in an equilibrium state; (c) it moves to find an equilibrium state on the hydrophilic plate, just at the boundary between the two regions.

two opposite sides ("uphill" or "downhill") of the drop. The required gradient in surface free energy was generated on the surface of a polished silicon wafer by exposing it to the diffusing front of a vapor of decyltrichlorosilane, Cl3Si(CH2)9CH3. The resulting surface displayed a gradient of hydrophobicity (with the contact angle of water changing from 97° to 25°) over a distance of 1 centimeter. When the wafer was tilted from the horizontal plane by 15°, with the hydrophobic end lower than the hydrophilic, and a drop of water (1 to 2 microliters) was placed at the hydrophobic end, the drop moved toward the hydrophilic end with an average velocity of approximately 1 to 2 millimeters per second.

On the other hand, it has been shown by Aaron Wheeler's group that a water droplet can go uphill under the action of electric forces [3]. It is shown in chapter 10 that the effect of electric forces is equivalent to a change of wettability. Hence, we obtain the same situation as that of Chaudhury.

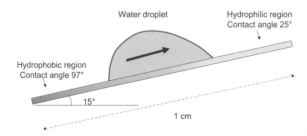

Figure 4.6 A water droplet, if sufficiently small, can run uphill towards the more hydrophilic region.

A similar motion can be obtained at a sudden wettability transition on a tilted plane. If a droplet is initially deposited over a wettability transition, it moves uphill from the lyophobic region (hydrophobic for a water droplet) towards the lyophilic region (hydrophilic for a water droplet). Figure 4.7 shows that the droplet is not stable when located on the transition line and moves uphill. The slope angle a micro-drop can move uphill depends on the size of the droplet (gravity acts in the opposite way) and the contact angles on both sides of the transition line.

An approximate calculation can be done to estimate the relative influence of the slope and

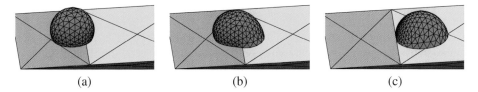

(a)   (b)   (c)

Figure 4.7 Droplet moving up a plane at a wettability transition (Surface Evolver): (a) the droplet is initially deposited over the hydrophilic/hydrophobic transition line (maintaining artificially a 120° contact angle on both regions); (b) after switching the contact angle to 80° on the hydrophilic region, the drop is not in an equilibrium state any more; (c) it moves up to find an equilibrium state on the hydrophilic plate, just at the boundary between the two regions.

that of the drop size. The capillary force F_C in the x-direction (direction of the maximum slope) is

$$F_C = e\gamma_{LG}(\cos\theta_a - \cos\theta_r) \tag{4.2}$$

where e is the approximate width of the droplet, θ_a and θ_r respectively the hydrophilic and hydrophobic contact angles. On the other hand, the gravitational force F_G is

$$F_G = -\rho g V \sin\alpha \approx -\rho g\left(\frac{2}{3}\pi e^3\right)\sin\alpha. \tag{4.3}$$

A comparison between these two forces is in figure 4.8. The gravitational force starts to balance the capillary force for a droplet of 1 to 2 mm radius. But very small drops can easily move uphill by capillarity.

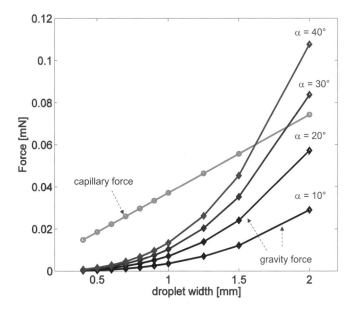

Figure 4.8 Capillary force and gravitational force as a function of the droplet size for four different values of the plane angle ($\theta_a = 80°$ and $\theta_r = 110°$).

4.2.3 Dynamic and Quasi-static Approach

At this stage, we stress the difference between the quasi-static and the dynamic approach: in the quasi-static approach presented above, only equilibrium or non-equilibrium status is determined. Of course, it indicates whether the droplet will move or not. But the real dynamic motion is more complex. Take the case of a droplet on an inclined plate with a uniform contact angle. The droplet is not at equilibrium because there is the resultant of the weight parallel to the plate, as shown in figure 4.9. However, depending on the droplet velocity, the dynamic and static shapes can be substantially different: first, in the dynamic motion, advancing and receding contact angles replace the static Young angle; second, the shape of the contact line can be very different [4]. If we denote by Ca the capillary number of the droplet,

$$Ca = \frac{\eta U}{\gamma}, \tag{4.4}$$

where U is the droplet velocity, then the advancing and receding contact angles are given by the Cox-Voïnov law[4],

$$\theta_{a,r}^3 = \theta_s^3 \pm A\vec{U} \cdot \vec{n}, \tag{4.5}$$

where θ_s denotes the static (Young) contact angle, $\vec{n}$ the unit vector normal to the triple line, and A is a constant characteristic of the fluid. The plus sign corresponds to the advancing angle and conversely the minus sign corresponds to the receding angle. We then have the relation $\theta_a > \theta_s > \theta_r$. When the droplet velocity decreases, the receding and advancing contact angles converge towards the static contact angle. On the other hand, above a critical capillary number Ca_C, the triple contact line has a singular shape at the back of the drop. The half-angle at the back of the drop is given by

$$\sin\phi = \min(\frac{Ca_C}{Ca}, 1). \tag{4.6}$$

Below the critical capillary number, there is no singularity of the profile.

4.2.4 Drop Moving Up a Step

Another interesting demonstration of the power of capillary forces at the microscale can be seen by making a drop move up a step. In such a case, a micro-drop of water is initially located on a step at the boundary of a hydrophilic region (on top of the step) and a hydrophobic region (at the base of the step). The calculation with the Evolver shows that, depending on the size of the droplet and on the height of the step, the drop may progressively move towards the hydrophilic region, even if this region is located at a higher level (fig 4.10). Again, in this example, capillary forces overcome gravity.

In biotechnology, wells, cusps and grooves are used to confine micro-drops. This confinement must be efficient. The preceding analysis has shown that attention should be given to the wettability of surfaces and that the combination of a non-wetting bottom of the well/groove and a wetting upper surface can result in poor confinement.

4.2.5 Drop Moving Over a Gradient of Surface Concentration of Surfactant

In the preceding sections, the capillary forces inducing droplet motion were linked to a local change of contact angle of the substrate. In this section, we show how chemical reactions

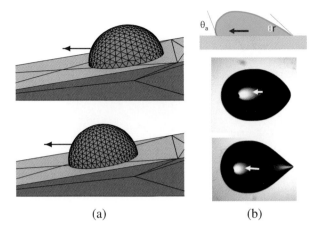

Figure 4.9 (a) Droplet sliding down an inclined plane (quasi-static Evolver approach); (b) the dynamic shape of the droplet with advancing and receding contact angles depends on the droplet velocity.

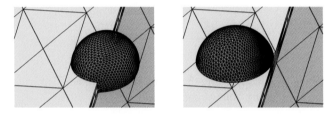

Figure 4.10 Motion of a drop up a step towards the hydrophilic plane (simulation made with Surface Evolver). Left: the droplet is located on the lower surface, with a part of it overlapping on the upper plane; the droplet is not in an equilibrium state, and moves up the step; right: motion continues until an equilibrium state is reached when the drop is entirely on the upper plate.

between the liquid of the droplet and the substrate can create droplet motion. To this extent the experiment of Dos Santos and colleagues [5] is characteristic. Suppose that a droplet of n-alkanes containing silane molecules is placed on a hydrophilic surface. It is well known that silane molecules form dense grafted monolayers on silicon or glass, rendering the surface hydrophobic. If we deposit such a droplet on a glass surface and initiate a motion by pushing it with a pipette, then the droplet continues to move on the substrate. It moves in nearly linear segments and changes its direction each time it encounters a hydrophobic barrier (figure 4.11). The droplet cannot cross its own tracks.

The motion is governed by an imbalance of capillary forces: the advancing contact line has a hydrophilic Young contact angle, whereas the receding line has a hydrophobic Young contact angle. After the initial push, the molecules of silane are concentrated at the vicinity of the receding contact line and stick to the substrate to form a hydrophobic path. The Young angle at the receding line is then a hydrophobic contact angle (fig. 4.12). As a result, the droplet moves in a straight line, except if it encounters surface defects – like an edge – or an existing hydrophobic trail.

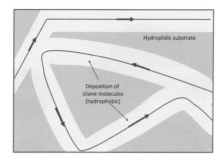

Figure 4.11 The Dos Santos experiment: tracks of a free running n-alkane droplet on a horizontal hydrophilic substrate; behind the droplet the deposition of silane molecules renders the substrate hydrophobic [5].

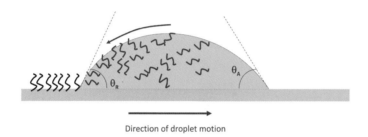

Direction of droplet motion

Figure 4.12 Advancing (θ_a) and receding (θ_r) contact angles are not identical. The hydrophobic monolayer changes the receding Young contact angle.

4.2.6   Conclusion

We have seen that capillary forces are often powerful enough to induce motion of micro-drops. However, we have considered perfectly smooth surfaces, and have neglected the effect of hysteresis – local change of contact angle – and its extreme form, which is droplet pinning – immobilization of a droplet due to defects of the surface. In reality, a drop does not move as soon as there is a gradient of wettability. It moves as soon as the gradient of wettability is sufficient for the capillary forces to dominate opposing forces, such as hysteresis and pinning. In the following section, we focus on the forces preventing droplet motion linked to the geometrical or chemical state of the surface.

4.3   Contact Angle Hysteresis

In the preceding section, the concept of advancing and receding contact angle has been introduced. In this section, we develop the notion of dynamic contact angle and that of contact angle hysteresis.

The concept of hysteresis for triple contact lines is difficult to define with accuracy. Indeed, it is not clear how to set the boundaries for such a concept. Let us start with a droplet sitting on a perfectly horizontal, clean, smooth plane. In such a case, as we have seen in the chapter 1, the static contact angle of the liquid is defined by Young's law

$$\cos\theta = \frac{(\gamma_{SG} - \gamma_{SL})}{\gamma_{LG}}, \tag{4.7}$$

where γ_{SG}, γ_{SL} and γ_{LG} are respectively the surface tension between the substrate S and the gas G, between the substrate S and the liquid L, and between liquid L and gas G.

However, no surface is ideally perfect; microscopic defects – morphological as well as chemical – are usually present. The concept of hysteresis is associated to local defects [6]. These local defects can be isolated point defects, surface defects, or line defects. A point defect can be an isolated roughness due to adherence of a micro or nanoparticle; a surface defect can be a stain of a chemical product; a line defect can be a long chain of polymer adhering to the surface, or the boundary between a hydrophobic and hydrophilic region.

Suppose we have a plane surface with dispersed local point defects. Let us deposit a droplet of water on the surface with a pipette in a first phase, and in a second phase, remove the water with the pipette (figure 4.13).

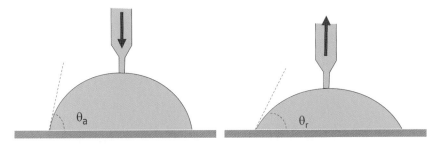

Figure 4.13 Contact angle is not the same when the droplet grows or recedes, due to hysteresis of the triple line.

We observe that the contact angle is not the same during the two phases. If θ_a denotes the contact angle during the first – advancing – phase and θ_r the contact angle during the second – receding – phase, we have

$$\theta_a > \theta_s > \theta_r. \tag{4.8}$$

Intuitively, the advancing front has a larger contact angle because it is locally "slowed down" by nano-inhomogeneities on the surface, whereas the receding front has a smaller contact angle because it is locally "pulled back" by these same inhomogeneities. Note that the same type of behavior is observed during the electrowetting process, with a similar method. More details will be given in Chapter 10.

Values of the change of contact angle are very difficult to predict, because they depend on many parameters. The Cox-Voinov law (4.5) is often used to describe the change of contact angle with the velocity of the liquid.

Cubaud and Fermigier [10] proposed a very instructive approach by showing that, in the presence of nano-defects at the solid surface, the contact line can be separated between a three-phase – far from the defect or heterogeneity – and a four-phase contact line at the contact of the defect. An image of local hysteresis at extremely low speed of motion is shown in figure 4.14.

The contact line Λ is the union of the set of the triple contact lines Λ_3 (index 3) and of the set of four-phase lines Λ_4 (index 4). This can be analytically described by

$$\Lambda = g(t)\Lambda_3 + (1 - g(t))\Lambda_4, \tag{4.9}$$

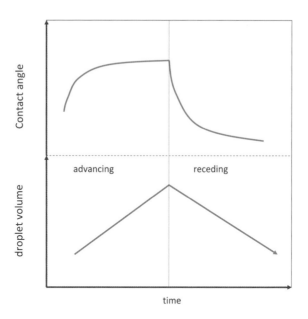

Figure 4.14 Typical time evolution of the contact angle during the growing and shrinking phase of a droplet on a flat substrate according to Wege e*et al.* [7], Lam et al. [8] and Tavana *et al.* [9].

where $g(t)$ is a function of time and surface geometry. For a quasi-static motion, with a small capillary number, the contact angle varies between the Young contact angle θ_3 of the triple line and the contact angle θ_4 corresponding to the defect. Figure 4.15 shows an enlarged view of the interface when it is stopped by surface defects. The deformation of the interface and the modified contact angle can be seen in the figure.

If we imagine now that there are many defects, we can easily understand that the global contact angle is an average between θ_3 and θ_4, resulting in a larger advancing contact angle because θ_4 is larger than θ_3.

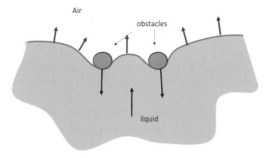

Figure 4.15 Sketch of a liquid stopped by pinning on obstacles. Far from the defect, the contact angle is equal to the Young contact angle θ_3; close to the defect the interface is twisted, and the contact angle varies between θ_3 and that of the defect θ_4 [10].

In the next section, we investigate droplet pinning, which is the extreme form of hysteresis where the droplet is totally blocked in its motion by defects on the solid surface.

4.4   Pinning and Canthotaxis

If the surface defects are sufficiently pronounced, the liquid may be stopped. This phenomenon is called pinning. We analyze successively the pinning on point, edge and line defects.

4.4.1   Droplet Pinning on a Surface Defect

Local chemical and/or geometrical defects locally modify the contact angle. If the defect is sufficiently important, or if there are sufficient numbers of defects, the droplet cannot move even if capillary – or electro-capillary – forces are applied on it. This phenomenon is called pinning. We show in figure 4.16 a numerical simulation of pinning of a droplet during its motion from a hydrophobic to a hydrophilic substrate.

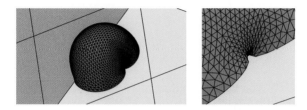

Figure 4.16 Left: Pinning of a drop moving from a hydrophobic area towards a hydrophilic surface due to a defect of the surface. Right: A close-up of the pinned spot.

Note that an important defect does not always pin the droplet. Depending on the relative location of the defect, the droplet may be pinned or just deviated in its motion, as is shown in figure 4.17.

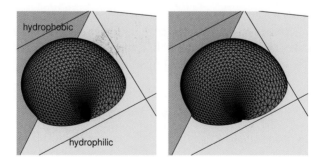

Figure 4.17 Depending on its location, a surface defect does not always prevent the water droplet from moving to the hydrophilic surface. The droplet can be pushed around the side of the defect.

4.4.2   Droplet Pinning on an Edge – Canthotaxis

Pinning – or attachment or anchoring – can also occur on an edge [11]. Consider a straight, perfect edge, and an interface coming to contact this edge (figure 4.18). The contact line on the angle stays fixed as long as the contact angle is not forced over the limit $\alpha + \theta$, where α is the angle between the two planes. The condition for pinning is then

$$\theta < \phi < \alpha + \theta, \tag{4.10}$$

where α is the interface angle, and θ the Young contact angle with the solid surfaces. In the case where the two planes have a different chemical surface, the Young contact angles can be denoted θ_1 and θ_2, and the condition (4.10) becomes

$$\theta_1 < \phi < \alpha + \theta_2. \tag{4.11}$$

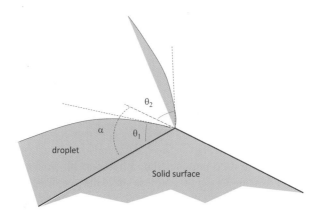

Figure 4.18 Droplet pinning on an edge: the droplet is pinned as long as the contact angle varies between the natural Young contact angle θ to the value $\theta + \alpha$. Above this value, the interface moves over the right plane and the droplet is released.

A simple reasoning based on the Young angle significance can explain canthotaxis. Suppose that the interface makes an angle ϕ with the wall, comprised in the limits of 4.10 (fig. 4.19). If the pinning is lost, the interface shifts a little from the edge. The resultant of the capillary forces along the plane is

$$F_x = \gamma_{SG} - \gamma_{SL} - \gamma_{LG} \cos \phi. \tag{4.12}$$

Using the Young law, the resultant F_x is

$$F_x = \gamma_{LG} (\cos \theta - \cos \phi). \tag{4.13}$$

When $\phi > \theta$, the resultant F_x is positive and pointing towards the edge, and the interface regains the edge. Conversely, when $\phi < \theta$, the resultant F_x is directed away from the edge and the pinning is lost. The same reasoning may be done on the other side of the edge, leading to the same conclusions (fig. 4.20).

4.4.3   Droplet Pinning at a Wettability Separation Line

Pinning may also occur at a transition line on a flat surface, between two regions with different chemical coatings, inducing a sharp transition of wettability [12]. When the contact line reaches

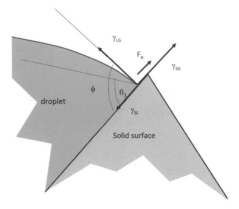

Figure 4.19 Sketch of the forces when the interface shifts from the edge: For $\phi > \theta_1$, the resultant F_x is directed towards the edge, restoring the pinning.

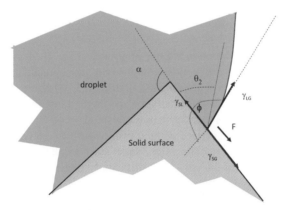

Figure 4.20 Sketch of the forces when the interface shifts to the other side of the edge: For $\phi < \alpha + \theta_2$, the resultant F_x is pointing towards the edge, restoring the pinning. For $\phi > \alpha + \theta_2$, the resultant F_x is pointing away from the edge and the pinning is broken.

the separation line, we have a four-phase contact line. The canthotaxis condition remains applicable if we set $\alpha = 0$. There is pinning as long as the contact angle is comprised between the Young angles on both sides, θ_1 and θ_2, as shown in figure 4.21 [13,14],

$$\theta_1 < \theta < \theta_2. \tag{4.14}$$

Equation 4.14 can be written in terms of surface energy,

$$\arccos\left(\frac{\gamma_{S_1 G} - \gamma_{S_1 L}}{\gamma_{LG}}\right) < \arccos\left(\frac{\gamma_{S_2 G} - \gamma_{S_2 L}}{\gamma_{LG}}\right), \tag{4.15}$$

where the indices S_1 and S_2 denote the left and right solid surfaces. If the external constraint is such that θ continues to increase, then the triple line is suddenly de-pinned and the liquid is released and invades the lyophobic surface.

Let us investigate in more detail the case of a sessile water droplet on a hydrophilic band located between two hydrophobic half-planes. The capillary forces acting on the part of the

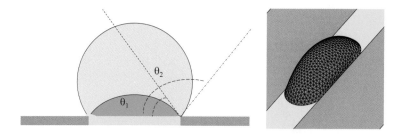

Figure 4.21 Quadruple contact line pinning on a wettability boundary. Left: vertical cross section of the droplet with the limiting contact angles θ_1 and θ_2. Right: an Evolver simulation of the droplet.

contact line located in the wetting region tend to stretch the droplet, while the capillary forces in the hydrophobic parts of the contact line tend to pinch the droplet. On the other hand, the surface tension force wants to keep the surface area of the droplet as small as possible. The question is: can a droplet be stretched by capillary forces to a point where it is completely resting on the wetting band?

This question has been answered in a series of articles [15-17], which have shown that four morphologies are possible, depending mostly on the lyophilic contact angle and the volume of the drop. These four morphologies are shown in figure 4.22. (I) If the liquid volume is small, the droplet has a spherical shape and is totally located on the lyophilic band. (II) If the volume is larger and the contact angle is smaller than a threshold value $\theta < \theta_{lim}$ the droplet spreads like a "band" along the lyophilic surface without overflowing on the lyophobic surface. (III) If $\theta > \theta_{lim}$ and the volume is moderate, then the droplet stays localized in a bulge state, i.e., does not spread. (IV)If $\theta > \theta_{lim}$ and the volume is sufficiently large, then the droplet spreads over the transition line onto the lyophobic surface.

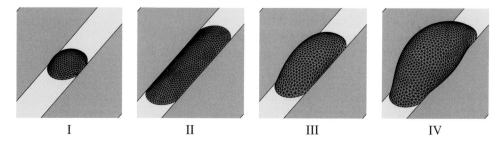

| I | II | III | IV |

Figure 4.22 Different shapes of a droplet on a lyophilic band: results of a numerical simulation.

The cross-sectional profiles of the droplet in the different types of morphologies are shown in figure 4.23. These profiles are circle arcs in all cases, with different curvatures. The curvature (and the internal pressure according to Laplace's law) is maximal in the bulge morphology III, where the ratio of the height of the drop to the base width is maximal.

In microfluidic systems, it is interesting to know if a volume of liquid deposited on a striped surface will spread on the lyophilic stripe or will remain in a bulge form. An investigation of the bifurcation between morphologies II and III has been done by Brinkmann *et al.* [17] and the

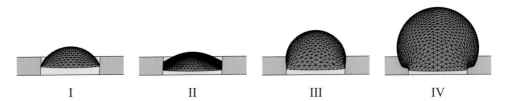

I   II   III   IV

Figure 4.23 Transverse shape of the droplet in the different morphologies: (I) the drop is a spherical cap with Young contact angle; (II) the drop has a circular transverse shape with radius of curvature about half that in the limit of case (I); (III) the droplet bulges over the hydrophobic band, but its contact surface lies within the hydrophilic band; (IV) curvature is reduced by liquid overflowing onto the lyophobic surfaces.

result is shown in figure 4.24. The two main parameters are the lyophilic contact angle θ and the dimensionless volume V/L^3, where L is the width of the band. It is easy to show that when V/L^3 is small, i.e. the contact line does not reach both edges of the strip, the morphology of the droplet is a spherical cap (number I in figure 4.23). For volumes in the next range there is a continuous smooth change from "channel" droplet to "bulge" droplet. Above the next critical volume, the transition between morphologies II and III is abrupt, akin to the Rayleigh-Plateau break-up of a long cylinder into droplets. It appears that there is a metastable region where the droplet can be in either one of the two morphologies II and III; this metastable region is comprised between the two curves denoted θ_{bu} and θ_{ch} in figure 4.24. It is particularly interesting to remark that the curve θ_{ch} is nearly vertical, meaning a large volume droplet deposited on the stripe totally spreads on the stripe (channel or band morphology) if the contact angle is smaller than the approximate value of 38°.

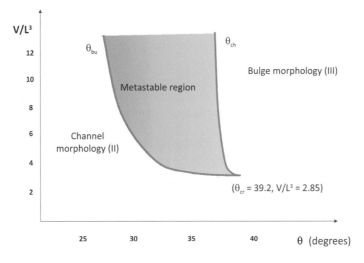

Figure 4.24 Bifurcation diagram for morphologies II and III.

Note that the value of 38° has a physical meaning: the interfacial free energy is

$$E = \gamma_{LG}S_{LG} - \gamma_{SG}S_{base} + \gamma_{SL}S_{base},$$ (4.16)

corresponding to the energy of the liquid-gas surface plus the energy of the base (where solid-gas contact has been replaced by solid-liquid contact). Using Young's law, we find

$$E = \gamma_{LG}(S_{LG} - \cos\theta S_{base}).$$ (4.17)

If we assume that the liquid in morphology II is a long cylinder of length L', and if we neglect the two ends, then we have the relations

$$S_{base} = LL',$$ (4.18)

$$S_{LG} = LL'\frac{\theta^*}{\sin\theta^*},$$ (4.19)

$$S_{cross-section} = \frac{L^2(\theta^* - \sin\theta^*\cos\theta^*)}{4\sin^2\theta^*},$$ (4.20)

where θ^* is the contact angle on the side of the cylindrical droplet. After substitution in (4.17) and introduction of the volume V, we deduce the expression of the surface energy as a function of the side contact angle θ^* and the Young contact angle θ,

$$E = \gamma_{LG}\left(\frac{4V}{L}\sin\theta^*\frac{\theta^* - \sin\theta^*\cos\theta}{\theta^* - \sin\theta^*\cos\theta^*}\right).$$ (4.21)

Stability of the interface is obtained by minimizing the surface energy relative to the angle θ^*, i.e., writing $\delta E/\delta\theta^* = 0$, and we find an implicit relation for θ^*:

$$\cos\theta = \frac{1}{2}\left(\frac{\theta^*}{\sin\theta^*} + \cos\theta^*\right).$$ (4.22)

The Laplace law indicates that the energy is always smaller than that obtained for the value of $\theta = 90°$, corresponding to a minimum curvature radius; and the maximum value of θ for having channel-type morphology is

$$\theta = \arccos\left(\frac{\pi}{4}\right) \approx 38°,$$ (4.23)

which is the value found by the numerical simulations. Note that 38° is quite a small value for a contact angle, and channel-type morphology requires very hydrophilic surfaces for water droplets.

So far we have been investigating the pinning along a straight line. An interesting investigation of the pinning along a sinusoidal line has been done by Ondarçuhu [12] using the theory of elasticity of the interface by de Gennes [6]. This approach is presented in Chapter 10 for the pinning of an interface on a jagged electrode boundary.

4.4.4 Pinning of an Interface by Pillars

Pinning may or may not be a desired feature in microsystems. Pinning can be a drawback when it wrongly stops a capillary flow, or it can be useful, for example when used to stabilize the interface between two immiscible liquids. Let us consider an example in which pillars are used to pin and maintain fixed an interface between two immiscible fluids [15,16] (fig. 4.25). This is the case of capillary valves, liquid-liquid extraction devices, etc.

Let us examine the case of triangular (or diamond shaped) pillars with hydrophobic surface. According to the Laplace theorem, the curvature radius of the interface is related to the pressure

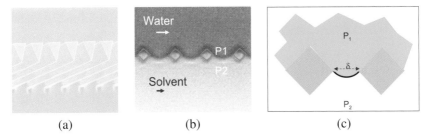

(a)   (b)   (c)

Figure 4.25 Sketch of an interface pinned by pillars. (a) Microfabricated pillars; (b) interface between two slowly flowing liquids stabilized by pillars, (c) blow up of the interface between two pillars. Depending on the pressure difference $P_1 - P_2$ the interface bulges more or less. If the pressure difference is too large, the interface breaks up. (photographs courtesy N. Sarrut, CEA-Leti).

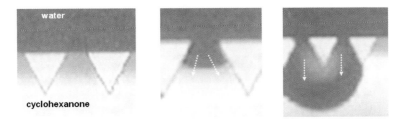

Figure 4.26 When the water pressure is increased, the interface is disrupted and water (dark, at the top) invades the solvent channel (photograph courtesy N. Sarrut, CEA-Leti).

difference across the interface. When the pressure on one side of the interface is increased, the curvature increases until the pinning limit is reached. Then the interface is disrupted and the high pressure liquid penetrates into the low pressure channel (fig. 4.26).

Take an interface pinned between the two facing edges of two similar micropillars, and suppose that the pressure P_1 in one liquid is progressively increased. Two conditions govern the pinning: the first condition is related to capillarity and to the phenomenon of canthotaxis [17], e.g. the pinning is effective if the condition

$$\theta < \theta_C \tag{4.24}$$

is met, and the interface does not slide on the facing pillar walls (fig. 4.27). In 4.24, θ_C is the (static) Young contact angle. Above this value, the interface slides irreversibly along the two facing walls of the pillars. The second condition corresponds to the minimum possible curvature of the interface. This curvature is obtained when the interface has the shape of a half-circle with a radius $\frac{\delta}{2}$ where δ is the distance between the two neighboring edges. In such a case $\theta = \alpha + 90°$. The second condition is then

$$\theta < \alpha + 90°. \tag{4.25}$$

Above this value, the interface cannot withstand the pressure difference and irreversibly slides along the walls. The general condition for pinning when the pressure P_1 is larger than P_2 is

$$\theta < \theta_{min} = min(\theta_C, \alpha + 90°). \tag{4.26}$$

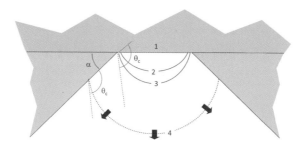

Figure 4.27 Under the effect of the pressure P_1, the interface bulges out. The interface stays pinned until $\theta > \theta_C$ (here $\theta_C < \alpha + 90°$). Position 1 corresponds to $P_1 = P_2$; position 3 is the canthotaxis limit, and position 4 shows the sliding of the interface.

The shape of the interface as a function of the pressure difference can be modeled with Evolver (fig. 4.28). In the case where 4.25 is the pinning condition, the interface bulges out until it takes a half-cylindrical shape, and then the calculation diverges because of the rupture of the interface.

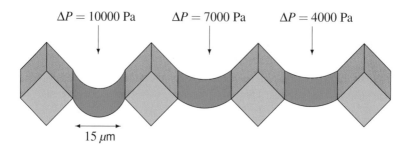

Figure 4.28 Evolver modeling of the bulging out of the interface as a function of the pressure difference (water/air interface).

4.5   Sessile Droplet on a Non-ideally Planar Surface

In biotechnological applications, it is always useful to have droplets aligned on a substrate. Alignment and ordering facilitate robotic operations. Hence, the question as to know if small reliefs or holes can be used to align the droplet is raised.

In order to answer such issues, this section deals with the following questions: is a sessile droplet placed on a small relief stable? Will it stay on the relief or move away from it? And conversely, is a sessile droplet placed in a V-shaped groove stable?

A reasoning based on the Laplace law furnishes an intuitive answer: the minimum of surface energy corresponds to a minimum pressure in the droplet, which in turn corresponds to the largest curvature radius. In the first case of a sessile droplet on a ridge, it is clear that the largest curvature radius is when the droplet stands aside from the ridge. In the second case of a droplet in a V-shaped groove, the largest curvature radius is when the droplet is at the bottom of the groove.

Modeling with Evolver confirms this reasoning (Figs. 4.29 and 4.30). Note that the energy difference between stable and unstable positions is small when the plane angle is close to 180° (this is the case of the figures). In such case, the motion occurs if the surfaces are perfect, else pinning may occur. In the following section we will deal in details with the notion of pinning.

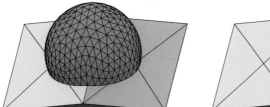

Figure 4.29 A droplet on a ridge (with zero gravity) is not stable, even if both sides have the same contact angle. It moves aside the ridge (case of a 48 μl water droplet and 90° contact angle with the substrate on both sides). Note that when the planes angle is close to 180°, the energy difference is small. The pressure in the left figure is 262.1 Pa whereas it is 259.2 Pa in the figure to the right. The approximate curvature radii are 275 and 278 μm.

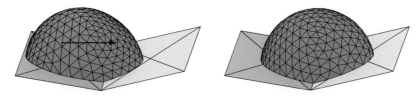

Figure 4.30 A droplet is stable when located symmetrically in a dip (case of a 43 μl water droplet and 90° contact angle with the substrate both sides, no gravity). Note that when the V angle is close to 180°, the energy difference is small. The pressure in the left figure is 241.5 Pa whereas it is 240.0 Pa in the stable position.

4.6   Droplet on Textured or Patterned Substrates

The reasoning used to establish Young's law in chapter 1 supposes a perfectly flat homogeneous surface. This is somewhat an abstraction, and surfaces – even when carefully microfabricated – have some roughness and may not be chemically homogeneous. The following sections deal with rough and chemically inhomogeneous surfaces. It will be shown that Young's law should be corrected to take into account the imperfections of the surface.

4.6.1   Wenzel's Law

It has been observed that roughness of the solid wall modifies the contact between the liquid and the solid. But the effect of roughness on the contact angle is not intuitive. It may seem a surprise – and also a very useful observation as we will see later on – that roughness amplifies the hydrophilic or hydrophobic character of the contact of an interface with a wall.

Suppose that θ^* is the contact angle with the rough surface and θ the angle with the smooth surface (in both cases, the solid, liquid, and gas are the same). One very important point here is that we make the implicit assumption that the size of the roughness is very small, so that the molecules of the liquid are macroscopically interacting with a plane surface but microscopically with a rough surface. This explains why we can define the unique angle of contact θ^*.

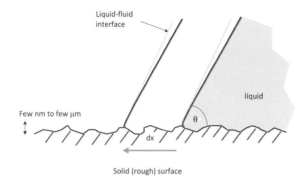

Figure 4.31 Sketch of the interface contact on a rough surface.

Suppose a very small displacement of the contact line (fig. 4.31). Then the work of the different forces acting on the contact line is given by

$$dW = \Sigma \vec{F} \cdot \vec{dl} = \Sigma F_x dx = (\gamma_{SL} - \gamma_{SG}) r \, dx + \gamma_{LG} \cos \theta^* dx, \qquad (4.27)$$

where r is the roughness, i.e. the ratio between the real surface area with the corresponding ideally smooth surface area, and $r\,dx$ is the real distance on the solid surface when the contact line is displaced by dx. Note that by definition, $r > 1$. Thus the change in energy is

$$dE = dW = (\gamma_{SL} - \gamma_{SG}) r \, dx + \gamma_{LG} \cos \theta^* dx. \qquad (4.28)$$

In fact, if we imagine that the drop finds its equilibrium state after the small perturbation dx, it finally stops at a position where its energy is minimum, so that

$$\frac{dE}{dx} = 0, \qquad (4.29)$$

and we obtain the relation

$$\gamma_{LG} \cos \theta^* = (\gamma_{SG} - \gamma_{SL}) r. \qquad (4.30)$$

If we recall that Young's law for a smooth surface is

$$\gamma_{SG} - \gamma_{SL} = \gamma_{LG} \cos \theta, \qquad (4.31)$$

then we obtain Wenzel's law

$$\cos \theta^* = r \cos \theta. \qquad (4.32)$$

Taking into account that $r > 1$, this relation implies that

$$|\cos \theta^*| > |\cos \theta|. \qquad (4.33)$$

We can deduce that if θ is larger than $90°$ (hydrophobic contact), then $\theta^* > \theta$ and the contact is still more hydrophobic due to the roughness. If θ is smaller than $90°$ (hydrophilic contact), then

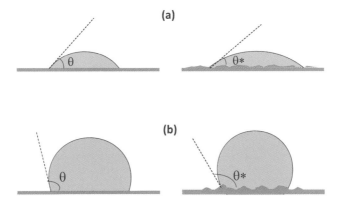

Figure 4.32 Contact of a liquid drop on a rough surface, comparison between smooth and rough surfaces: (a) hydrophilic (lyophilic) case, (b) hydrophobic(lyophobic) case.

$\theta^* < \theta$ and the contact is still more hydrophilic. In conclusion, surface roughness intensifies the wetting characteristic (fig. 4.32).

An important observation at this stage is that the scale of the roughness on the solid surface is small compared to that of the drop. Indeed, if not, it would not be possible to define a unique contact angle; for, example if the characteristic dimension of the asperities of the substrate is of the same order than the droplet size, the drop position would change depending where the drop is placed (fig. 4.33).

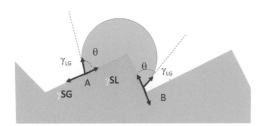

Figure 4.33 Large scale roughness: schematic view of a drop located on an angle of the solid surface. The drop might not be stable and moves to equilibrate the capillary forces.

4.6.2 Cassie-Baxter Law

The same analysis of Wenzel was done by Cassie and Baxter for chemically inhomogeneous solid surfaces. As for Wenzel's law, the same requirement of small size heterogeneities compared to interaction size between liquid and solid wall applies. For simplicity we analyze the case of a solid wall constituted by microscopic inclusions of two different materials. If θ_1 and θ_2 are the contact angles for each material at a macroscopic size, and f_1 and f_2 are the surface fractions of the two materials (fig. 4.34), then the energy to move the interface by dx is

$$dE = dW = (\gamma_{SL} - \gamma_{SG})_1 f_1 dx + (\gamma_{SL} - \gamma_{SG})_2 f_2 dx + \gamma_{LG} \cos \theta^* dx. \tag{4.34}$$

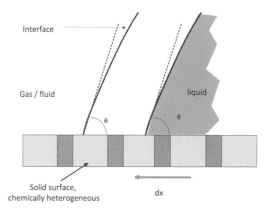

Figure 4.34 Displacement of the contact line of a drop on an inhomogeneous solid surface.

The equilibrium is obtained by taking the minimum of E

$$dE = dW = (\gamma_{SL} - \gamma_{SG})_1 f_1 dx + (\gamma_{SL} - \gamma_{SG})_2 f_2 dx + \gamma_{LG} \cos\theta^* dx = 0, \qquad (4.35)$$

and by comparison with Young's law, we obtain the Cassie-Baxter relation

$$\cos\theta^* = f_1 \cos\theta_1 + f_2 \cos\theta_2. \qquad (4.36)$$

This relation may be generalized to a more inhomogeneous material

$$\cos\theta^* = \Sigma_i f_i \cos\theta_i. \qquad (4.37)$$

Note that

$$f_1 + f_2 = 1 \qquad \text{or} \qquad \Sigma_i f_i = 1. \qquad (4.38)$$

The Cassie-Baxter relation shows that the cosine of the contact angle on a microscopically inhomogeneous solid surface is the barycenter of the cosine of the contact angles on the different chemical components of the surface.

The Cassie-Baxter law explains some unexpected experimental results. Sometimes – if not enough care was taken during micro-fabrication – a micro-fabricated surface may present chemical inhomogeneity and the wetting properties are not those that were intended. For example, if a uniform layer of Teflon is deposited on a substrate, the surface should become hydrophobic. However, if the layer is too thin, the Teflon layer may be porous and the coating inhomogeneous; the wetting properties are then modified according to the Cassie-Baxter law and the gain in hydrophobicity may not be as large as expected.

As for Wenzel's law, an important remark at this stage is that the scale of change of the different chemical materials of the solid surface is small compared to that of the drop. Indeed, if not, it would not be possible to define a unique contact angle any more. This latter type of inhomogeneity is related to drop pinning or to drop motion depending on the morphology of the inhomogeneity, as we have seen in the preceding sections.

4.6.3   Contact on Microfabricated Surfaces: the Transition Between the Wenzel and Cassie Laws

4.6.3.1   Introduction

We have seen in the preceding sections that the contact angle of a liquid on a solid surface depends on the roughness and the chemical homogeneity of the surface; this contact angle is given by the Wenzel or Cassie-Baxter laws. In particular, the Wenzel law states that asperities of the surface increase the hydrophobic or hydrophilic characteristics of a solid surface. Hence it could be interesting to artificially grow asperities on a substrate to increase its wetting properties in order to reach super-hydrophilicity or super-hydrophobicity. These two properties are approximately defined by the value of the contact angle. Super-hydrophilicity corresponds approximately to a contact angle

$$\theta^* < 5° \tag{4.39}$$

and super-hydrophobicity approximately to

$$\theta^* > 160°. \tag{4.40}$$

Note that, in biotechnology, the materials used are limited – PDMS, polystyrene, Teflon, SU8, glass, silicon, gold. Plastics (Teflon, polystyrene, SU8, PDMS) are generally hydrophobic or neutral, whereas glass and metals are hydrophilic and silicon neutral.

A way to make a solid surface hydrophobic or hydrophilic is to chemically deposit a hydrophobic or hydrophilic coating on top of the substrate. For example, Uelzen and Mueller [18] have developed a technique that consists of growing microscopic tin crystals on a silicon substrate – in order to increase the roughness (fig. 4.35) – followed by the deposition of a hydrophobic coating of TFE or a hydrophilic coating of gold by CVD (Chemical Vapor Deposition). The result is an enhanced hydrophobic or hydrophilic contact.

(a)   (b)

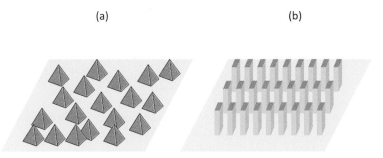

Figure 4.35 (a) A very hydrophobic surface obtained by growing tin crystals on a silicon substrate and covered with a hydrophilic or hydrophobic coating by CVD, after [18]; (b) micropillars are commonly used to increase the roughness of a substrate.

CFx (fluorocarbon) coatings are also often used in micro and nanotechnology to make a surface superhydrophobic. For example, a coating of CF4-H2-He achieved by successive plasma discharges increases considerably the contact angle with water [19]. The asperities of the surface are very small, of the order of 10 nm in this case.

Another way to increase the hydrophilic or hydrophobic character of a surface is to artificially create patterned microstructures in order to increase the roughness of the surface – by creating micro-pillars or grooves. Figure 4.35 shows an example of patterning a silicon surface with micro-pillars [20].

Using Wenzel's law, it is expected that the hydrophilic or hydrophobic character will be increased by the microfabricated roughnesses. So the question is: can Wenzel's formula, taking into account a roughness based on the shape of the micro-structures, be used to derive the contact angle? The answer is not straightforward. In the hydrophobic case, it has been observed that the droplet does not always contact the bottom plate and sometimes stays on top of the pillars – a phenomenon which is called the "fakir effect" (fig. 4.36). In such a case, should not the Cassie law, based on a juxtaposition of solid surface and air, be used? And what is the limit between a Wenzel droplet and a Cassie droplet? All these questions are discussed next.

Figure 4.36 Left: Young contact angle (slightly hydrophobic case); middle: Wenzel droplet touching the substrate textured with micropillars; right: Cassie droplet sitting on top of the pillars ("fakir effect"). θ_W is the Wenzel contact angle and θ_C is the Cassie contact angle.

4.6.3.2   Contact Angle on a Microfabricated Substrate. Case of Hydrophobic Contact

The contact angle of a sessile drop sitting on microfabricated pillars has been the subject of many investigations recently. As we have seen previously, Young's law defines the contact angle on the substrate material,

$$\cos\theta = \frac{\gamma_{SG} - \gamma_{SL}}{\gamma_{LG}}. \tag{4.41}$$

If the drop penetrates between the pillars, one can write the Wenzel angle as

$$\cos\theta_W = r\cos\theta, \tag{4.42}$$

where θ_W is the "Wenzel" contact angle and r the roughness of the surface. If the drop stays on top of the pillars, one can write the Cassie law in the form

$$\cos\theta_C = f\cos\theta + (1-f)\cos\theta_0, \tag{4.43}$$

where θ_C is the "Cassie" contact angle, θ_0 the contact angle with the layer of air, and f the ratio of the contact surface (top of the pillars) to the total horizontal surface. If the pillars are not too far from each other, the value of θ_0 is roughly $\theta_0 = 180°$ (see figure 4.37).

Then equation 4.43 becomes

$$\cos\theta_C = -1 + f(1 + \cos\theta). \tag{4.44}$$

It is usual to plot the two relations 4.42 and 4.44 in the diagram of figure 4.38 [21-23]. In such a representation, the two equations correspond to two straight lines, the first one with a slope r, and the second one with a slope f.

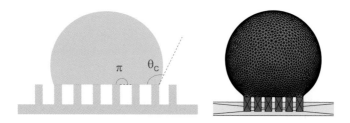

Figure 4.37 Left: Sketch of a Cassie drop (fakir effect). The interface between the pillars is roughly horizontal. Right: Evolver simulation showing a nearly flat interface between the pillars ($\theta_{Young} = 105°$, roughness $r = 3.55$, Cassie ratio $f = 0.44$).

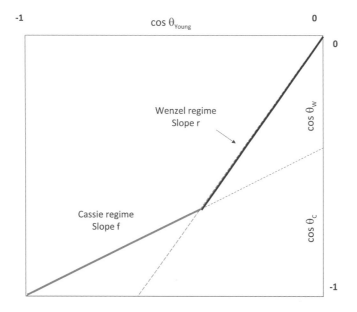

Figure 4.38 Plot of the Wenzel and Cassie laws for a (water) sessile droplet sitting on a hydrophobic surface textured with micro-pillars. Using the Laplace law, it can be seen that the energy of the drop is lower when the contact angle is the smallest. The physical situation then corresponds to the continuous lines, with a Wenzel drop at small Young contact angle and a Cassie drop for large Young contact angle.

The two lines intersect, because

$$r = \frac{S_{total}}{S_{horizontal}} > f = \frac{S_{top}}{S_{horizontal}}. \tag{4.45}$$

The two lines intersect at a Young contact angle θ_i defined by $\theta_C = \theta_W$, so that

$$\cos\theta_i = \frac{f-1}{r-f}. \tag{4.46}$$

In the diagram of figure 4.38, for a given Young angle, there are two contact angles. The question is now: which one is the real one? From energy considerations – for example by using Laplace's law – it can be deduced that the real contact angle is the smaller one, so when the Young contact angle is not very hydrophobic ($\theta < \theta_i$), the contact corresponds to a Wenzel regime and the drop wets the whole surface. When the Young contact angle is more hydrophobic ($\theta > \theta_i$), the drop is in a Cassie regime and sits on top of the pillars.

4.6.3.3   Example of Square, Hydrophobic Micropillars

Consider the case of a hydrophobic surface textured with square pillars. This problem has been investigated by Patankar [23] and Zhu *et al.* [24]. In the first hypothesis of a Wenzel contact, if the pillars have a square section $a \times a$, and a height H, with a spacing b (fig. 4.39), the roughness is determined by

$$r = \frac{S_{total}}{S_{horizontal}} = \frac{(a+b)^2 + 4aH}{(a+b)^2} = 1 + \frac{4A}{\left(\frac{a}{H}\right)}, \qquad (4.47)$$

where

$$A = \left(\frac{a}{a+b}\right)^2. \qquad (4.48)$$

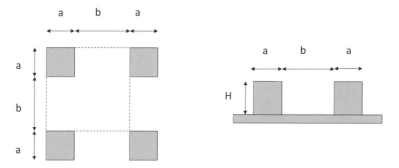

Figure 4.39 Top and cross-sectional schematic sketches of a substrate with square pillars.

On the other hand, the Cassie coefficient f is given by

$$f = \frac{S_{top}}{S_{horizontal}} = \frac{a^2}{(a+b)^2} = A. \qquad (4.49)$$

The Wenzel and Cassie formulas for square pillars disposed in a square "motif" are deduced from the expressions of r and f. The Wenzel contact angle is given by

$$\cos\theta_W = \left[1 + \frac{4A}{\left(\frac{a}{H}\right)}\right]\cos\theta, \qquad (4.50)$$

and, assuming that the interface is horizontal between the pillars, the Cassie contact angle is

$$\cos\theta_C = A\cos\theta + (1-A)\cos(180°) = -1 + A(1+\cos\theta), \qquad (4.51)$$

where the indices W and C respectively stand for the Wenzel and Cassie laws.

The Young contact angle θ_i for which $\theta_C = \theta_C$ is given by

$$\cos\theta_i = \frac{-1}{1 + \frac{4aH}{b(2a+b)}}. \tag{4.52}$$

Let us estimate the influence of the parameters a, H, and b in 4.52. If the height of the pillars decreases, i.e. $H \to 0$, using 4.47 we derive $r \to 1$, meaning that the roughness induced by the pillars disappears and, using 4.52 $\theta_i \to 180°$. Referring to figure 4.38, reducing the height of the pillars increases the transition angle between Cassie and Wenzel regimes and favors the Wenzel regime. Indeed, in this case, the bulging down of the interface between the pillars makes the liquid contact the flat plate supporting the pillars: this phenomenon can be clearly seen in the numerical simulation (fig. 4.40).

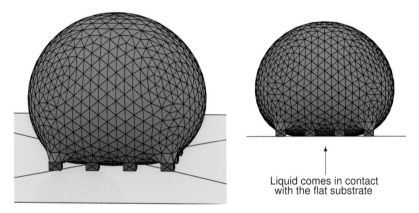

Liquid comes in contact
with the flat substrate

Figure 4.40 When H becomes small compared to the other dimensions a and b, the "fakir effect", i.e. the Cassie regime, is lost because the droplet comes into contact with the flat substrate. Contact angle 115°.

This property is similar to that observed with membranes pierced with very small holes: water does not penetrate in the holes, except if forced by pressure. Figure 4.41 shows a simulation done with Evolver of a sessile droplet on a flat plate with a hole.

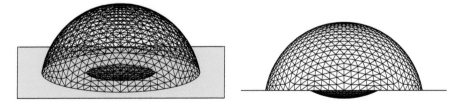

Figure 4.41 A sessile droplet sitting on a plate pierced by a hole does not bulge out much if the hole dimension is sufficiently small.

When the pillars are thin, $a \to 0$, and using 4.52, we find $\theta_i \to 180°$. The regime is then again the Wenzel regime. Thin pillars or small pillar heights are favorable to the Wenzel regime. In this case the liquid sinks between the pillars and contacts the flat substrate, as shown in figure 4.42.

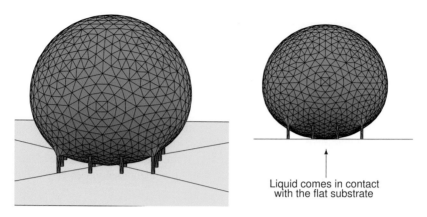

Figure 4.42 When *a* becomes small compared to the other dimensions *H* and *b*, the "fakir effect", i.e. the Cassie regime, is lost because the droplet comes into contact with the flat substrate. Contact angle 145°.

On the other hand, when the pillars are very close to one another, $b \to 0$, using 4.52 we find that $\theta_i \to 0$ and we observe a Cassie regime and the liquid does not contact the flat substrate, as shown in figure 4.43.

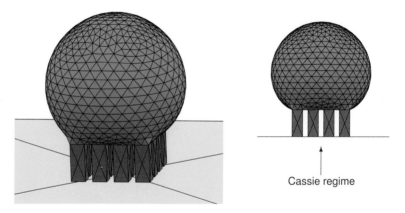

Figure 4.43 When *b* is small compared to the other dimensions *a* and *H*, the Cassie regime is likely to occur; here $b/a = 1/2$. Contact angle 155°.

4.6.3.4 Wenzel-Cassie Hysteresis

Note that the situation we have just described does not always correspond to reality. It happens that a droplet is not always in its lowest energy level and that droplets are sometimes in meta-stable regimes. One example was given by Bico *et al.* [21-22]. It is that of a drop deposited by a pipette on a pillared surface. Even if it should be in a Wenzel regime and the droplet should penetrate between the pillars, it stays on top of the pillars. It needs an impulse, mechanic, electric, or acoustic, to enter the expected Wenzel regime.

Transition from a Cassie droplet to a Wenzel droplet can be obtained by electrowetting actuation. Krupenkin *et al.* [25] have shown that a droplet sitting on pillars in a Cassie regime, i.e., sitting only on the top of the pillars, sinks down as soon as electrowetting actuation is turned on (fig. 4.44).

Figure 4.44 Sessile droplet sitting on micropillars: (a) with no electric actuation, the drop sits on top of the pillars; (b) electrode actuated with 22 V electric potential: the droplet wets the flat part of the substrate. The wire at the top is the zero potential electrode. Reprinted with permission from [25], ©EPL,1999.

4.6.3.5   Superhydrophobicity

A surface is said to be super-hydrophobic when the contact angle of aqueous liquid is larger than 160°. In nature, some tree leaves in wet regions of the globe have super-hydrophobic surfaces in order to force water droplets to roll off the leaves, preventing rotting of the leaves, as shown in figure 4.45, taken from Barthlot, *et al.* [26].

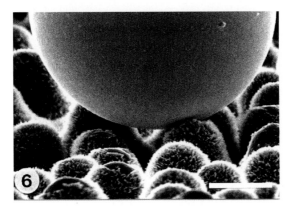

Figure 4.45 Mercury droplet on the papillose adaxial epidermal surface of Colocasia esculenta demonstrating the effect of roughness on wettability. Due to the decreased contact area between liquid and surface, air is enclosed between the droplet and the leaf, resulting in a particularly strong water-repellent surface. The bar is 20 μm. Reprinted with permission from [26], ©ACS, 2004.

Super-hydrophobicity requires either a super-hydrophobic Young contact angle ($\theta_{Young} > 160°$) or a Cassie regime ($\theta_C > 160°$). Ideally, super-hydrophobicity corresponds to $\cos\theta_{effective} = -1$ or $\theta_{effective} = 180°$. By looking at the diagram 4.38, one sees that super-rhydrophobicity cannot be obtained in the Wenzel regime (except if $r \approx 1$ and $\theta_{Young} \approx 180°$). Otherwise, it can only be achieved in a Cassie regime when the value of f is very small. Using 4.49, for a pillar lattice, f can be cast in the form

$$f = \frac{1}{\left(1 + \frac{b}{a}\right)^2}.$$

(4.53)

Hence f is small when $b/a \gg 1$; but if we want that the liquid to not touch the substrate, the pillar height H should be sufficient to obtain the "fakir effect". In such a case, the pillars are thin and the contact area of the droplet with the solid substrate is very small, it is only that of the top of the pillars. In other words, the best situation for super-hydrophobicity for a geometrically textured surface is f as small as possible and r as large as possible (figure 4.46). Using the approximation $b/a \gg 1$ in the expression of the roughness r, we find $r = 1 + \frac{4aH}{b^2}$. This latter relation shows that if H is sufficiently high, r will be large.

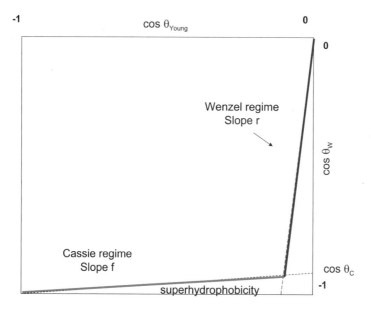

Figure 4.46 Super-hydrophobicity requires a Cassie/Wenzel diagram with a very small coefficient f and a large coefficient r.

4.6.3.6   Irregularly Textured and Fractal Surfaces

In the introduction of this section, we have seen that an irregular coating was obtained by growing crystals on a surface. In this case, the height of the relief is not large enough and the Wenzel regime is the regime for a water droplet on this surface. In this section, we analyze the situation when the rugosities are irregular and sufficient to obtain the Cassie regime.

The first example is that of Fan $et\ al.$ [27], who made pillars grow from a surface (fig. 4.47) using a GLAD technique. GLAD (glancing angle deposition) is a physical vapor deposition technique along a selected angle.

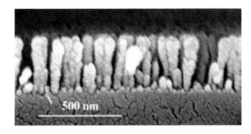 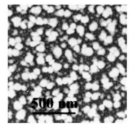

Figure 4.47 Irregular pillars obtained by Fan*et al*, from [27]; ©IOP 2004; reprinted with permission.

In this case, the contact of the water with the pillars can be sketched as in figure 4.48. Typically this is a Cassie droplet. However, due to the roughness of the tops of the pillars, the Cassie contact angle can then be defined by an extension of relation (4.44), using the Wenzel formulation for the contact of the liquid on the top of the pillars

$$\cos\theta_C = r * f\cos\theta + (f-1). \tag{4.54}$$

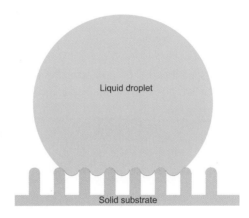

Figure 4.48 Sketch of droplet sitting on rounded pillars.

In the case where the tops of the pillars are spherical, Bico *et al* [22] have shown that 4.54 becomes

$$\cos\theta_C = -1 + f(1+\cos\theta)^2. \tag{4.55}$$

This situation combines the effects of Cassie and Wenzel regimes. Note that equation 4.55 shows that the Cassie regime is not linear in $\cos\theta$ but quadratic.

Similarly, Onda *et al* and Shibuichi *et al* [28,29] have fabricated fractal surfaces using AKD (alkylketene dimer). These surfaces may be called biomimetic because they resemble the structure of hydrophobic leaf surfaces, which are also fractal. Using fractal dimensions they have deduced a relation similar to the Wenzel law,

$$\cos\theta_W = \left(\frac{L}{l}\right)^{D-2}\cos\theta, \tag{4.56}$$

where L and l are the largest and smallest fractal size limits of surface and D is the fractal dimension ($D = 2.3$ in their case). It is typically a case of super-hydrophobicity since the coefficient $r = (\frac{L}{l})^{D-2}$ is relatively large and the value of f is very small.

4.6.3.7   Hydrophilic Case

In the preceding sections, we have dealt with non-wetting textured surfaces. Here we examine the case of a hydrophilic (wetting) textured surface. This case refers to the theory of impregnation: a droplet on a rough wetting surface has a smaller contact angle than the Young contact angle, according to Wenzel's law. However, it has been observed that in some cases, spreading occurs, i.e., a part of the liquid forms a film on the substrate.

If we keep the same notations, r the roughness of the surface and f the Cassie ratio, the Wenzel law is still written as

$$\cos \theta_W = r \cos \theta. \tag{4.57}$$

On the other hand, the "Cassie" regime corresponds this time to total wetting of the substrate

$$\cos \theta_C = 1 - f + f \cos \theta. \tag{4.58}$$

The critical contact angle corresponds to $\theta_W = \theta_C$, and is given by

$$\cos \theta_{crit} = \frac{1 - f}{r - f}. \tag{4.59}$$

If the Young contact angle θ is such that $\theta < \theta_{crit}$, then the liquid wets the surface, i.e., a liquid film spreads on the surface. The two possible morphologies are shown in figure 4.49.

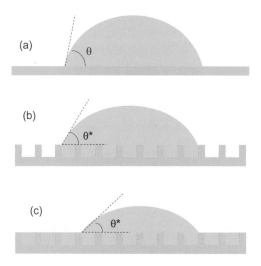

Figure 4.49 The possible morphologies of a droplet on a wetting textured surface: (a) flat surface; (b) Wenzel droplet; (c) Cassie droplet.

For a flat surface $r \rightarrow 1$ and the surface is wetted only if the Young contact angle $\theta \rightarrow 0$. For a porous medium (r infinite), equation (4.51) indicates that wetting occurs when $\theta < 90°$.

This result is the same as that of capillary inside a microchannel, which is detailed in chapter 7. More generally, equation 4.59 defines a critical angle comprised between 0 and 90°.

This equation shows that the presence of a film improves the wetting ($\theta^* < \theta$), i.e. the part of the droplet that bulges above the pillars has a contact angle $0 < \theta^* < \theta$, but it is not possible to induce a wetting transition (total wetting) by texturing a solid surface: equation 4.58 shows that complete wetting $\theta^* = 0$ requires $\theta = 0$.

4.6.3.8 Simultaneous Superhydrophobic and Oleophobic Surfaces – the Concept of Omniphobicity

Very low energy surfaces are of utmost importance in materials science, biotechnology, biomedical devices, fuel transport and even architecture. Such surfaces have the great advantage that liquids, both aqueous and organic, do not wet them, hence do not deposit particles and chemical substances that they transport. In the special case of microfluidics, these surfaces have the advantage to induce less drag [33].

When speaking of water only, the concept is that of superhydrophobicity: water droplets do not wet the surface and they easily roll away. A superhydrophobic surface has a large contact angle with aqueous liquids, a low roll-off angle and a low contact angle hysteresis.

But a more general and useful concept is that of omniphobicity, which encompasses both superhydrophobicity and superoleophobicity. Note that a superoleophobic surface repels extremely low-surface tension liquids, such as alkanes and organic liquids. As no naturally oleophobic surface exists, superoleophobicity is a challenge.

At first look, the situation is paradoxical. A simple hydrophobic surface is easily wetted by organic liquids. A pillared hydrophobic surface – like that described earlier in this chapter – may reach superhydrophobicity, but it will behave as an oleophilic, or even a superoleophilic surface. From a general standpoint, one can only make use of two parameters to design liquid-repellence: the surface energy and the surface roughness [34].

Two types of solutions have been found. The first one – and the most documented in the literature – uses air trapping between the liquid and the solid surface (Cassie effect). The second solution, which has appeared more recently [35], uses low energy liquid trapping in a microporous substrate.

The first solution has been found by analyzing the topology of the lotus leaf: it is well known that water droplets roll off these leaves; however the contact angle of water on the leaf (flat) surface is of the order of 74°. The surprising superhydrophobicity of such plant leaves is expected to be a consequence of reentrant surface texture, that is, the surface topography cannot be described by a simple univalued function $z = h(x, y)$, and a vector projected normal to the xy plane intersects the texture more than once. Hence the concept of reentrant surfaces provides a solution to superoleophobicity [34, 36-38]. Such an observation can be used to design a oleophobic surface.

Consider a textured surface like that shown in figure 4.50. If the Young contact angle θ is smaller than the composite surface angle ψ, then the liquid penetrates between the pillars. Indeed, the Laplace law indicates that the pressure is negative in the vicinity in the liquid at the vicinity of the pillars, whereas the overall pressure of the droplet is positive. In the opposite case, an air layer may be stably trapped below the liquid. The conditions for air trapping have been derived by Nosonovsky [39]. So, if the composite angle ψ is sufficiently small, even a moderately wetting liquid may stay in the Cassie configuration and the composite surface is then both superhydrophobic and superoleophobic. Hence surfaces with reentrant shapes like that shown in figure 4.51 are good candidates for superhydrophobicity and superoleophobicity.

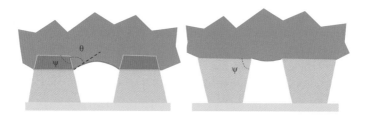

Figure 4.50 Left: when $\theta < \psi$, the liquid is not at equilibrium and penetrates between the pillars. Right: the liquid is at equilibrium and stays on top of the pillars.

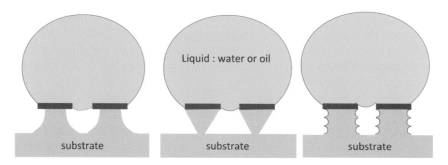

Figure 4.51 Typical omniphobic surfaces.

However, if the liquid is pushed downwards by either inertial forces, or pressure or electric forces, it may penetrate between the pillars. So another, more recent, approach has been proposed, based on the "lepenthes effect" [35]. Lepenthes leaves have a porous surface which traps water. Insects falling on the leaves slip and fall into the corolla, where the acids of the plant digest them. This observation has inspired the concept of SLIPS (for Slippery Liquid Infused Porous Surfaces) using microporous surfaces wicked by a very low surface tension liquid. Instead of trapping air as in the lotus effect, a liquid is trapped in this new design.

In order to have at the same time a superhydrophobic and superoleophobic surface, the trapped (lubrificating) liquid must repel at the same time aqueous and most organic liquids. If the trapped (lubrificating) liquid is denoted B and the liquid that should be repelled is denoted A, as in figure 4.52, we must have the conditions

$$\Delta E_1 = E_A - E_1 > 0, \tag{4.60}$$

$$\Delta E_2 = E_A - E_2 > 0. \tag{4.61}$$

Then

$$E_A - E_1 = r(\gamma_{SA} - \gamma_{SB}) - \gamma_{AB} > 0, \tag{4.62}$$

$$E_A - E_2 = r(\gamma_{SA} - \gamma_{SB}) + \gamma_A - \gamma_B > 0, \tag{4.63}$$

where r is the roughness. Introducing the Young relation, we obtain the two conditions

$$E_A - E_1 = r(\gamma_B \cos\theta_B + \gamma_A \cos\theta_A) - \gamma_{AB} > 0, \tag{4.64}$$

$$E_A - E_2 = r(\gamma_B \cos\theta_B + \gamma_A \cos\theta_A) + \gamma_A - \gamma_B > 0. \tag{4.65}$$

Solutions given in [35] are a solid constituted of an array of nanoposts functionalized with a low-surface-energy polyfluoroalkyl silane, and a random network of Teflon nanotubes. The lubrificating liquid is a low-surface-tension perfluorinated liquid.

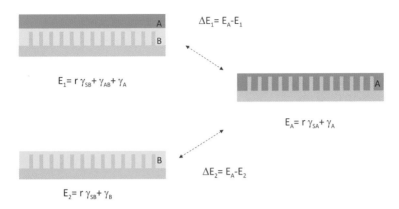

Figure 4.52 The different possibilities of wetting: Liquid B should be chosen so that it wets preferentially the solid. The advantage of SLIPS is that they can withstand very high pressures and that they are self repairing: an impact may temporarily displaced a small amount of lubrificating liquid, but it will quickly comes back in its initial position.

4.6.3.9   Conclusion

A complete diagram of wetting transitions is shown in figure 4.53 [30,31].

In the hydrophobic situation, if the Young angle θ is such that $\theta > \theta_i$ where θ_i is defined by $\cos\theta_i = (1 - f)/(r - f)$ (equation 4.46), the droplet stays on the pillar tops (fakir effect, i.e. Cassie droplet) producing a super-hydrophobic situation. If $90° < \theta < \theta_i$ the droplet is in the Wenzel regime, completely in contact with the surface of the pillars, with a contact angle larger than the Young contact angle.

In the hydrophilic situation, if the Young contact angle θ is such that $90° > \theta > \theta_{crit}$ the droplet is in the Wenzel regime, with a real contact angle θ^* smaller than θ. For $\theta < \theta_{crit}$, where θ_{crit} is defined by equation 4.58, the liquid partly spreads between the pillars and leaves a droplet above the pillars with a very small contact angle.

Figure 4.53 summarizes the different cases.

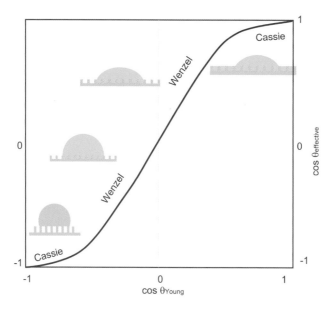

Figure 4.53 Plot of the relation between $\cos(\theta_{effective})$ and $\cos(\theta_{Young})$ for patterned surfaces.

4.7 **References**

[1] N. Moumen, Subramanian R.S., J. McLMaughlin, "Experiment on the motion of drops on a horizontal solid surface due to a wettability gradient," *Langmuir* **22**, pp. 2682–2690, 2006.

[2] M.K. Chaudhury, G.M. Whitesides, "How to make water run uphill," *Science* **256**, pp. 1539–1541, 1992.

[3] A. R. Wheeler, "Putting electrowetting to work," *Science* **322** (24), pp. 539–540, 2008.

[4] T. Podgorski, J.-M. Flesselles and L. Limat, "Corners, cusps and pearls in running drops," *Phys. Rev. Lett.* **87**, 036102–036105, 2001.

[5] D. Dos Santos, F. and T. Ondarçuhu, "Free-running droplets," *Phys. Rev. Letters* **75**(16), pp. 2972–2975, 1995.

[6] de Gennes, P-G., F. Brochard-Wyart, D. Quèrè, *Capillarity and wetting phenomena: Drops, bubbles, pearls, waves*, Springer, New-York, 2004.

[7] H.A. Wege, J.A. Holgado-Terriza, J.I. Rosales-Leal, R. Osorio, M. Toledano, M.A. Cabrerizo-Vilchez, "Contact angle hysterisis on dentin surfaces measured with ADSA on drops and bubbles," *Colloids and Surfaces, A: Physicochemical and Engineering Aspects* **206**, pp. 469–483, 2002.

[8] C.N.C Lam, R. Wu, D. Li, M.L. Hair, A.W. Neumann, "Study of the advancing and receding contact angles: liquid sorption as a cause of contact angle hysteresis," *Advances in Colloid and Interface Science* **96**, pp. 169–191, 2002.

[9] H. Tavana, A.W. Neumann, "On the question of rate-dependence of contact angles," *Colloids and Surfaces, A: Physicochemical Engineering Aspects* **282-283**, pp. 256–262, 2006.

[10] T. Cubaud, M. Fermigier, "Advancing contact lines on chemically patterned surfaces," *Journal of Colloid and Interface Science* **269**, pp. 171–177, 2004.

[11] A. Buguin, L. Talini, P. Silberzan, "Ratchet-like topological structures for the control of

microdops," *Appl. Phys. A.* **75**, pp. 207–212, 2002.

[12] T. Ondarçuhu, "Total or partial pinning of a droplet on a surface with chemical discontinuity," *J. Phys. II France* **5**, pp. 227–241, 1995.

[13] J. Bico, C. Marzolin, D. Quèrè, "Pearl drops," *Europhys. Lett.* **47**(2), pp. 220–226, 1999.

[14] J. Bico, C. Tordeux, D. Quèrè, "Rough wetting," *Europhys. Lett.* **55**(2), pp. 214–220, 2001.

[15] P. Lenz, R. Lipowsky, "Morphological transitions of wetting layers on structured surfaces," *Phys. Rev. Letters* **80**(9), pp. 1920–1923, 1998.

[16] H. Gau, S. Herminghaus, P. Lenz, R. Lipowsky, "Liquid morphologies on structured surfaces: From microchannels to microchips," *Science* **383**, pp. 46–49, 1999.

[17] M. Brinkmann, R. Lipowsky, "Wetting morphologies on substrates with striped surface domains," *Journal of Applied Physics* **92**(8), pp. 4296–4306, 2002.

[18] W. Ehrfeld, C. Gärtner, K. Golbig, V. Hessel, R Konrad, H. Löwe, T. Richter, C. Schulz, "Fabrication of components and systems for chemical and biological microreactors," *Microreaction Technology, Proceedings of the 1st International Conference on Microreaction Technology IMRET* **1**, pp.72–90, Berlin, 1997.

[19] J. Berthier, Van-Man Tran, F. Mittler, N. Sarrut, "The physics of a coflow micro-extractor: interface stability and optimal extraction length," *Sensors and Actuators A* **149**, pp. 56–64, 2009.

[20] J. Berthier, F. Loe-Mie, V.-M. Tran, S. Schoumacker, F. Mittler, G. Marchand, N. Sarrut, "On the pinning of interfaces on micropillar edges," *Journal of Colloid and Interface Science* **338**, pp. 296–303, 2009.

[21] Th. Uelzen, J. Müller, "Wettability enhancement by rough surfaces generated by thin film technology," *Thin Solid Films* **434**, pp. 311–315, 2003.

[22] Seong H. Kim, Jeong-Hoon Kim, Bang-Kwon Kang, Han S. Uhm, "Superhydrophobic CFx coating via in-line atmospheric RF plasma of He-CF4-H2," *Langmuir* **21**, pp. 12213–12217, 2005.

[23] Liang Zhu, Yanying Feng, Xiongying Ye, Zhaoying Zhou, "Tuning wettability and getting superhydrophobic surface by controlling surface roughness with well-designed microstructures," *Sensors and Actuators, A Physical* **130-131**, pp.595–600, 2006.

[24] J. Bico, U. Thiele, D. Quèrè, "Wetting of textured surfaces," *Colloids and Surfaces A: Physicochemical and Engineering Aspects* **206**, pp. 41–46, 2002.

[25] J. Bico, C. Marzolin, D. Quèrè, "Pearl drops," *Europhys. Lett.* **47**(2), pp. 220–226, 1999.

[26] N.A. Patankar, "Transition between superhydrophobic states on rough surfaces," *Langmuir* **20**, pp. 7097–7102, 2004.

[27] L. Zhu, Y.Y. Feng, X.Y. Ye, Z.Y. Zhou, "Tuning wettability and getting superhydrophobic surfaces by controlling surface roughness with well-designed microstructures," *Transducers '05, 13rd International Conference on Solid-State Sensors, Actuators and microsystems*, Seoul, Korea, June 5-9, 2005.

[28] T.N. Krupenkin, J. Ashley Taylor, Tobias M. Schneider, Shu Yang, "From Rolling Ball to Complete Wetting: The Dynamic Tuning of Liquids on Nanostructured Surfaces," *Langmuir* **20**, pp. 3824–3827, 2004.

[29] W. Barthlott, C. Neinhuis, "Purity of the sacred lotus, or escape from contamination in biological surfaces," *Planta* **202**, pp. 1–8, 1997.

[30] J-G Fan, X-J Tang and Y-P Zhao, "Water contact angles of vertically aligned Si nanorod arrays," *Nanotechnology* **15**, pp. 501–504, 2004.

[31] T. Onda, S. Shibuichi, N. Satoh, K. Tsujii, "Super water-repellent surfaces," *Langmuir* **12**(9), pp. 2125–2127, 1996.

[32] S. Shibuichi, T. Onda, N. Satoh, and K. Tsujii, "Super water-repellent surfaces resulting from fractal structures," *J. Phys. Chem.* **100**, pp. 19512–19517, 1996.

[33] B. Bhustan, "Biomimetics inspired surfaces for drag reduction and oleophobicity/philicity," *Beilstein J. Nanotechnol.* **2**, pp 64–68, 2011.

[34] A.Tuteja, Wonjae Choi, G.H. McKinley, R.E.Cohen, M.F. Rubner, "Design parameters for superhydrophobicity and superoleophobicity," *MRS Bulletin* **33**, August 2008.

[35] Tak-Sing Wong, Sung Hoon Kang, Sindy K.Y. Tang, E.J. Smythe, B.D. Hatton, A. Grinthal, J. Aizenberg, "Bioinspired self-repairing slippery surfaces with pressure-stable omniphobicity," *Nature* **477**, pp. 443–447, 2011.

[36] L. Zhai, M.C. Berg, F.C. Cebeci, Y. Kim, J.M. Milwid, M.F. Rubner, R.E. Cohen, "Patterned superhydrophobic substrates: toward a synthetic mimic of the Namid desert beetle," *Nano Lett.* **6** (6), p. 1213, 2006.

[37] L. Cao, H.H. Hu, D. Gao,"Design and fabrication of micro-textures for inducing a super-hydrophobic behavior on hydrophilic materials," *Langmuir* **23** (8), p.4310, 2007.

[38] Maesoon Im, Hown Im, Joo-Hyung Lee, Jun-Bo Yoon and Yang-Kyu Choi, "A robust superhydrophobic and superoleophobic surface with inverse-trapezoidal microstructures on a large transparent flexible substrate," *Soft Matter* **6**, pp. 1401–1404, 2010.

[38] M. Nosonovsky, "Multiscale Roughness and Stability of Superhydrophobic Biomimetic Interfaces," *Langmuir* **23**, pp. 3157–3161, 2007.

5

Droplets Between Two Non-parallel Planes: From Tapered Planes to Wedges

5.1 Abstract

In this chapter, we focus on the behavior of a droplet or a liquid plug placed between two non-parallel plates. It is assumed that the droplet is sufficiently small that gravitational forces can be neglected (droplet Bond number smaller than 1). The case of tapered plates is first investigated. It is shown that, if the walls are perfectly smooth, e.g. there is no pinning hysteresis, a wetting droplet is not at equilibrium in such a geometry. The conventional Hauksbee's formulation for wetting walls is presented, and a generalization for any wall wetting property (two wetting walls, two non-wetting walls, wetting and non-wetting walls) is derived. In a second part, the focus is on the Concus-Finn relations for a wedge, which governs the location of the droplet in the wedge. A generalization to different wall wetting properties is presented.

5.2 Droplet Self-motion Between Two Non-parallel Planes

It was first observed by Hauksbee [1,2] that a water plug limited by two non-parallel wetting plates – hydrophilic for a water droplet or lyophilic for any liquid – moves towards the narrow gap region. A sketch of the plug is shown in figure 5.1.

It was shown in Chapter 2 that the minimum energy configuration is a spherical droplet with the proper contact angles, which can only happen if the plates are hydrophobic enough, or else the liquid goes to the corner of the wedge. This section takes a force-based approach to arrive at the same conclusions, and although the arguements here are not as rigorous, they do give insight on the dynamics of what happens.

In this section, it is demonstrated how Laplace's law furnishes a clear explanation of droplet motion. Moreover, the analysis of self-motion of a droplet between two non-parallel plates is extended to any contact angle. It is shown that there exist a critical contact angle θ^* larger than $\pi/2$ above which the droplet stays away from the corner. Below this critical angle, the droplet moves towards the corner, in agreement with Hauksbee's analysis. It is shown also that if the

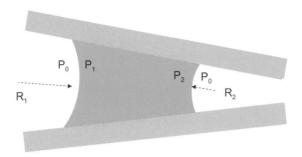

Figure 5.1 Sketch of a liquid plug between two wetting plates. The contact angle is $\theta < 90°$ and the plug is not at equilibrium: it moves in the direction of the smaller cross-section. Note that the Laplace pressures are negative in such a case.

Young contact angles of the liquid with the plates are different, it is their average value that has to be taken into account.

5.2.1 Identical Young Contact Angle with Both Plates

Let us assume a geometry where the two plates form a wedge of half-angle α in the vertical plane. The different possible geometries, depending on the concavity or convexity of the interface are shown in figure 5.2.

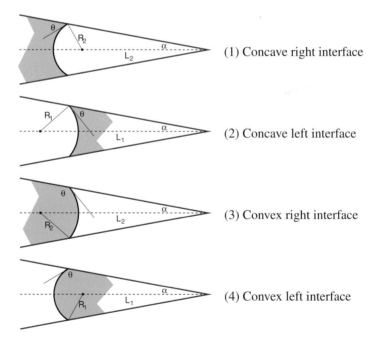

Figure 5.2 The different cases of the interfaces. α is the wedge half-angle, L is the distance from the wedge vertex to the center of curvature, R is the radius of curvature, and θ is the liquid contact angle.

In our notations, R is always positive. L is the distance from the dihedral to the center of curvature. Let us write Laplace's law for the left interface (index 1). The curvature of the interface is determined by the two curvature radii in two perpendicular planes. Let us consider the vertical and horizontal planes, with the curvature radii R_1 and R_H

$$P_0 - P_1 = \gamma \left[\frac{c_1}{R_1} + \frac{1}{R_H} \right], \tag{5.1}$$

where c_1 has the value 1 or -1 depending whether the interface is concave or convex relative to the liquid phase. In case (2) of figure 5.2, the interface is concave and $c_1 = 1$, and in case (4), the interface is convex and $c_1 = -1$. Using the Law of Sines we find

$$\frac{R_1}{\sin \alpha} = \frac{L_1}{\sin(\theta + \pi/2)} \quad \text{or} \quad R_r = \frac{L_1 \sin \alpha}{\cos \theta}. \tag{5.2}$$

Likewise, case (4) gives

$$\frac{R_1}{\sin \alpha} = \frac{L_1}{\sin(\theta - \pi/2)} \quad \text{or} \quad R_r = \frac{L_1 \sin \alpha}{-\cos \theta}. \tag{5.3}$$

Independently of the concavity, relation (5.1) becomes

$$P_0 - P_1 = \gamma \left[\frac{\cos \alpha}{L_1} + \frac{1}{R_H} \right]. \tag{5.4}$$

For the right side interface (index 2)

$$P_0 - P_2 = \gamma \left[\frac{c_2}{R_2} + \frac{1}{R_H} \right], \tag{5.5}$$

Because the droplet has a circular shape in the horizontal plane, the curvature radius R_H is the same for both interfaces. The same reasoning as done for the left side leads to

$$P_0 - P_2 = \gamma \left[\frac{\cos \alpha}{L_2} + \frac{1}{R_H} \right]. \tag{5.6}$$

where c_2 has the same definition as c_1.

By subtracting the two relations (5.4) and (5.6), the horizontal curvature radius R_H vanishes, and the pressure difference is

$$P_1 - P_2 = \gamma \left[\frac{\cos \alpha}{L_2} - \frac{\cos \alpha}{L_1} \right] = \gamma \cos \alpha \left[\frac{1}{L_2} - \frac{1}{L_1} \right]. \tag{5.7}$$

Inspection of (5.7) immediately shows that if the centers of curvature coincide, $L_1 = L_2$, then the droplet is in equilibrium. This obviously agrees with the result from Chapter 2 that a sphere is a stable configuration. Clearly this can happen only for hydrophobic planes; for hydrophilic planes, the centers of curvature are outside the drop on opposite sides, $L_1 > L_2$, and the force is always towards the narrow part of the wedge. Returning to the hydrophobic case, if the drop is flatter than a sphere, then $L_1 > L_2$, and the net force is towards the opening of the wedge, recalling $\cos \alpha < 0$ here. If the drop is stretched from a sphere, the force is narrowward.

It can be shown that when the droplet moves towards the corner it accelerates. As the droplet moves to the right, L_1 will decrease much faster than L_2 as the drop stretches out, and the force increases.

5.2.2 Different Young Contact Angles

What happens if the two planes have a different contact angle with the liquid? Let us denote by θ_1 and θ_2 the two contact angles. The behavior turns out to be the same as with a common contact angle, but the common contact angle θ has to be replaced by the average $\theta = (\theta_1 + \theta_2)/2$.

There are three possible stable configurations: a spherical bridge between the planes that does not wet the dihedral (high contact angles), a spherical wedge that does wet the dihedral (medium contact angles), and a long filament (low contact angles).

The dividing line between the bridge and wetting wedge comes when the sphere of the droplet is just tangent to the junction of the planes, as shown in figure 5.3. A little geometry with some isosceles triangles defined by some radii shows that the angles satisfy

$$2\alpha = \left(\theta_1 - \frac{\pi}{2}\right) + \left(\theta_2 - \frac{\pi}{2}\right), \tag{5.8}$$

or

$$\frac{\theta_1 + \theta_2}{2} = \alpha + \frac{\pi}{2}. \tag{5.9}$$

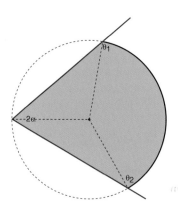

Figure 5.3 Critical contact angles for a spherical drop wetting the dihedral line of a wedge.

The dividing line between the wetting wedge and filaments is the Concus-Finn criterion, which happens when the liquid/air surface is planar, so a cross-section is triangular, as shown in figure 5.4. The critical condition may be written as the standard Euclidean angle sum of a triangle:

$$2\alpha + \theta_1 + \theta_2 = \pi, \tag{5.10}$$

or

$$\frac{\theta_1 + \theta_2}{2} = \frac{\pi}{2} - \alpha. \tag{5.11}$$

The contact angles here necessarily make it so the total of the plane contact energy and the liquid/air surface energy is exactly zero. This can most directly be seen by dropping an altitude from the wedge vertex to the planar surface. If the wetted plane lenghts are b_1 and b_2 respectively, and c the length of the planar surface, then projecting b_1 and b_2 down to c gives

$$c = b_1 \cos\theta_1 + b_2 \cos\theta_2, \tag{5.12}$$

which is exactly the condition for the total surface tension energy to be zero:

$$c - b_1 \cos\theta_1 - b_2 \cos\theta_2 = 0 \tag{5.13}$$

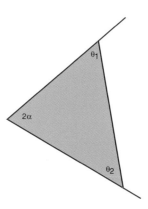

Figure 5.4 Critical Concus-Finn contact angles for the formation of filaments; the dihedral angle and the contact angles add to π, the liquid surface is flat, and there is zero pressure difference across the surface.

For contact angles smaller that the Concus-Finn critical condition, the reason filaments grow arbitrarily may be deduced from the fact that the total surface contact energy is negative. The liquid/air surface must be concave (since the Concus-Finn critical condition is that it is flat), which means the liquid has negative pressure. Pressure being the rate of change of energy with respect to volume, that means the surface energy must be negative since the surfaces increase as volume increases. Now consider a fixed volume of liquid contemplating spreading into a filament. Narrowing into a filament greatly increases the surface-to-volume ratio, and since the surface energy is negative, spreading into a filament is favored.

Another perspective on the two critical conditions is to consider the tangent plane where a wetting wedge surface meets the dihedral line. For the wedge to exist, it must be possible to have a single tangent plane that meets both the contact angle conditions simultaneously. This turns out to be possible exactly between the two critical conditions found above, so the wedge can wet for

$$\frac{\pi}{2} - \alpha < \frac{\theta_1 + \theta_2}{2} < \frac{\pi}{2} + \alpha. \tag{5.14}$$

Figure 5.5 shows a graphical representation of the various possibilities.

5.2.3   Numerical Simulation – 2D and 3D Cases

Consider two plates making a half-angle of $2.85°$, and consider the two following cases: (1) water droplet located between two hydrophilic solid plates ($\theta = 80°$), and (2) water droplet located between two hydrophobic solid plates ($\theta = 130°$). We can use Surface Evolver, and show that the results agree with the theoretical analysis of the preceding sections, i.e. the droplet moves towards the corner in the first (hydrophilic) case, and stays away from the corner in the second case (hydrophobic).

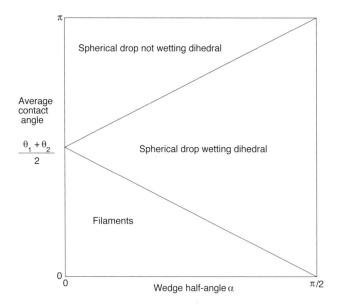

Figure 5.5 Phase plane for the different types of stable behaviors of a drop between nonparallel planes.

The motion of the droplet is illustrated in the figures 5.6 to 5.9, depending on the values of the contact angles. The contact angles are $80°/80°$ in figure 5.6, $130°/130°$ in figure 5.7, $80°/130°$ in figure 5.8 and $105°/140°$ in figure 5.9.

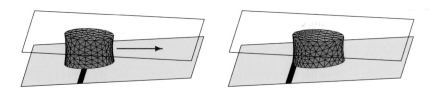

Figure 5.6 Evolver simulation of a wetting droplet moving towards the smaller cross-section area: the black line is a fixed mark for visualization and the upper plate has been dematerialized for better visibility.

5.2.4   A Reciprocal to the Hauksbee Problem

In the chapter 11 (Microelectronics), we present an analysis of the tilt of a chip placed on a sessile water droplet deposited on a hydrophilic pad on a wafer [3]. It is shown that the chip slowly tilts until a wedge is formed. The analysis is closely related to the Hauksbee problem. In fact it can be seen as a reciprocal to the Hauksbee problem (fig. 5.10). In the first case, it has been demonstrated here that a wetting droplet moves towards the dihedral edge of the two fixed planes. In the second case, the upper plate tilts on top of the fixed droplet until a liquid wedge is formed. These two reciprocal properties are linked by the observation that the minimal free surface area has a wedge shape [4].

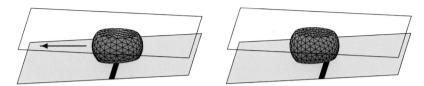

Figure 5.7 Evolver simulation of a non-wetting droplet moving towards the larger cross-section area.

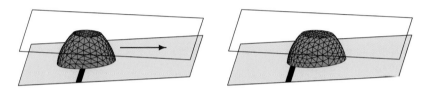

Figure 5.8 Evolver simulation of a droplet with different Young contact angles with both planes ($\theta_{lower} = 80°$, $\theta_{upper} = 130°$); in this case, the initial droplet is stretched, so it moves towards the smaller cross-section area.

5.2.5   Example of Tapered System for Passive Pumping in Fuel Cells

Recently, Litterst *et al.* and Paust et al. have given an interesting application of the Hauksbee analysis [5,6]. Their application concerns passive pumping for methanol-powered fuel cells. In electronics, direct micro methanol fuel cells (μMFCs) are attractive due to the high density of methanol, fast refueling, and extremely long shelf life. Therefore, they are promising candidates for powering small electronic devices. However, a significant challenge is the removal of carbon dioxide bubbles from the anode. At the anode, CO_2 bubbles are generated as a reaction product of the methanol oxidation reaction; these bubbles may block channels and cause the μMFC to malfunction. Hence degassing is required. Using tapered channels – as shown in figure 5.11 – bubbles can be removed from the cell; at the same time the extraction of the bubbles from the system contributes to the pumping of the fuel into the cell. A close-up on the tapered plate area is shown in figure 5.12.

The physics behind the pumping system is that described earlier in this chapter, except for the fact that the droplet volume is not constant. The walls are treated to be hydrophilic so that

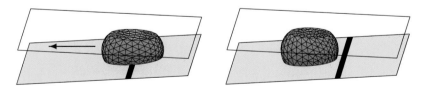

Figure 5.9 Evolver simulation of a droplet with different Young contact angles with both planes ($\theta_{lower} = 105°$, $\theta_{upper} = 140°$); in this case, the initial droplet is compressed, so it moves towards the larger cross-section area.

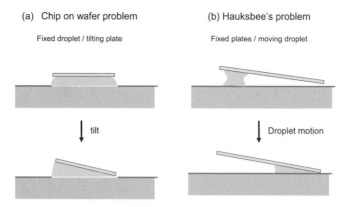

Figure 5.10 Sketches of the "Chip on wafer" tilt instability (a), and of Hauksbee's problem (b). In both cases, the minimal surface for the droplet is reached when the droplet is located in the dihedral corner.

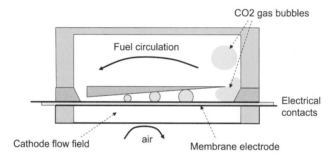

Figure 5.11 Sketch of the fuel cell with the tapered plate.

the contact angle of the CO_2 with the wall is sufficiently above $\pi/2$ to obtain the motion away from the dihedral (the planes' angle is of the order of 1.5° to 3°). At first a bubble grows into the tapered channel (t_1 on figure 5.12); it touches the upper channel wall (t_2); the bubble right interface grows towards the increasing cross-section of the channel while its left interface stays pinned at the channel entrance (t_3); at time t_4 the Laplace pressure difference is sufficient to trigger de-pinning and the release of the bubble. The Laplace pressure difference increases because the length L_2 in (5.7) is constant and L_1 increases: the first, negative term increases (in absolute value) while the positive term decreases, and the absolute value of pressure difference

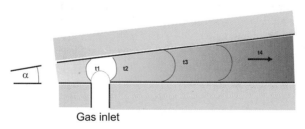

Figure 5.12 Sketch of the passive pumping of CO_2 bubbles in a fuel cell.

$|P_1 - P_2|$ increases. Once the droplet has evacuated the channel, the same sequence then restarts.

In their approach, Paust *et al.* have taken into account a receding and an advancing contact angle with the walls. In such a case, equations (5.3) and (5.6) become

$$P_0 - P_1 = \gamma \frac{\cos(\theta_{adv} + \alpha)}{d_1} \tag{5.15}$$

and

$$P_0 - P_2 = \gamma \frac{\cos(\theta_{rec} + \alpha)}{d_2}. \tag{5.16}$$

However, at the low flow velocities characteristic of this microsystem, the contact angle hysteresis is small and does not significantly affect the droplet motion.

5.2.6 Discussion

We have demonstrated that a droplet between two non-parallel plates is not at equilibrium. There exists a critical angle θ^* larger than $\pi/2$ below which the droplet moves towards the corner ($\theta < \theta^*$) and above which the droplet moves away from the corner ($\theta > \theta^*$). Moreover, a droplet moving towards the corner progressively accelerates while a droplet moving away from the corner decelerates and stops.

Bouasse [7] has remarked that the same type of motion applies for a cone, where the plug moves towards the tip of the cone. In reality, Bouasse used a conical frustum (slice of cone) in order to let the gas escape during plug motion. More recently, Renvoisé *et al.* have performed a same type of experiment using a narrowing tube with its tapered end pointing upwards, analyzing the balance between capillary force, gravity and shape of the tube [8].

In this section, the wedge half-angle α was supposed relatively small, so that the constant volume can move along a noticeable length between the two planes. In the following section we address the case of a droplet located very close to the groove line. We shall see that the position of the droplet is predicted by the Concus-Finn relations, which indicate the conditions for a droplet to spread like a filament in the corner, or to detach from the corner. In particular, we will show that a sufficiently wetting droplet in a small angle wedge (tapered planes) moves towards the corner, wets the corner and spreads laterally as a filament.

5.3 Droplet in a Corner

In this section we investigate the behavior of a droplet placed very close to the edge of the wedge. Microfluidic channels and chambers are usually etched in silicon, glass, or plastic, and liquids are either aqueous or organic. Hence contact angles can be anywhere in the interval $[0,\pi]$. We show that the shape that a droplet takes in a wedge depends on the wedge angle and on the Young contact angle of the liquid with the walls.

5.3.1 Dimensions of the Droplet and Effect of Gravity

Let us investigate first the effect of a corner – or a wedge – on the droplet interface. Take the case of a 90° wedge. The shape of the droplet is shown in figure 5.13, depending on the Bond number $Bo = \rho g R^2 / \gamma$, where R is a characteristic dimension of the droplet, which can be scaled as the 1/3 power of the value of the volume of liquid. A small volume droplet of 0.125 μl tends to take the form of a portion of sphere despite the different contact angles on the two planes (the Bond number is of the order of 0.04). A larger droplet of 1.25 μl – Bond number of 4 – is flattened by gravity. In the rest of this chapter we consider droplets sufficiently small not to be affected by gravity.

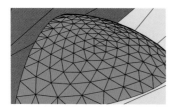 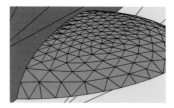

Figure 5.13 Shape of a liquid drop in a 90° wedge. Left: a small volume droplet of 0.125 μl tends to take the form of a sphere despite the different contact angles on the two planes (the Bond number is of the order of 0.04). Right, a larger droplet of 1.25 μl – Bond number 4 – is flattened by gravity.

5.3.2   Concus-Finn Relations

It has been observed that liquid interfaces in contact with highly wetting solid walls forming a wedge tend to spread in the corner. This behavior results from the fact that the interface curvature is strongly reduced in the corner (in figure 5.14, the vertical curvature radius is small): the Laplace pressure is low in the corner and liquid tends to spread in the corner. Concus and Finn (1969 and 1974) [9] have investigated this phenomenon and they have derived a criterion for capillary motion in the corner of the wedge. If θ is the Young contact angle on both planes and α the wedge half-angle, the condition for capillary self-motion is

$$\theta < \frac{\pi}{2} - \alpha. \tag{5.17}$$

This case corresponds to wetting walls. Conversely, when the walls are non-wetting the condition for de-wetting of the corner is

$$\theta > \frac{\pi}{2} + \alpha. \tag{5.18}$$

In figure 5.15, the Concus-Finn relations have been plotted in a (θ, α) coordinate system. One verifies that, for a flat angle ($\alpha = 0°$), the Concus-Finn relations reduce to the usual capillary analysis.

5.3.3   Numerical Approach

The Concus-Finn relations can be numerically verified using the Surface Evolver software. Figure 5.16 shows the shape taken by a wetting droplet in a 90° wedge according to the contact angle with the walls. Spreading of the liquid in the corner occurs when condition (5.15) is met. If condition (5.15) is not met, the droplet takes a nearly spherical shape.

The decrease of the Laplace pressure in the corner mentioned earlier can be checked by plotting the stretching length (length of the groove line in contact with the liquid) versus the contact angle (fig. 5.17).

The second Concus-Finn relation can also be checked numerically. Figure 5.18 shows the shape taken by a non-wetting droplet in a 90° wedge according to the contact angle with the

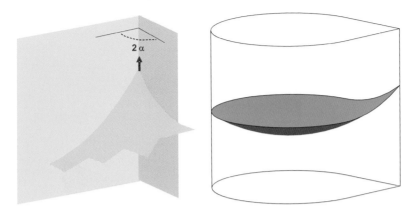

Figure 5.14 A liquid interface is lifted in the corner of a wedge made of two wetting plates. This phenomenon is due to the need for a common tangent plane making the proper contact angles on the two walls. The tube-with-wedge on the right has a wedge angle of $2\alpha = 90°$ and contact angle of $60°$.

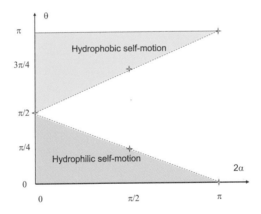

Figure 5.15 Plot of the domains of self-motion in a corner, according to the Concus-Finn relations.

walls. If the contact angles on both planes are larger than $135°$, the drop detaches from the corner, and does not wet the corner any more.

5.3.4   Example of a Liquid in a Micro-beaker

An illustrative example of the Concus-Finn criterion is that of water in a micro-beaker with a square base (fig. 5.19). If the contact angle with the walls is larger than $45°$, the free surface has the curvature indicated in the left figure. When the contact angle is decreased below the Concus-Finn limit, filaments spread upward in the corners. Their extent is limited by their weight.

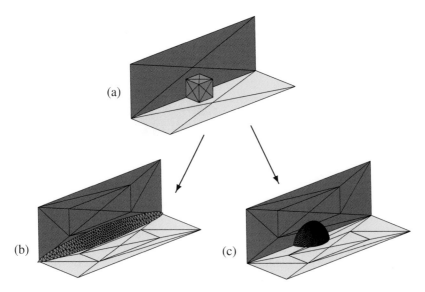

Figure 5.16 Droplet in a 90° wedge: (a) initial (non-physical) droplet; (b) the droplet spreads in the corner when the contact angles satisfy the Concus-Finn condition (here θ = 45°); (c) the droplet adopts a nearly spherical shape when the Concus-Finn condition is not met (here θ = 80°).

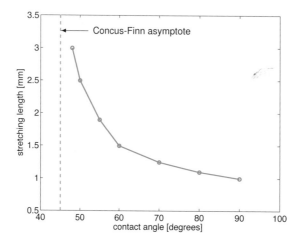

Figure 5.17 Stretching length of the droplet as a function of the contact angle: the Concus-Finn limit appears as an asymptote.

5.3.5   Extended Concus-Finn Relation

A generalization of the Concus-Finn relation has been proposed in [10] derived from the work of K. Brakke [4]. When the two planes do not have the same wettability (contact angles θ_1 and

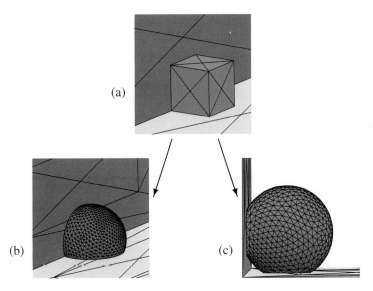

Figure 5.18 Droplet in a hydrophobic wedge: (a) initial (non-physical) droplet; (b) the droplet at equilibrium still wets the corner if the second Concus-Finn relation is not satisfied (here $\theta = 120°$); (c) if the contact angle is larger than $135°$ (here $\theta = 150°$), the droplet detaches from the corner, in accordance to the second Concus-Finn relation.

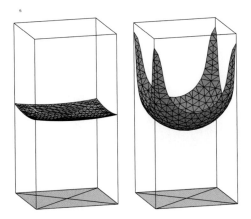

Figure 5.19 Free surface of water in a micro-beaker: filaments spread upwards when the contact angle is below the Concus-Finn limit of $45°$. Left, contact angle $70°$; right, contact angle $40°$.

θ_2), the relation (5.15) becomes

$$\frac{\theta_1 + \theta_2}{2} < \frac{\pi}{2} - \alpha. \tag{5.19}$$

Remember that α is the wedge half-angle. Usually microfluidic systems are etched in silicon, glass or plastic with a cover plate on top, glued or fixed by direct bonding. Hence the upper corners frequently have different contact angles with the liquid (fig. 5.19). An important consequence of relation (5.17) applies to trapezoidal microchannels (fig. 5.20).

When a glass cover is sealed on top of the system, the upper corners may form an angle of

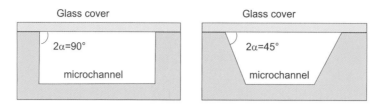

Figure 5.20 Cross section of rectangular and trapezoidal microchannels.

45° and the extended Concus-Finn condition indicates the following limit

$$\frac{\theta_1 + \theta_2}{2} < \frac{\pi}{2} - \alpha = 67.5°. \tag{5.20}$$

The glass cover may be quite hydrophilic, say $\theta_1 \sim 60°$, and if the channel is also hydrophilic, say $\theta_2 \sim 70°$, then $(\theta_1 + \theta_2)/2 \sim 65°$ and the liquid spreads in the upper corners, leading to unwanted leakage.

The same generalization applies to the non-wetting case. The extended Brakke-Concus-Finn de-wetting condition for a corner is

$$\frac{\theta_1 + \theta_2}{2} > \frac{\pi}{2} + \alpha. \tag{5.21}$$

Figure 5.20 shows two shapes of a droplet in a groove for two couples of contact angle: $(\theta_1, \theta_2) = (45°, 60°)$ and $(\theta_1, \theta_2) = (145°, 160°)$. In the first case, relation (5.17) is verified and the droplet spreads; in the second case, relation (5.18) is verified and the droplet detaches from the groove line.

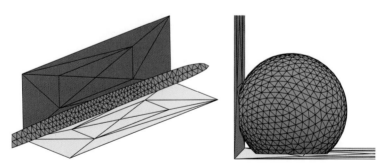

Figure 5.21 Two shapes of droplet in a 90° wedge. Left, filament when $(\theta_1, \theta_2) = (45°, 25°)$. Right, detachment when $(\theta_1, \theta_2) = (140°, 160°)$.

5.3.6   Droplet in a Wetting/Non-wetting Corner

In this section, the behavior of a droplet in a 90° wedge is investigated, in the case where one side of the wedge is wetting and the other non-wetting. It is expected that, if the contact angle on the wetting side is small and that on the non-wetting side is large, the drop will be positioned on the wetting side. It is intuitive to think that this will happen if the difference between the non-wetting contact angle and the wetting contact angle is large. By referring to [4], a criterion for the drop to be positioned on the wetting side only is

$$\theta_1 - \theta_2 > \pi - 2\alpha, \tag{5.22}$$

α being the wedge half-angle. Relation (5.20) can be verified by numerical simulation as shown in figure 5.21.

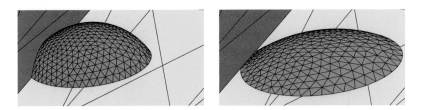

Figure 5.22 Droplet in a corner with wetting and non-wetting sides, depending on the surface properties. Left: the droplet stays attached to the corner for contact angles $60°, 130°$. Right: When the floor contact angle is reduced to $30°$, the drop migrates entirely to the floor.

The results of sections 5.3.4 and 5.3.5 can be shown on the same plot in a (θ_1, θ_2) coordinate system for a given groove angle α (fig. 5.22). Along the line $\theta_1 = \theta_2$, we have the Concus-Finn relations. The droplet spreads as a filament below the limit defined by $\theta_1 + \theta_2 < \pi - 2\alpha$; the droplet detaches from the groove line above the limit defined by $\theta_1 + \theta_2 > \pi + 2\alpha$. If $\theta_1 - \theta_2 > \pi - 2\alpha$, the droplet migrates onto the θ_1 plane, and in the opposite case, onto the θ_2 plane. In the domain ($\theta_1 + \theta_2 > \pi - 2\alpha$; $\theta_1 + \theta_2 < \pi + 2\alpha$; $|\theta_1 - \theta_2| < \pi - 2\alpha$) the droplet stays attached to the groove line.

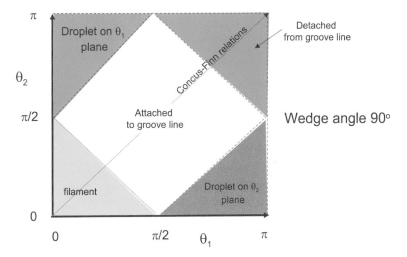

Figure 5.23 The different possible droplet location depending on the values of θ_1 and θ_2 ($2\alpha = 90°$).

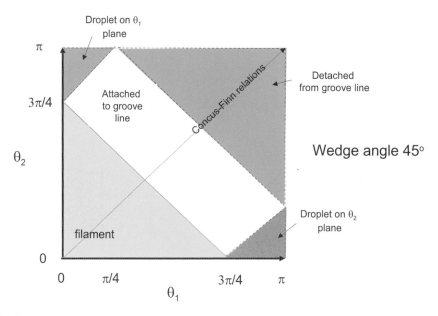

Figure 5.24 The different possible droplet location depending on the values of θ_1 and θ_2 ($2\alpha = 45°$).

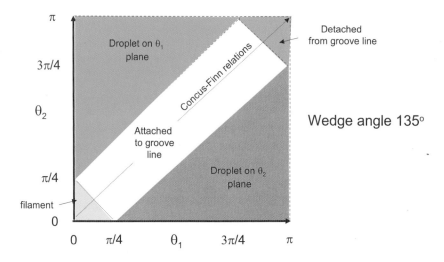

Figure 5.25 The different possible droplet location depending on the values of θ_1 and θ_2 ($2\alpha = 135°$).

5.3.7 Discussion

In a general way, the surface tension does not appear in the Concus-Finn relations. Thus, these relations also apply for two-phase liquids, i.e. droplets of water in organic continuous phase, or droplets of organic liquid in an aqueous phase. Relation (5.15) shows that a droplet of oil surrounded by water is likely to form in hydrophilic channels, because the water spreads along the solid wall, and relation (5.16) shows the opposite for hydrophobic channels, i.e., oil spreads and a water droplet is formed. This is why flow focusing devices (FFDs), used to produce calibrated microdrops of aqueous phase in an organic media, must have their interior walls hydrophobic, a state usually obtained by silanization.

In microtechnology, wedges and corners often form a 90° angle, so that a droplet disappears in the form of filaments if the wetting angle on both planes is smaller than 45°. One must be wary that, when coating the interior of microsystems with a strongly wetting layer, in order to have a very hydrophilic (wetting) surface, droplets may disappear; they are transformed into filaments in the corners.

An interesting confirmation of the Concus-Finn relations has been done by Khare *et al.* in triangular shape wedges. They have shown that filament spreading can be triggered by the use of an electric field: it is known that electrowetting voltage reduces the contact angle and, once the contact angle is below the Concus-Finn limit, the droplet spreads along the groove line as a filament [11].

5.4   Conclusion

In this chapter the behavior of a microdrop located between two non-parallel plates has been investigated. In the case of tapered plates, a droplet is not, in general, at equilibrium. The generalized Hauksbee law governs the motion of the droplet, which is either towards the wedge line or away from it depending on the values of the contact angles of the liquid with each plate and the.

In the case of a droplet located very close to the groove line, the generalized Concus-Finn relations govern the equilibrium morphology of the droplet: depending on the liquid contact angles with the solid plates, it can spread along the groove line as a filament, detach from the groove line, migrate onto one of the two plates or stay attached to the groove line. Figures 5.23, 5.24, and 5.25 summarize the possibilities for a selection of wedge angles.

5.5   References

[1]   F. Hauksbee, *Philos. Trans.* **27**, p. 395, 1712.

[2]   A.A. Darhuber, S.M. Troian, "Principles of microfluidic actuation by modulation of surface stresses," *Ann. Rev. Fluid Mech.* **37**, pp. 425–455, 2005.

[3]   J. Berthier, K. Brakke, F. Grossi, L Sanchez, L. Di Cioccio, "Silicon die self-alignment on a wafer: stable and unstable modes," Sensors and *Transducers Journal* **115**(4), p. 135, 2010.

[4]   K. Brakke, "Minimal surfaces, corners, and wires," *J. Geom. Anal.* **2**, pp. 11–36, 1992.

[5]   C. Litterst, S. Eccarius, C. Hebling, R. Zengerle, P. Koltay, "Increasing μDMFC efficiency by passive $CO2$ bubble removal and discontinuous operation," *J. Micromech. Microeng.* **16**, pp. 248–253, 2006.

[6]   N. Paust, C. Litterst, T. Metz, M. Eck, C. Ziegler, R. Zengerle, P. Koltay, "Capillary-driven pumping for passive degassing and fuel supply in direct fuel cells," *Microfluid. Nanofluid.* **7**, pp. 531–543, 2009.

[7]   H. Bouasse. *Capillarité et Phénomènes Superficiels.* Librairie Delagrave, Paris, 1924.

[8]   P. Renvoisé, J. W. M. Bush, M. Prakash and D. Quèrè, "Drop propulsion in tapered tubes," *EPL* **86**, p. 64003, 2009.

[9]   P. Concus, R. Finn, "On the behavior of a capillary surface in a wedge," *PNAS* **63**(2), pp. 292–299, 1969.

[10] J. Berthier. *Microdrops and Digital Microfluidics.* William-Andrew Publishing, 2008.

[11] K. Khare, M. Brinkmann, B.M. Law, S. Herminghaus, and R. Seemann, "Switching wetting morphologies in triangular grooves," *Eur. Phys. J. Special Topics* **166**, pp. 151–154, 2009.

6

Microdrops in Microchannels and Microchambers

6.1 Abstract

This chapter is dedicated to the study of droplets and plugs (large droplets contacting the walls) in the geometry of microsystems, i.e. in microchannels and micro-chambers. In micro-wells, the droplet is at rest, and a static approach can be performed. On the other hand, in microchannels, fluid velocities are most of the time small; the capillary number is then much smaller than unity and a quasi-static approach is valid. However, we will discuss when needed the role of the drag force on the droplet. We shall assume gravity is negligible.

In the first part, we focus on the behavior of liquids in micro-wells by analyzing the shape of the surface, which we show depends on evaporation; in the second part, we deal with the shapes and arrangements of droplets in microchannels. In microchannel geometry, we will call the "dispersed" phase the droplets or plugs and the "continuous" phase the carrier fluid.

6.2 Droplets in Micro-wells

Micro-wells are largely used in biotechnology as containers for performing chemical and bio-chemical reactions. Droplets are usually delivered in the wells by hand or by robots (fig. 6.1). In this section, the behavior of liquids in micro-wells is investigated depending on the wetting properties of the walls.

6.2.1 Shape of the Liquid Surface in a Micro-well

The walls of a micro-well are of the same nature, either hydrophilic or hydrophobic, depending on the material and fabrication. Without treatment, glass and silicon oxide walls are hydrophilic and plastic (PDMS, polystyrene, parylene) walls are hydrophobic. Let us examine first the case of a hydrophilic micro-well. As shown in figures 6.2 and 6.3a, the liquid interface is concave, and if the well has a polygonal shape, the liquid surface peaks upwards in the corners depending on the value of the corner angle. At rest, e.g. at equilibrium, the curvature of the interface is the same everywhere. In sharp corners, the interface points upwards in order to satisfy the contact

161

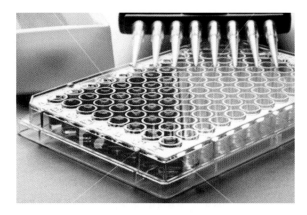

Figure 6.1 A micro-well plate.

angles. This phenomenon increases with the sharpness of the corner angle. This observation was first made by Concus and Finn in the year 1969 [1]; further theoretical developments are in [2]. If the well is a regular polygon, then the liquid surface is exactly spherical (recall we are assuming gravity is negligible). A hydrophobic micro-well may be viewed as an inverted hydrophilic micro-well, with gas and liquid exchanged.

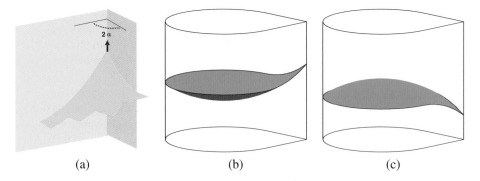

(a)   (b)   (c)

Figure 6.2 (a) A liquid interface is lifted by the corner between two wetting plates. (b) In the hydrophilic case (contact angle 60°), the surface points upwards; (c) while in the hydrophobic case (contact angle 120°) it points downwards. The wedge angle is $2\alpha = 90°$.

The shape of the liquid surface in a square well is shown in figure 6.3. In the case of a hydrophilic well, the surface of the aqueous liquid is concave and dips in its center; conversely, in the case of a hydrophobic well, the liquid surface has a convex shape and bulges up.

6.2.2   Evaporation of Liquid in a Micro-well

When working with microdrops or small amounts of liquid in an open space, evaporation is a phenomenon that systematically occurs if no precaution is taken. This section investigates how liquid recedes during evaporation.

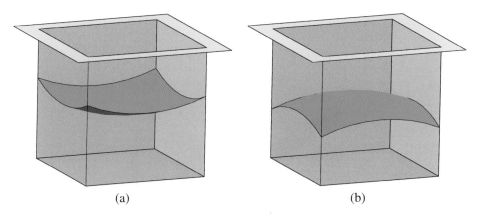

(a)   (b)

Figure 6.3 Aqueous liquid in a micro-well: (a) hydrophilic case ($\theta = 70°$); (b) hydrophobic case ($\theta = 110°$).

6.2.2.1 Hydrophilic Case

First, it is assumed that the walls and floor of the micro-well are hydrophilic (the liquid is aqueous) with the same contact angle. The shape of the interface is shown in figure 6.4 at different stages of the evaporation process. The center of the well de-wets first, and a ring of liquid remains longer along the bottom corners. This result has been experimentally observed by microscopic imagery (fig. 6.5).

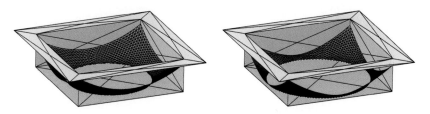

Figure 6.4 Left: in an evaporating lyophilic (hydrophilic for water) micro-well, the surface is concave and soon a central region at the bottom of the micro-well is de-wetted. Right: the liquid progressively retreats, the corners keeping traces of liquid for some time. Contact angle 60°.

If the wetting properties of the wall are chosen so that the liquid meniscus is not too curved, evaporation can be used to achieve coating of the walls. For example, Yu-Ying Lin *et al.* have obtained a uniform monolayer of 490 nm polystyrene beads by controlling the evaporation in a micro-well containing 2 μl of deionized water with 10wt% polystyrene beads [3]. Deegan *et al.* have shown that the evaporation of a droplet on a planar surface leads to the deposition of particulate rings because the evaporation rate is not uniform on the droplet surface, leading to a strong convective internal motion [4]. By using a micro-well with adequate wetting properties, the evaporation rate is much more uniform, the suspension is homogeneous and a uniform layer of beads on the floor is obtained (fig. 6.6).

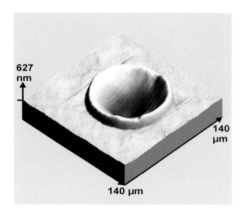

Figure 6.5 Microwell formation after evaporation of a droplet of toluene on polystyrene (PS): The image shows an Atomic Force Micrograph (AFM). (Image processing with WSxM from http://www.nanotec.es)

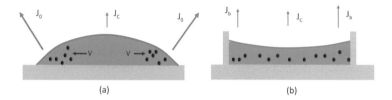

Figure 6.6 (a) Sessile droplet (Deegan *et al.* analysis), (b) droplet in a micro-well (Yu-Yinh Lin analysis): the evaporation rate being more uniform in a micro-well geometry, the suspension of microbeads is more homogeneous and a uniform coating of the floor is obtained after complete evaporation.

6.2.2.2 Hydrophobic Case

Secondly, it is assumed that the walls of the micro-well are hydrophobic (the liquid is water). In this case, the side walls and the bottom corners de-wet first, and a sessile droplet in the center of the well remains until the last stages of evaporation (fig. 6.7). But if there is the slightest asymmetry in the de-wetting, the liquid will wind up in one spherical shape attached to two walls in a corner.

Note the hydrophobic case, with positive pressure in the liquid, wants the liquid to gather in one compact shape, while the hydrophilic case, with negative pressure, wants the liquid to be shared equally among the corners.

6.2.2.3 Controlling Evaporation in Microsystems

In biotechnology, when working with droplets, evaporation must be avoided as much as possible because evaporation time is much smaller than the time needed for performing the biologic protocol [5]. Closed atmospheres on one hand, and sacrificial droplets on the other hand are often used to maintain a constant vapor pressure and limit evaporation (fig. 6.8). The purpose of sacrificial droplets is to saturate the closed atmosphere with water vapor, preventing further evaporation of the droplet of interest.

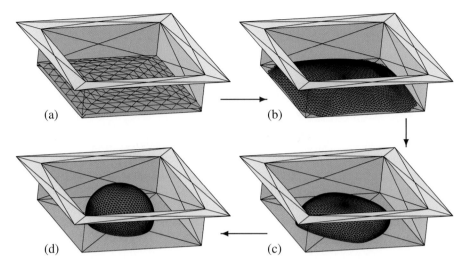

Figure 6.7 (a) In a hydrophobic micro-well, the surface of the liquid is convex; (b,c) evaporation results first in the de-wetting of the bottom corners, and of all the vertical walls; (d) as soon as the triple line is totally on the bottom surface, the liquid moves to the center to form a sessile droplet, which continues to evaporate.

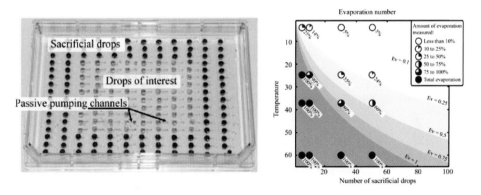

Figure 6.8 Left: evaporation from sacrificial droplets maintains the water vapor pressure in a closed box and prevents noticeable evaporation of droplets of interest (4-inch Petri dish). Right: the evaporation from droplets of interest is less than 10% when the evaporation number is smaller than 0.1. Reprinted with permission from [6], ©RSC, 2008, http://dx.doi.org/10. 1039/B717423C.

An evaporation number (denoted Ev) has been established; it indicates whether the quantity of water contained in the sacrificial droplets is sufficient to limit the evaporation from the droplets of interest [6]. More specifically, recalling that the evaporation rate of a micro-drop is proportional to its radius and inversely proportional to the liquid density [7,8], the relative evaporation of droplets of interest compared to the total relative evaporation is given by

$$\mathrm{Ev} = \frac{\dfrac{\Sigma E_i}{V_i}}{\dfrac{\Sigma E}{V_{tot}}} = \frac{\dfrac{\Sigma R_i}{\rho_i V_i}}{\dfrac{\Sigma R}{\rho_i V_{tot}}} = \frac{\dfrac{\Sigma R_i}{\rho_i V_i}}{\dfrac{\Sigma R}{\rho_{sat} V_a}} = \frac{V_a \Sigma R_i \rho_{sat}}{V_i \Sigma R \rho_i}, \tag{6.1}$$

where the index i indicates the liquid of interest (not summation), R_i the radii of the droplets of interest, V_i and E_i the initial total volume of liquid of interest and its evaporation rate, V_{tot} the total liquid volume that is evaporated, V_a the container volume and θ_{sat} the density of the vapor phase. Figure 6.8 shows the evaporation amount as a function of temperature and number of sacrificial droplets in a closed 4-inch Petri dish. The evaporation number Ev depends on the mass ratio of liquid of interest and sacrificial liquid, container volume and temperature.

6.2.3   Filling a Micro-well

Feeding liquid into a micro-well is usually done with a pipette, manually or robotically. It is important that the liquid detaches from the tip of the pipette, so that the robot can move the pipette horizontally to the next micro-well. It has been experimentally observed that the droplet detaches or not, depending on the capillary forces on the solid surfaces and on the vertical elevation of the pipette. In other words, it depends on the droplet volume delivered by the pipette, the contact angles in the well (bulging of the surface) and the vertical elevation of the pipette. This observation can be simulated with Evolver by modifying the data file corresponding to liquid droplet in a microwell with the introduction of the tip of the pipette (fig. 6.9).

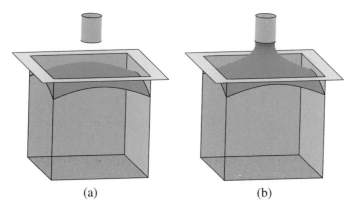

(a)   (b)

Figure 6.9 Filling a micro-well with a pipette: the droplet detaches from the tip of the pipette depending on the droplet volume, contact angles and vertical position of the pipette; (a) droplet detaches; (b) in the case of a larger liquid volume, the liquid surface stays pinned to the pipette.

Note that not all the liquid available in the pipette is dispensed, due to the formation of a neck at the onset of droplet detachment, separating the liquid in the micro-well from the liquid attached to the pipette (fig. 6.10).

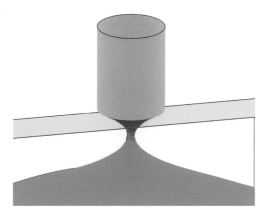

Figure 6.10 Necking at the onset of droplet detachment.

6.3   Droplets in Microchannels

Three-dimensional microfluidics addresses the physics of droplets and plugs at rest or circulating in microchannels. The applications are numerous, from flow-focusing devices (FFDs) for encapsulation purposes, to the production of emulsions, and to T-junctions for chemical and biochemical reactions [8-12].

6.3.1   Capillary, Weber and Bond Numbers

In droplet microfluidics, droplets and plugs circulate inside etched microchannels. Several dimensionless numbers describe the relative importance of various features.

When the droplet or plug velocity is not very high, the surface tension forces are predominant over inertia. The capillary number Ca represents the ratio between inertia and surface tension:

$$Ca = \frac{\eta U}{\gamma},\qquad(6.2)$$

where η is the liquid viscosity, U the average liquid velocity and γ the surface tension. Note that two capillary numbers can be defined, the internal capillary number corresponding to the viscosity η_i and velocity U_i of the dispersed phase, and the external capillary number corresponding to the viscosity η_e and velocity U_e of the continuous phase. Usually the external flow is dominant and the external capillary number indicates the relative importance of inertia and surface tension. It is very seldom that inertia dominates: for example, consider a typical flow of water ($\eta \approx 10^{-3}$ Pa.s) at a velocity of less than 10 mm/s carrying oil plugs ($\gamma_{oil-water} \approx 30$ mN/m). The external capillary number is of the order of $Ca_e \approx 3 \times 10^{-4}$.

Another dimensionless number, the Weber number We, is used to predict the disruption of an interface under the action of inertial forces working against the surface tension force. More specifically, the Weber number is the ratio of inertial forces to surface tension forces:

$$We = \frac{\rho U^2 R}{\gamma} = \frac{\rho U^2}{\frac{\gamma}{R}}.\qquad(6.3)$$

The numerator corresponds to a dynamic pressure and the denominator to a capillary pressure. A strong surface tension maintains the droplet as a unique microfluidic entity, with a convex interface. If the inertia forces are progressively increased, the interface is first deformed by waves, becomes locally concave, and is finally disrupted [13]. However, that happens seldom in microfluidics, since typical values of the Weber number are of the order of 10^{-3} to 10^{-1}, magnitudes obtained for the typical values $\rho \approx 1000$ kg/m^3, $V \approx 1$ cm/s, $\gamma \approx 1\text{-}30$ mN/m, and $R \approx 100\text{-}1000$ μm.

Finally, the relative importance of gravity and surface tension forces is given by the Bond number Bo:

$$Bo = \frac{g\Delta\rho R^2}{\gamma}, \qquad (6.4)$$

where g is the gravitational constant, $\Delta\rho$ the buoyancy term and R a characteristic dimension of the droplet or plug. Typically, $R < 1$ mm, $\Delta\rho < 1000$ kg/m^3 and $\gamma > 10$ mN/m, so that the Bond number is smaller than 0.1. There are a few interesting experiments using liquids with very low surface tension, $\gamma \approx 1$ mN/m, but in this case the density difference is small, of the order of $\Delta\rho \approx 200$ kg/m^3 and the Bond number is still small ($Bo \approx 0.2$). Hence, most of the time in microfluidic systems, it is the capillary forces and surface tension that govern the shape of the droplet, gravity and inertia having very small effects.

6.3.2 Non-wetting Droplets and Plugs

In this section, the behavior of droplets and plugs in non-wetting channels is investigated. First we describe the very useful silanization treatment, which renders a solid surface hydrophobic; then different configurations of droplets/plugs are investigated.

6.3.2.1 Silanization Surface Treatment

When it is wished that droplets and plugs not adhere to a wall, it is usual to perform a specific surface treatment at the end of the microfabrication process. Silanization is commonly used to render a wall hydrophobic. There exist different silanization processes. A common process for silicon substrates consists of grafting on the wall a monolayer of silane, i.e. FDTS perFluoro-DecylTricloro-Silane, CF3(CF2)7(CH2)2 - SiCl3. In such a case silanization is realized at 35°C in a vacuum chamber (MVD-100). The first step consists of wetting the surface to produce Si-OH bonds, and the second step creates Si-O-Si bonds as shown in figure 6.11. Many different silanization protocols can be found in the literature [14].

6.3.2.2 Rectangular Microchannels

When the droplet or plug does not wet the channel walls, a film of the continuous phase separates the dispersed phase from the walls. The images of figure 6.12 and the modeling results of figure 6.13 show typical morphologies of droplets in non-wetting channels. In the following sections, an analysis of droplets in different geometrical configurations is made.

6.3.2.3 T-shaped Microchannels – Abacus Groove

Rectangular grooves – or troughs – etched at the bottom of a rectangular channel can be used as railings to guide droplets, and to perform operations on droplets such as droplet coalescence [15,16]. Such a feature is also called an Abacus groove. The principle is sketched in figure 6.14. This principle has been proved to work well by designing an S-shaped Abacus groove in a Hele-Shaw microchamber (fig. 6.15) [16]. Let us recall here that a Hele-Shaw cell is a "flat" channel

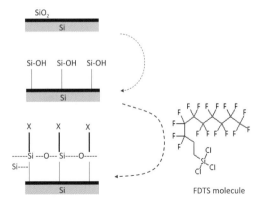

Figure 6.11 Principle of gaseous-phase silanization by X-SiCl3.

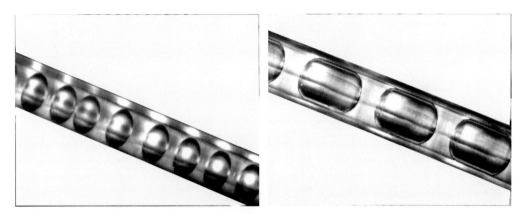

Figure 6.12 Microdrops and plugs in a capillary tube: water in oil.

whose width is much larger than its depth [17]. Note the drag force that shifts the droplets slightly sideways when the Abacus groove makes an angle with the carrier flow. If the Abacus groove is set perpendicular to the carrier flow, then a droplet may be stopped until the next

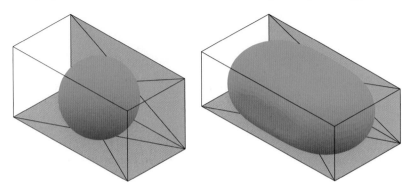

Figure 6.13 Plug in a rectangular microchannel as calculated with Evolver.

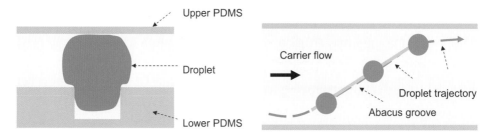

Figure 6.14 Principle of the Abacus groove for droplet guiding. Left: channel cross-section showing the deformation of the droplet in the groove. Right: deflection of the trajectory of a droplet by the Abacus groove.

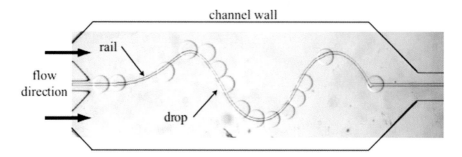

Figure 6.15 Abacus groove guiding droplets: the drag effect due to the carrier flow velocity slightly shifts the droplets in the diagonal sections of the groove. The flow velocity is of the order of 1 to 2 mm/s and the channel height 50 μm. Reprinted with permission from [15], ©Socété Hydrotechnique de France, Paris, 2010.

incoming droplet arrives and coalesces, as shown in figure 6.16 [15]. Numerical simulations show the deformation of the droplet by the Abacus groove, depending on the volume of the droplet (fig. 6.17). From the modeling results, it is seen that the guiding force depends on the anchoring of the droplet in the groove. Intuitively, the anchoring depends on the relative

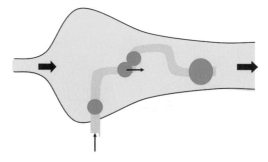

Figure 6.16 If the Abacus groove is perpendicular to the direction of the carrier flow, the droplet stops; then the next incoming droplet arrives and merges (coalesces) with the immobilized droplet. The resulting droplet is large enough to experience a sufficient drag and is transported out of the system.

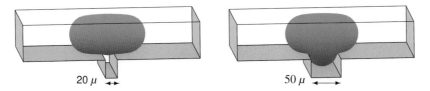

Figure 6.17 Surface Evolver results: the guiding of a droplet depends on the relative width of the rail: left, narrow rail (20 μm); right, larger rail (50 μm).

dimension of the droplet (radius R), main channel height h and width of groove w; for example, a droplet cannot penetrate into a groove of too small width, leading to poor guiding ability. As seen in figure 6.16, the drag force exerted by the carrier fluid may be sufficient to override the anchoring; the worst case being when the Abacus groove is perpendicular to the flow. Following Dangla *et al.*, an analysis of the guiding force can be done by considering the balance between two forces: on one hand, the drag force given by [18]

$$F_{drag} = 24\pi\mu\frac{R^2}{h}U_f, \tag{6.5}$$

where μ is the carrier fluid viscosity, h the depth of the Hele-Shaw cell, R the radius of the droplet, and U_f the velocity of the carrier fluid. It has been assumed that the droplet has a pancake shape and is not too deep into the trough. On the other hand, there is the surface tension force. If E is the surface energy, then

$$E = \gamma S, \tag{6.6}$$

where S is the surface area and γ the surface tension. Hence the capillary force is

$$\vec{F}_{cap} = -\gamma\vec{\nabla}S, \tag{6.7}$$

A droplet is trapped by the trough as long as $F_{drag} < F_{cap}$. The pancake-shaped droplet having a large surface energy, the indentation caused by the presence of the trough decreases the surface energy by allowing the droplet to partially enter into the trough. The trough – or Abacus groove – can be seen as an energy well. As long as $R \gg w$, the energy decrease of the droplet sitting on the trough is independent of R and scales as

$$\Delta E \approx -\gamma w^2, \tag{6.8}$$

and the capillary force then scales as

$$F_{cap} \approx -\gamma w. \tag{6.9}$$

From (6.5) and (6.8), the maximum velocity of the carrier fluid for which the droplet stays anchored by the trough is

$$U_f = \frac{\gamma}{\mu}\frac{hw}{R^2}. \tag{6.10}$$

Relation (6.9 indicates that the larger the trough width w, the better the anchoring. According to (6.9), the best anchoring would be for $h \approx w \approx R$. This situation can be reached if the depth

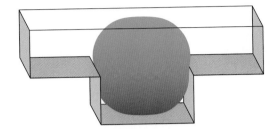

Figure 6.18 Left: a droplet can be guided by a larger trough without being deformed; right: using Abacus grooves the size of the droplet, droplets are forced to follow the system of rails. Reprinted with permission from [15], ©Socété Hydrotechnique de France, Paris, 2010.

of the trough is also of the order of R (fig. 6.18). In such a case, the droplet is almost spherical and trapped into the trough. Using (6.9) with $h \approx w \approx R$, one finds $U_f \approx \frac{\gamma}{\mu}$, which states that the droplet detaches from the trough above the critical capillary number. Besides the extremely good anchoring, another advantage of the geometry where the cross-section dimensions of the trough are of the order of the droplet radius, is that the droplet is not deformed. This is important when the droplets are capsules carrying cells, which are very sensitive biologic objects that should not be submitted to high mechanical stresses.

6.3.2.4   Plugs Slowed Down by Pillars

In the preceding section, we have seen how a trough (Abacus groove) could be used as an energy well for the guiding of droplets. Here we investigate the opposite situation, where surface energy is increased by local deformations of a liquid plug. The situation is that described by Chung *et al.* [19], where a liquid plug is wending its way between two rows of pillars under the drag force of a microflow (fig. 6.19). We study here the quasi-static behavior of a plug between pillars. The deformation of the interface between the pillars increases the surface area and results in an energy increase. Hence the plug is slowed down by the presence of the pillars. This principle can be used to force droplets or plugs to merge together, as shown by Niu *et al.* in figure 6.20 [20].

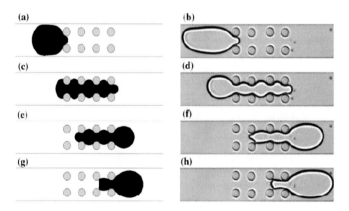

Figure 6.19 Sketch and images of a liquid plug pushed between two rows of pillars (and slowed down by the deformation imposed by the pillars) [19].

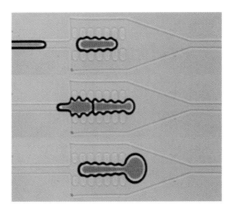

Figure 6.20 A plug is slowed down by the two rows of pillars, until a second plug arrives and merges. The merged plug is then large enough to be expelled by the hydrodynamic forces. Reprinted from [19], ©Springer, 2010.

The capillary situation can be analyzed numerically with Evolver: let us consider a plug initially squished between the two rows of pillars, with a head outside of the rows of pillars (fig. 6.21). The situation is not stable, as the high pressure at the narrow end of the plug pumps liquid into the large low-pressure bulge outside the pillars, and plug moves outside the pillar row under the action of the surface tension forces.

Figure 6.21 Numerical simulation of a plug between two rows of pillars with top and bottom nonwetting planes: in order to minimize the surface energy, the "head" increases and the "tail" progressively disappears.

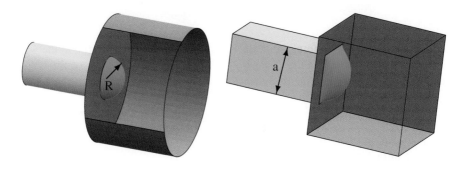

Figure 6.22 Sketch of a capillary burst valve: the liquid interface is pinned at the enlargement until the pressure exceeds a threshold value. In the case of a cylindrical channel, the threshold is $2\gamma/R$; while in the case of a square channel, it is of the order of $2\gamma/(a/2)$.

6.3.3   Wetting Droplets and Plugs

It is not always wished that the dispersed phase constituted by the droplets and plugs be isolated from the walls by a film of the continuous liquid. It is sometimes useful to have plugs touching the walls; for example plugs can be used to make a microflow stop, as is the case in capillary valving [21,22] (fig. 6.22). In the wetting case, the "dispersed" fluid interface makes an angle with the walls which is the Young contact angle if the velocity of the fluid is sufficiently small. When the velocity of the carrier fluid is important, advancing and receding contact angles must be taken into account. The Cox-Voinov relation is usually used to predict the values of the advancing and receding contact angles as functions of the static (Young) contact angle [23]:

$$\theta_{a,r}^3 = \theta_s^3 \pm AU, \tag{6.11}$$

where θ_s denotes the static (Young) contact angle, U the fluid velocity and A is a constant characteristic of the fluid. The plus sign corresponds to the advancing angle and conversely the minus sign corresponds to the receding angle. We then have the relation $\theta_a > \theta_s > \theta_r$. When the droplet velocity decreases, the receding and advancing contact angles converge towards the static contact angle. In the following we suppose that the fluid velocity is small enough to approximate the contact angles by their static values.

6.3.3.1   General Case: Contact Angle Between $45°$ and $135°$

Figure 6.23 shows the morphologies of two plugs in a rectangular channel, depending on the wetting properties of the liquid. In case of lyophobic contact (hydrophobic for water) the plug surface is convex, whereas it is concave in the lyophilic case.

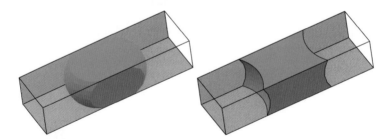

Figure 6.23 Left: a water droplet in a hydrophobic channel (with no film at the wall). Right: a water droplet in a hydrophilic channel. In both cases, the velocity of the carrier fluid is considered small enough so that the contact angles are the Young contact angles.

6.3.3.2   Contact Angle Smaller Than $45°$

In the particular case where the contact angle is smaller than $45°$, the Concus-Finn relation states that, for a dihedral with a $90°$ angle, the droplet progressively extends as a filament in the corner (fig. 6.24a). This is why sometimes a droplet can totally (and mysteriously) disappear from microscope view. The continuous phase is then separated from the walls in the corner regions (and flows in the middle).

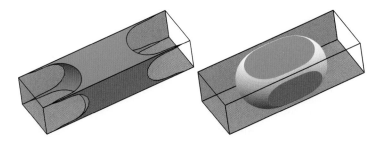

Figure 6.24 Left, filaments extends progressively when the contact angle is such that $\theta < 90° - \alpha/2$, where α is the dihedral angle between the walls. Right, droplet detaches from the corners when $\theta > 180° - \alpha/2$.

6.3.3.3   Contact Angle Larger Than $135°$

The second Concus-Finn relation states that, for a dihedral with a $90°$ angle, if the contact angle of the liquid with the wall is larger than $135°$, the droplet detaches from the corner (fig. 6.24b). Hence, tunnels are formed along the corner. If the continuous phase is a liquid, it forms a filament between the droplet and the corners.

6.3.4   Trains of Droplets – Compound Droplets

Droplets are often not isolated in a microchannel; usually there are many droplets circulating at the same time in the channel. As long as the droplets are separated by the carrier liquid, the situation is that described in the preceding sections. But droplets can come into contact. They might merge if the liquids are miscible (and without the presence of surfactants); conversely, in the case of immiscible liquids, they form a train, as shown in figure 6.25.

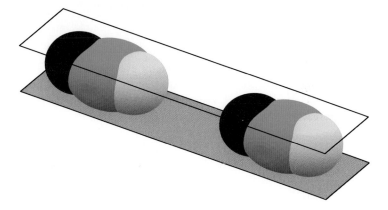

Figure 6.25 Two trains of droplets in a microchannel.

6.3.4.1   Trains Moving Inside a Microchannel

Droplet-based microfluidics has become important in today's microfluidics because it can produce micro-emulsions [9,12], encapsulate chemical compounds and even live cells [24], or simply act as micro-reactors where the droplets function not only as isolated reaction flasks, but are also capable of on-drop separation and sensing [24-26]. In their approach, Barikbin *et al.* have chosen three immiscible liquids: fluorinated oil as the continuous carrier phase, aqueous sample containing a specific target, and ionic liquid as the sensing fluid [26]. Different type of ionic liquids exist whose chemical and physical properties can be tailored in task-specific fashion. These ionic liquids are called TSIL, for Task Specific Ionic Liquids. Using a flow focusing device (FFD) like that of figure 6.26, compound droplets can be produced. Extraction and sensing of the target present in the aqueous phase is done by the TSIL. The morphology of the compound depends on three surface tensions: the surface tension γ_{WIL} between water and ionic liquid, γ_{FOW} between water and fluorinated oil, and γ_{ILFO} between ionic liquid and fluorinated oil.

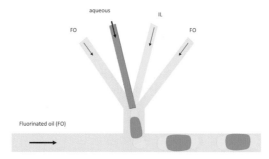

Figure 6.26 Schematic of a multiple FFD for the formation of compound droplets of water-ionic liquid in fluorinated oil [24] (IL= Ionic Liquid, W=water, and FO=Fluorinated Oil).

Note that the sketch of figure 6.26 is not always true: engulfment may occur depending on the values of these three surface tensions: the condition for stability of plugs in contact (no engulfment) is given by the balance of the surface tension forces at the triple line (Neumann's construction)

$$\gamma_{ILFO} + \gamma_{FOW} + \gamma_{WIL} = 0 \tag{6.12}$$

Relation (6.10) can be satisfied only if the magnitude of every force is smaller than the sum of the magnitudes of the other two forces. This statement can be easily verified by remarking that if the magnitude of a force is larger than the sum of the magnitudes of the two others, equilibrium cannot be reached (fig. 6.27) [8].

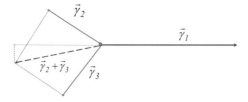

Figure 6.27 Assuming that γ_1 is larger than $\gamma_2 + \gamma_3$, the resultant of the forces projected on the direction of γ_1 cannot equilibrate.

With the three liquids mentioned above, there is no engulfment and the compounds have the morphology shown in figure 6.28.

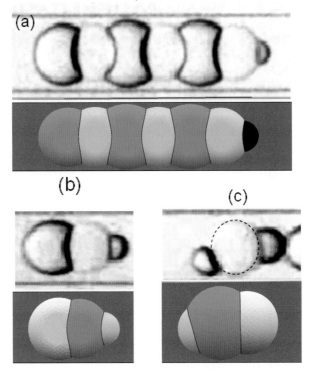

Figure 6.28 Trains of immiscible droplets: (a) train constituted of ionic liquid and water; (b) and (c) according to their size, small trains can be transported aligned or not with the flow direction. (upper images reprinted with permission from [26], ©RSC 2011, http://dx.doi.org/10.1039/C004853D)

6.3.4.2   Packing of Droplets

Micro-emulsions are constituted by an extremely high number of droplets, usually stabilized by the addition of surfactants. Due to the number of droplets, compacting emulsions is a necessity. The principle is to compress the droplets inside the microchannel. Packing patterns in flat microchannel have been investigated by Fleury *et al.* [27]. For given rectangular channel dimensions, two parameters govern the packing: first, the relative size a of the droplet defined by the ratio of the surface area A of the droplet (pancake surface) to the corresponding channel surface area. If d is the depth of the channel and w its width ($d << w$) then the volume of the droplet is

$$V_{drop} \approx Ad = \pi R^2 d, \qquad (6.13)$$

and the corresponding channel volume

$$V_{chan} \approx w^2 d. \qquad (6.14)$$

Then the size ratio is defined by

$$a = \frac{A}{w^2}. \qquad (6.15)$$

The shape of the droplets changes during the compression. But the droplet being incompressible, its volume V_{drop} is constant, and also its surface area A. Hence the coefficient a only depends on the initial size of the droplets.

The second parameter governing the arrangement of the droplets is the volume fraction of droplets, noted ϕ. This volume fraction is defined by the ratio of the total volume of droplets to the available volume of the channel. The available volume of channel depends on the compression exerted on both ends of the channel. Figure 6.29 shows the observed arrangement patterns as a function of these two parameters. In the figure, the non-dimensional compressive force f is the normalized pressure exerted on the droplets from both ends of the channel. Note that for small droplets ($a < 0.475$), when increasing the compressive force f, the arrangement suddenly jumps from a zig-zag configuration to a "staircase" configuration.

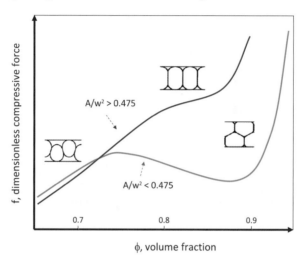

Figure 6.29 Diagram of droplet arrangement patterns in a microchannel depending on the parameters $a = A/w^2$ and ϕ. Reprinted with permission from [27], ©Socété Hydrotechnique de France, Paris, 2010.

A numerical analysis with Evolver produces similar arrangement patterns. For small compressive force, the zig-zag pattern is the unique arrangement of the droplets (assuming that the width w is not too large compared to the size of the droplet), as shown in figure 6.30. The droplets adopt an approximate triangular shape.

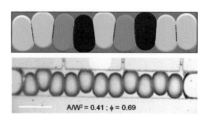

Figure 6.30 Comparison between Evolver results and experimental results showing a zig-zag arrangement at small compressive force. Reprinted with permission from [27], ©Socété Hydrotechnique de France, Paris, 2010.

If the compressive force is increased, the arrangement pattern depends on the coefficient $a = A/w^2$. If the relative droplet size a is larger than 0.475, the droplet are squished parallel (one-row bamboo structure), as shown in figure 6.31.

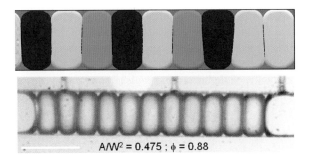

Figure 6.31 Comparison between Evolver results and experimental results showing a parallel arrangement for $a = 0.475$ and $\phi = 0.88$. Reprinted with permission from [27], ©Societe Hydrotechnique de France, Paris, 2010.

On the other hand, if the relative droplet size a is smaller than 0.475, a two-row staircase structure is formed (fig. 6.32).

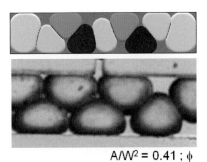

Figure 6.32 Comparison between Evolver results and experimental results showing a two-row staircase structure for a=0.41 and $\phi = 0.9$. Reprinted with permission from [27], ©Societe Hydrotechnique de France, Paris, 2010.

In fact, the entire value interval for the relative droplet size was not covered in Fleury's experiments. Using the model, it is shown that if a is sufficiently small, then two rows of droplets are obtained upon compression (fig. 6.33).

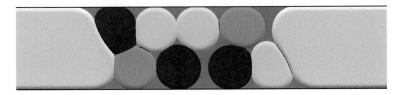

Figure 6.33 If $a = A/w^2$ is sufficiently small, the droplets rearrange in two rows. Here, $a = 0.32$.

6.4   Conclusion

In this chapter, the behavior of droplets in micro-wells and microchannels has been investigated. It has been shown that many arrangements of droplets in the geometry of microsystems are governed by capillary forces and inertia, and gravity are often negligible. In such cases, the droplet or interface morphology is governed by surface tension away from the walls, and capillary forces exerted on the triple lines at the walls. Situations such as micro-wells, capillary valving and droplets circulating in channels can then be described with a quasi-static approach.

6.5   References

[1] P. Concus, R. Finn, "On the behavior of a capillary surface in a wedge," *PNAS* **63**(2), pp. 292–299, 1969.

[2] K. Brakke, "Minimal surfaces, corners, and wires," *J. Geom. Anal.* **2**, pp. 11–36, 1992.

[3] Yu-Ying Lin, Fan-Gang Tseng,Da-Jeng Yao, "A Large Uniform Monolayer Area Obtained by Droplet Evaporation in Microwells," *Proceedings of the 2010 5th IEEE International Conference on Nano/Micro Engineered and Molecular Systems*, January 20-23, Xiamen, China.

[4] R. D. Deegan, O. Bakajin, T. F. Dupont, G. Huber, S. R. Nagel, and T. A. Witten, "Capillary flow as the cause of ring stains from dried liquid drops," *Nature* **389**(6653), p. 827, 1997.

[5] R. J. Jackman, D. C. Duffy, E. Ostuni, N. D. Willmore and G. M. Whitesides, "Fabricating large arrays of microwells with arbitrary dimensions and filling them using discontinuous dewetting," *Analytical Chemistry* **70**, pp. 2280–2287, 1998.

[6] E. Berthier, J. Warrick, H. Yu, DJ. Beebe, "Managing evaporation for more robust microscale assays. Part 1. Volume loss in high throughput assays," *Lab Chip* **8**(6), pp. 852–859, 2008.

[7] K.S. Birdi, D.T. Vu, A. Winter, "A study of the evaporation rate of small water drops placed on a solid surface," *J. Phys. Chem.* **93**, pp. 3702–3703, 1989.

[8] J. Berthier. *Microdrops and Digital Microfluidics*, William Andrew Publishers, 2008.

[9] S.L. Anna, N. Bontoux, and H.A. Stone, "Formation dispersions using "flow focusing" in microchannels," *Appl. Phys. Lett.*, **82**(3), pp. 364–366, 2003.

[10] J. Atencia, D.J. Beebe, "Controlled microfluidic interfaces," *Nature* **437**, pp. 648–655, 2005.

[11] P. Garstecki, H. A. Stone, George M. Whitesides, "Mechanism for flow-rate controlled breakup in confined geometries: a route to monodisperse emulsions," *Phys. Rev. Lett.* **94**, pp. 164501-1-4, 2005.

[12] A.M. Gañán-Calvo, J.M. Gordillo, "Perfectly monodisperse microbubbling by capillary flow focusing," *Phys. Rev. Lett.* **87**(27), pp. 274501-1-4, 2001.

[13] Yu Pan, Kazuhiko Suga, "A numerical study on the breakup process of laminar jets into a gas," *Physics of Fluids* **18**, p. 052101, 2006.

[14] www.bcm.edu/microarray/Silanization.pdf.

[15] Eujin Um and Je-Kyun Park, "A microfluidic abacus channel for controlling the addition of droplets," *Lab Chip* **9**, pp. 207–212, 2009.

[16] R. Dangla, Sungyon Lee, C. Baroud, "Anchors and rails: trapping and guiding drops in 2D," *Proceedings of the 2nd European Microfluidic Conference*, Toulouse 7-9 December 2010.

[17] http://en.wikipedia.org/wiki/Hele-Shaw_flow.

[18] S.R.K. Maruvada and C.W. Park, "Retarded motion of bubbles in Hele-Shaw cells," *Physics of Fluids* **8**, p. 3229, 1996.

[19] Changkwon Chung, Misook Lee, Kookheon Char, Kyung Hyun Ahn, Seung Jong Lee, "Droplet dynamics passing through obstructions in confined microchannel flow," *Microfluidics and Nanofluidics Journal* **9**, pp. 1151–1163, 2010.

[20] X. Niu, S. Gulati, J.B. Edel, A.J. deMello, "Pillar-induced droplet merging in microfluidic circuits," *Lab Chip* **8**, pp. 1837–1841, 2008.

[21] D. Irima, *Encyclopedia of Microfluidics and Nanofluidics*, ed. D. Li, Springer, 2008.

[22] Hansang Cho, Ho-Young Kim, Ji Yoon Kang, and Tae Song Kim, "Capillary passive valve in microfluidic systems," *Technical Proceedings of the 2004 NSTI Nanotechnology Conference and Trade Show*, Volume 1, 7-11 March 2004, Boston, USA.

[23] T. Podgorski, J.-M. Flesselles and L. Limat, "Corners, cusps and pearls in running drops," *Phys. Rev. Lett.* **87**, pp. 036102-036105, 2001.

[24] Choong Kim, Seok Chung, Young Eun Kim, Kang Sun Lee, Soo Hyun Lee, Kwang Wook Oh, and Ji Yoon Kang, "Generation of core-shell microcapsules with three-dimensional focusing device for efficient formation of cell spheroid," *Lab Chip* **11**, pp. 246–252, 2011.

[25] H. Song, D.L. Chen, and R.F. Ismagilov, "Reactions in droplets in microfluidics channels," *Angewandte Chemie* **45**(44), pp. 7336-7356, 2006.

[26] Zahra Barikbin, M Taifur Rahman, Pravien Parthiban, Anandkumar S. Rane, Vaibhav Jain, Suhanya Duraiswamy, S. H. Sophia Lee and Saif A. Khan, "Ionic liquid-based compound droplet microfluidics for 'on-drop' separations and sensing," *Lab Chip* **10**, pp. 2458–2463, 2011.

[27] J-B Fleury, O. Caussen, S. Herminghaus, M. Brinkman and R. Seemann, "Topological transition in a stack of water droplets," *Proceedings of the 2nd European Microfluidic Conference*, Toulouse 7-9 December 2010.

7

Capillary Effects: Capillary Rise, Capillary Pumping, and Capillary Valve

7.1   Abstract

In this chapter, we investigate the effect of capillarity forces on liquid rise, liquid pumping and valving. In the first part, we analyze how capillary forces oppose gravity to trigger capillary rise in and around solid structures such as plates, hollow pillars and bundle of solid pillars. In the second part, the theory of capillary pumping is developed and examples of capillary pumps are given. In the third part, the effectiveness of capillary valves is analyzed. Note that capillary pumping and valving are wide-spread for small scale microsystems because of their simplicity. For example, in biotechnology, most point of care (POC) microsystems use capillarity and wetting for filling of components with liquid samples.

7.2   Capillary Rise

In the introduction of this book, we have shown that capillary forces are usually negligible at the macroscale compared to other forces such as gravity, and inertia, but become important as the geometric scales are reduced. They are often dominant at the microscale and nearly always at the nanoscale. In this first section, we investigate how capillary forces can make liquids rise against gravitational forces. It is shown that the height that the liquid can reach is considerable.

7.2.1   Cylindrical Tubes: Jurin's Law

When a capillary tube is plunged into a volume of wetting liquid, the liquid rises inside the tube under the effect of capillary forces (figure 7.1). It is observed that the height reached by the liquid is inversely proportional to the inner radius of the tube. This property is usually referred to as Jurin's law. Using the principle of minimum energy, one can conclude that the liquid goes up in the tube if the surface energy of the dry wall is larger than that of the wetted wall. If we define the impregnation criterion I by

$$I = \gamma_{SG} - \gamma_{SL},$$ (7.1)

then the liquid rises in the tube if $I > 0$. Upon substitution of the Young law in (7.1), the impregnation criterion can be written in the form

$$I = \gamma_{SG} \cos \theta.$$ (7.2)

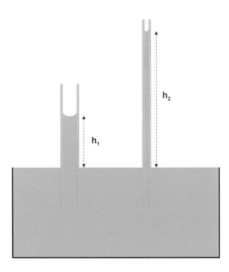

Figure 7.1 Capillary rise is inversely proportional to the capillary diameter.

When the liquid rises in the tube, the system gains potential energy – because of the elevation of a volume of liquid – and loses capillary energy – due to the reduction of the surface energy. The balance is [1]

$$E = \frac{1}{2}\rho g h V_{liquid} - S_{contact} I = \frac{1}{2}\rho g h (\pi R^2 h) - 2\pi R h I = \frac{1}{2}\rho g \pi R^2 h^2 - 2\pi R h \gamma_{SG} \cos \theta.$$ (7.3)

Note that the detailed shape of the meniscus has not been taken into account in (7.3). The interface stabilizes when

$$\frac{\delta E}{\delta h} = 0,$$ (7.4)

which results in

$$h = \frac{2\gamma \cos \theta}{\rho g R}.$$ (7.5)

Equation (7.4) is termed the Jurin law: the capillary rise is inversely proportional to the tube radius. It can also be applied to the case where the liquid level in the tube decreases below the outer liquid surface; this situation happens when $\theta > 90°$. The maximum possible height that a liquid can reach corresponds to $\theta = 0$: $h = 2\gamma/\rho g R$. In microfluidics, capillary tubes of 100 μm diameter are currently used; if the liquid is water ($\gamma = 72$ mN/m), and using the approximate value $\cos\theta \approx \frac{1}{2}$, the capillary rise is of the order of 14 cm, which is quite important at the scale of a microcomponent.

What is the capillary force associated to the capillary rise? As stated first by Princen [2,3], at equilibrium, the capillary force balances the weight of the liquid in the tube. This weight is given by

$$F = \rho g \pi R^2 h. \tag{7.6}$$

Replacing h by its value from equation (7.4), we find the capillary force

$$F = 2\pi R \gamma \cos\theta. \tag{7.7}$$

The capillary force is the product of the length of the contact line $2\pi R$ times the line force $f = \gamma\cos\theta$. This line force is sketched in figure 7.2.

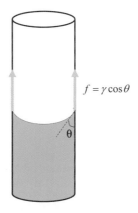

Figure 7.2 Sketch of the capillary force of a liquid inside a tube.

For $f > 0$ the liquid goes up in the tube and for $f < 0$ the liquid goes down. Note that figure 7.1 is not quite exact. There are also capillary forces on the outside of the tube, as shown in figure 7.3.

To derive the expression of the capillary rise inside the tube, we have considered a control volume corresponding to the liquid volume inside the tube. We now can consider a control volume defined by a pipette (figure 7.4). The vertical force oriented upwards to maintain the pipette is

$$F = P - P_A + P_{c,i} + P_{c,e}, \tag{7.8}$$

where P is the weight of the pipette, P_A the floatation force and $P_{c,i}$ and $P_{c,e}$ are respectively the interior and exterior capillary forces exerted on the solid. Thus

$$F = P - P_A + 2\pi R_i \gamma \cos\theta + 2\pi R_e \gamma \cos\theta. \tag{7.9}$$

Note that this force is a function of the surface tension γ.

7.2.2 Capillary Rise in Square Tubes

The same arguments of the Princen approach apply for square tubes [2,3]. The capillary force pointing upwards is

$$F_{cap} = 4a\gamma\cos\theta, \tag{7.10}$$

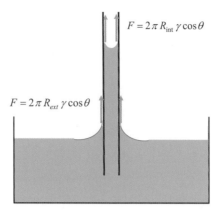

Figure 7.3 Capillary forces act also on the tube exterior, raising the level of the liquid around the tube.

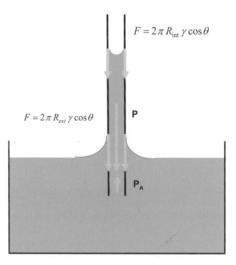

Figure 7.4 Forces acting on the pipette.

where a is the side of the square section. We can approximate the liquid column weight by

$$F_{grav} = \rho g a^2 h, \tag{7.11}$$

where h is the height of the liquid column (in the middle of the channel). Equating the two forces results in

$$h = \frac{4\gamma\cos\theta}{\rho g a}. \tag{7.12}$$

Introducing the capillary length $\kappa^{-1} = \sqrt{\frac{\gamma}{\rho g}}$, relation (7.12) becomes

$$h = \frac{4\cos\theta\kappa^{-2}}{a}. \tag{7.13}$$

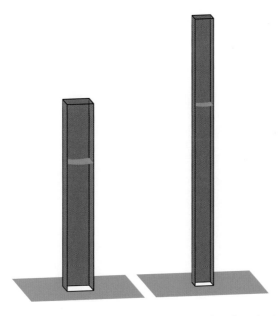

Figure 7.5 Capillary rise in vertical square tubes (Evolver) showing the height of the liquid in hollow square tubes.

Again, it is found that the elevation of the liquid column is inversely proportional to the characteristic dimension of the cross-section of the tube (figure 7.5).

A more accurate approach for the liquid weight would be to take into account the horizontal curvature of the meniscus. This approach brings a noticeable correction only if filaments occupy the corners; this happens when the Concus-Finn relation is not respected, i.e. $\theta + \alpha/2 < \pi/2$ for corner dihedral angle α. Taking into account the filaments in the corners, the total weight of the liquid can be written (figure 7.6)

$$F_{grav} = \rho g a^2 h + 4\rho g \int_h^\infty \left(1 - \frac{\pi}{4}\right) R^2(z) dz. \tag{7.14}$$

Bico and Quèrè [4] have made an interesting simplification by neglecting the curvature in the vertical plane (x, z) in favor of the curvature $1/R$ in the horizontal plane ($z = $ constant). Then the Laplace law becomes

$$\frac{\gamma}{R} = \rho g z. \tag{7.15}$$

Substitution of (7.15) in (7.14) followed by integration yields a second order polynomial in h. If for simplicity we assume that $\theta = 0$, the polynomial is

$$h^2 - 4\kappa^{-2} \frac{h}{a} + 4 \left(1 - \frac{\pi}{4}\right) \frac{\kappa^{-4}}{a^2} = 0. \tag{7.16}$$

The solution is

$$h = (2 + \sqrt{\pi}) \frac{\kappa^{-2}}{a}. \tag{7.17}$$

This expression is very similar to (7.13), with a coefficient $2 + \sqrt{\pi}$ instead of 4.

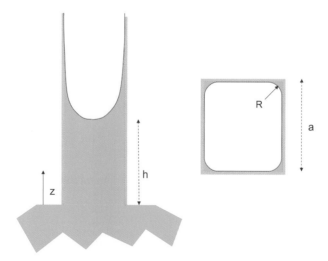

Figure 7.6 Sketch of the capillary rise inside a square tube for a contact angle satisfying the Concus-Finn relation (filament extending to infinity in the corners).

7.2.3   Capillary Rise on a Vertical Plate – Surface Tension Measurement by the Wilhelmy Method

In section 7.2.1 we have seen that the force exerted on a solid dipped in a liquid depends on the surface tension of the liquid with the surrounding air. A Wilhelmy plate is a thin plate that is used to measure equilibrium surface or interfacial tension at an air-liquid or liquid-liquid interface [5]. The plate is oriented perpendicular to the interface, and the force exerted on it is measured. The Wilhelmy plate consists of a thin plate, usually on the order of a few centimeters square. The plate is attached to a scale or balance via a thin metal wire. The force on the plate due to wetting is measured via a tensiometer or microbalance and used to calculate the surface tension using the equation

$$\gamma = \frac{F}{L\cos\theta},$$
(7.18)

where L is the wetted perimeter $(2w + 2d)$ of the Wilhelmy plate and θ is the Young contact angle between the liquid phase and the plate. In practice the contact angle is rarely measured, instead either literature values are used, or complete wetting ($\theta = 0$) is assumed.

In the Wilhelmy method, it is important that the position of the plate is correct, meaning that the lower end of the plate is exactly on the same level as the surface of the liquid. Otherwise the buoyancy effect must be calculated separately. Figure 7.7 shows a numerical simulation of the Wilhelmy method with Surface Evolver.

A broader scope of surface tension measurement can be found in [6,7].

7.2.4   Capillary Rise Between Two Parallel Vertical Plates

The same balance between capillary forces and gravitational forces can be done for a meniscus between two parallel plates (figure 7.8) separated by a distance $d = 2R$. The approach leads to a capillary force given by

$$F_{cap} = 2w\gamma\cos\theta,$$
(7.19)

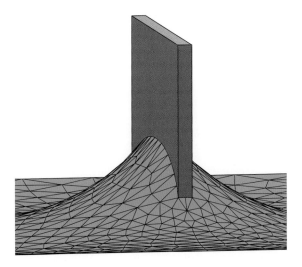

Figure 7.7 Capillary rise on a Wilhelmy plate (Evolver).

where w is the width of the plates, and an approximate gravitational force

$$F_{grav} \approx \rho g(2R)wh. \tag{7.20}$$

Note that the relation (7.20) is only approximate because the weight of the volume of water above the $z = 0$ plane is slightly different than that predicted by (7.48) due to the rise of liquids beyond the rectangle comprised between the two plates (figure 7.9). At equilibrium the two forces balance each other, yielding a capillary rise

$$h \approx \frac{\gamma \cos \theta}{\rho g R}. \tag{7.21}$$

By substituting in (7.21) the capillary length defined by

$$\kappa^{-1} = \sqrt{\frac{\gamma}{\rho g}}, \tag{7.22}$$

one obtains

$$h = \kappa^{-2} \frac{\cos \theta}{R}. \tag{7.23}$$

Figure 7.9 shows the capillary rise between two identical, vertical parallel plates calculated with Evolver. A comparison of the height reached by the liquid as a function of the inter-plate distance is shown in figure 7.10.

Note that the expressions for the two geometries (cylinder and two parallel plates) are similar. If we use the coefficient c, with $c = 2$ for a cylinder and $c = 1$ for parallel plates, we obtain

$$h = c\kappa^{-2} \frac{\cos \theta}{R}, \tag{7.24}$$

where R is either the radius of the cylinder or the half-distance between the plates.

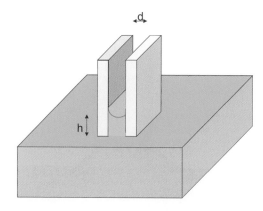

Figure 7.8 Sketch of the capillary rise between two parallel vertical plates.

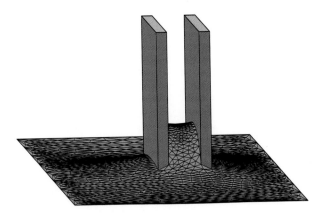

Figure 7.9 Capillary rise between two parallel vertical plates (Evolver).

7.2.5   Capillary Rise in a Dihedral

When a corner formed by two solid plates is put in contact with a wetting liquid, a meniscus generally rises inside the corner. The questions that can be raised are: what are the conditions

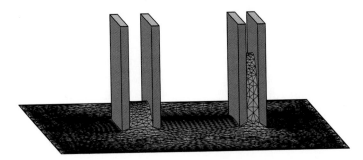

Figure 7.10 Capillary rise is inversely proportional to the distance between the two plates ($\gamma =$ 72 mN/m and $\theta = 60°$).

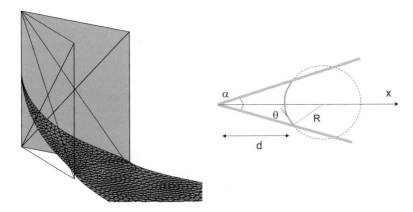

Figure 7.11 Left: Shape of the interface obtained with Evolver; right: cross cut in a horizontal plane: R is the horizontal curvature radius, α the dihedral angle and θ the contact angle.

for observing a liquid rise, and what is the shape of the meniscus.

Bico and Quèrè [4] have made an interesting approach by neglecting the curvature in vertical planes in favor of the curvature in the horizontal plane ($z = $ constant). A schematic view of the interface is shown in figure 7.11. The Laplace law then yields

$$\frac{\gamma}{R} = \rho g z, \tag{7.25}$$

where R is the horizontal curvature radius. Geometrical considerations in figure 7.11 lead to

$$(d + R) \sin \frac{\alpha}{2} = R \cos \theta, \tag{7.26}$$

which can be written in the form

$$d = R \left(\frac{\cos \theta}{\sin \frac{\alpha}{2}} - 1 \right). \tag{7.27}$$

The distance d is positive (which means that the liquid rises) if

$$\cos \theta > \sin \frac{\alpha}{2}, \tag{7.28}$$

which is the Concus-Finn relation

$$\theta + \frac{\alpha}{2} < \frac{\pi}{2}. \tag{7.29}$$

After substitution of (7.25) in (7.27), we obtain

$$d = \frac{\gamma}{\rho g z} \left(\frac{\cos \theta}{\sin \frac{\alpha}{2}} - 1 \right), \tag{7.30}$$

showing that the vertical profile in the xz-plane is a hyperbola. Upon introducing the capillary length $\kappa^{-1} = \sqrt{\frac{\gamma}{\rho g}}$, the expression of the vertical profile is

$$d \propto \frac{\kappa^{-2}}{z} \tag{7.31}$$

It has been shown recently that the addition of the vertical curvature radius does not modify notably the shape of the interface [8].

7.2.6   Capillary Rise in an Array of Four Vertical Square Pillars

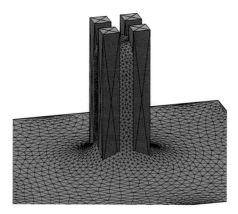

Figure 7.12 Capillary rise triggered by four square pillars of 50 μm, with 40 μm spacing: case of water with a Young contact angle of 60° with the pillars.

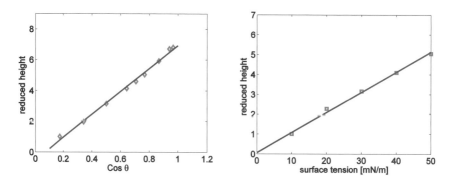

Figure 7.13 As expected, the capillary rise is proportional to $\cos\theta$ and to the surface tension γ.

Now we turn to the capillary rise between parallel, vertical pillars. First, we perform a numerical analysis with Evolver for a bundle constituted of four square pillars (figure 7.12) with spacing δ and pillar width a. It is verified that the capillary height is proportional to the cosine of the contact angle θ and to the surface tension γ (figure 7.13).

On the other hand, the variation of the height of the capillary rise as the inverse of the capillary characteristic dimension $h \approx 1/\delta$ that has been observed previously for ducts or between plates is still valid in the case of pillars. Figure 7.14 shows the relation between the liquid height h and the pillar spacing δ for a bundle of four pillars. The different computational results align along a straight line in a $1/\delta$ coordinate system, except for large values of δ, where the capillary rise is almost constant: as long as the contact angle is less than 90°, there is some capillary rise along the pillars.

In conclusion, it is found that, for small pillar distance ($\delta < a$) the capillary rise in a bundle

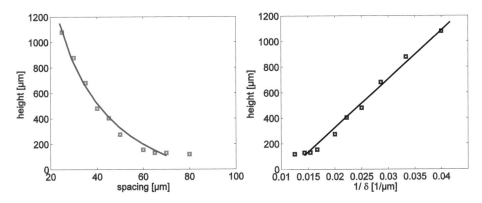

Figure 7.14 Height of capillary rise h versus pillar spacing δ: h varies as $1/\delta$ except at large spacing where the capillary rise is nearly constant ($\theta = 60°$ and $\gamma = 72$ mN/m).

of pillars satisfies the relation

$$d \propto \frac{\gamma \cos \theta}{\rho g \delta}.$$ (7.32)

7.2.7 Comparison of Capillary Rise Between Wilhelmy Plate and Pillars

In this section, it is shown that a densely packed bundle of pillars produces a larger capillary rise than a simple Wilhelmy plate.

Let us compare a Wilhelmy plate of cross-section $a = 350$ μm, $b = 200$ μm with a bundle of six square pillars delimiting the same rectangle as the Wilhelmy plate cross-section (figure 7.15). The pillar side is $c = 50$ μm, and pillar spacing is $\delta = 100$ μm. Hence, $a = 7c$ and $b = 4c$.

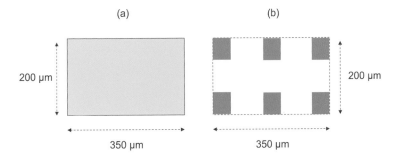

Figure 7.15 Schematic of the cross-sections: (a) Wilhelmy plate, (b) pillars.

A comparison of the capillary forces yields:

$$F_1 = 2(a+b)\gamma \cos \theta$$ (7.33)

and

$$F_2 = 6(4c)\gamma \cos \theta.$$ (7.34)

The ratio of these two forces is

$$\frac{F_1}{F_2} = \frac{a+b}{12c} = \frac{11}{12} \approx 1. \tag{7.35}$$

Thus, in the two cases, the capillary forces are of the same order. The Evolver calculation indicates comparable liquid rise, with $h = 170\,\mu$m in the first case and $h = 220\,\mu$m in the second case (figure 7.16).

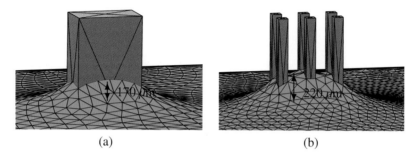

(a)   (b)

Figure 7.16 (a) Capillary rise for a Wilhelmy plate and (b) a bundle of pillars.

Now, if we consider a more densely packed bundle, the capillary rise can be substantially higher. Consider a bundle of pillars of $c' = 20\,\mu$m with a spacing of $\delta = 10\,\mu$m. The same cross-section will then be delimited by 12×7 pillars. In this case, and $a = 17.5c'$, $b = 10c'$. The capillary force is

$$F' = 84(4c')\gamma\cos\theta. \tag{7.36}$$

The capillary force ratio is now

$$\frac{F_1}{F'} = \frac{a+b}{168c'} \approx 0.16. \tag{7.37}$$

The capillary force is now six times larger than that of the Wilhelmy plate. In consequence, for liquids with a very small surface tension, a bundle of pillars will still produce a noticeable rise compared to a Wilhelmy plate.

7.2.8   Oblique Tubes – Capillary Rise in a Pipette

The analysis of the capillary rise – or descent – in cylindrical tubes or between two parallel plates has been recently extended by Tsori [9] to the case where the walls are not parallel, as, for instance, for a conical pipette (figure 7.17).

Mechanical equilibrium states that the Laplace pressure is balanced by the hydrostatic pressure,

$$P_0 + \frac{c\gamma}{R} = P_0 - \rho g h, \tag{7.38}$$

where c is the index defined previously, $c = 2$ for cones and $c = 1$ for wedges, and r is the curvature radius of the meniscus. Note that the depth h is counted negatively from the surface. The curvature radius is expressed by

$$r(h) = -\frac{R(h)}{\cos(\theta + \alpha)} \tag{7.39}$$

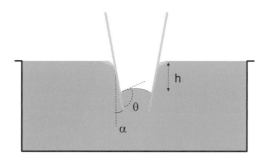

Figure 7.17 Capillary descent inside a hydrophobic pipette.

with

$$R(h) = R_0 + h \tan \alpha, \tag{7.40}$$

where R_0 is the internal radius of the pipette at $h = 0$ and $R(h)$ the horizontal radius at the height h. Substituting (7.40) in (7.39) and then in (7.38), we obtain

$$\cos(\theta + \alpha) = \frac{1}{c}(R_0 + h \tan \alpha)\frac{\rho g}{\gamma}. \tag{7.41}$$

Using capillary length $\kappa^{-1} = \sqrt{\frac{\gamma}{\rho g}}$ to scale the variables, we obtain the dimensionless variables $\bar{R} = \kappa R$ and $\bar{h} = \kappa h$. Equation (7.41) then becomes

$$\tan \alpha \, \bar{h}^2 + \bar{R}_0 \bar{h} - c \cos(\theta + \alpha) = 0. \tag{7.42}$$

This is a quadratic equation in $\bar{h}$. The discussion of this equation is somewhat complex. Depending on the values of α and θ, the meniscus may be stable or not stable; in this latter case, the meniscus jumps to the top or the bottom of the pipette, where it gets stabilized by pinning (anchoring to an angle). The diagram of figure 7.18 summarizes the meniscus behavior.

The important information here is that there are two domains where the meniscus "jumps" inside the pipette until it finds a pinning edge (figure 7.18). The first case is that of a cone/wedge angle α larger than a critical value α^* (and contact angle θ sufficiently large): the meniscus stays pinned at the bottom of the pipette, and no liquid penetrates the pipette, unless a negative pressure is established. On the other hand, when the angles α and θ are sufficiently small (α smaller than a negative critical value $-\alpha^*$), the liquid jumps to the top of the cone/wedge. The critical values depend on the internal radius R_0 of the pipette at $h = 0$. In the case of figure 7.17, the cone angle α is positive, and there is a large set of conditions where the meniscus will stay at the bottom.

In conclusion, a cone-shaped micro-pipette dipped into a liquid does not always have the expected behavior, i.e., there might not be the expected capillary rise/descent. However, if the absolute value of the angle of the cone or wedge is smaller than a critical value α^*, there will be a continuous rise or descent similar to that observed for parallel walls.

7.3 Capillary Pumping

Passive microfluidics systems are interesting for their simplicity and stand-alone solutions. For such micro- and nano-systems, capillary pumping is most of the time the best approach. In

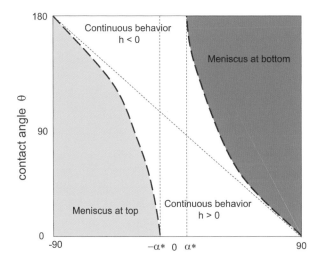

Figure 7.18 Diagram of meniscus behavior in the (θ, α) plane. The angle α^* is defined by $\sin \alpha^* = \bar{R}_0^2 / 4c$. Depending on the contact angle, if the angle of a pipette is sufficiently large, whether it is hydrophobic or hydrophilic, the meniscus will stay at the bottom. No liquid will penetrate the pipette unless a negative pressure is established in the pipette [5].

this section, we deal with the principle of capillary pumping and present some designs for microsystems.

7.3.1   Principles of Capillary Pumping

There are two ways to look at the physics of capillary pumping: either by investigating the capillary forces exerted on the triple contact line, or by calculating the capillary pressure.

On one hand, the capillary line force (per unit length) is

$$f = \gamma \cos \theta. \tag{7.43}$$

On the other hand, the capillary pressure can be calculated by two different approaches. The first one is the Laplace law. Assuming a 2D situation, it is easy to show that the curvature radius of the interface is

$$R = \frac{\delta}{2 \cos \theta}, \tag{7.44}$$

and the capillary pressure is

$$P = 2 \frac{\gamma \cos \theta}{\delta}. \tag{7.45}$$

The second approach uses the energy variation with a change of the liquid volume [10]:

$$P = -\frac{\delta E}{\delta V} = \gamma_{SG} \frac{\delta A_{SG}}{\delta V} + \gamma_{SL} \frac{\delta A_{SL}}{\delta V} + \gamma_{LG} \frac{\delta A_{LG}}{\delta V}. \tag{7.46}$$

Considering that any change of A_{SL} is at the expense of A_{SG},

$$\delta A_{SG} = -\delta A_{SL}, \tag{7.47}$$

and using the Young equation,

$$\gamma_{LG}\cos\theta = \gamma_{SG} - \gamma_{SL}, \tag{7.48}$$

equation (7.46) collapses to

$$P = \gamma_{LG}(\gamma_{SG} - \gamma_{SL}). \tag{7.49}$$

If the channel has a uniform cross-section (area A, perimeter p), then

$$\frac{\delta A_{SL}}{\delta V} = \frac{p\,dx}{a\,dx} = \frac{p}{a} = \frac{2h}{\delta h} = \frac{2}{\delta}, \tag{7.50}$$

$$\frac{\delta A_{SG}}{\delta V} = 0. \tag{7.51}$$

Substitution in (7.49) yields

$$P = \frac{2\gamma_{LG}\cos\theta}{\delta}, \tag{7.52}$$

which is the same as relation (7.45).

7.3.2 Capillary Pumping and Channel Dimensions

The effectiveness of capillary pumping as a function of the channel dimensions is shown in figure 7.19. Consider a given volume of liquid which is wanted to be moved by capillary forces. In the large channel, the capillary force acting on the fluid is

$$F_1 = 2\gamma_{LG}h\cos\theta, \tag{7.53}$$

where h is the width of the contact line. If the large channel is divided by a number n of auxiliary walls, the capillary force is

$$F_1 = 2\gamma_{LG}h\cos\theta + 2n\gamma_{LG}h\cos\theta = 2(n+1)\gamma_{LG}h\cos\theta. \tag{7.54}$$

Therefore

$$\frac{F_2}{F_1} = n+1. \tag{7.55}$$

Hence the partition of the main channel into subchannels of smaller dimensions increases considerably the effectiveness of capillary pumping. This is the reason for the different designs of capillary pumps presented in the next subsection.

7.3.3 The Dynamics of capillary Pumping: Horizontal Microchannel

The next two sections deal with the dynamics of capillary pumping. Two approaches are presented. The first one is based on the Washburn law for a flat microchannel, the second one is more general and can take into account conical channels – channels whose cross-section slightly increases.

Consider a microchannel placed horizontally instead of vertically (figure 7.20), with a constant inlet pressure, and suppose the pressure of the liquid overbalances the capillary force so a continuous flow is set up that lasts as long as there is liquid available in the entry port. Let us

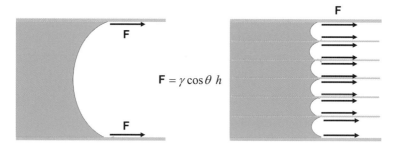

Figure 7.19 Capillary pumping is more efficient when the dimensions of the channel are reduced by the use of secondary walls.

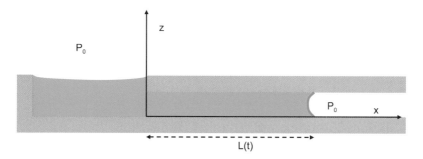

Figure 7.20 Principle of capillary pumping: liquid at right progresses under the action of capillary forces.

assume that the reservoir is large so that the curvature of the reservoir interface can be neglected (for example the size of the reservoir is larger than twice the capillary length). Assuming that gravity is negligible in the small depth of the reservoir, the pressure at $x - 0$ is then P_0, the atmospheric pressure. Let us assume also that the channel is rectangular (width w, depth d, with $d \ll w$).

Following Bruus' approach [11], which directly derives from the Washburn law [12], the continuous flow in the horizontal channel is supposed to have a Poiseuille-Hagen profile, and the flow rate can be symbolically written

$$\Delta P = R_V V = \frac{\eta L}{f(w,d)} V = \frac{\eta L}{f(w,d)} \frac{Q}{wd},$$ (7.56)

where f is a function of the cross dimensions of the channel and its aspect ratio, L is the length of the wetted channel, V the average velocity of the flow, Q is the flow rate and η the fluid viscosity. A currently used formula to calculate the laminar pressure drop in a rectangular microchannel of aspect ratio $\varepsilon = \min(\frac{w}{d}, \frac{d}{w}) = \frac{d}{w}$ is the expression [13,14]

$$\Delta P = \frac{4\eta LV}{\min(w,d)^2 q(\varepsilon)} = \frac{4\eta LV}{d^2 q(\varepsilon)},$$ (7.57)

with

$$q(\varepsilon) = \frac{1}{3} - \frac{\varepsilon \tanh(\frac{\pi}{2\varepsilon})}{\frac{64}{\pi^5}}.$$ (7.58)

This formula, derived from an approximated velocity profile, produces very accurate values of the pressure drop, and its validity regarding the aspect ratio has been checked numerically in [15]. Hence, in (7.56), the function f is of the form

$$f = \min(w,d)^2 q(\varepsilon) = d^2 \frac{1}{3} - \frac{\varepsilon \tanh\left(\frac{\pi}{2\varepsilon}\right)}{\frac{64}{\pi^5}}. \tag{7.59}$$

On the other hand, the velocity V of the flow is given by

$$V = \frac{\delta L}{\delta t}. \tag{7.60}$$

And, because the channel depth d is much smaller than the width w, the driving pressure ΔP can be written

$$\Delta P = \frac{2\gamma \cos\theta}{\delta}. \tag{7.61}$$

After substitution of (7.60) in (7.57) and using (7.61), the length L of the liquid in the channel is the solution of the differential equation

$$2L\Delta L = \frac{d\gamma \cos\theta q(\varepsilon)}{\eta} \delta t, \tag{7.62}$$

which can be readily integrated, yielding

$$L(t) = \sqrt{\frac{d\gamma \cos\theta q(\varepsilon)}{\eta} t}. \tag{7.63}$$

Expression (7.63) is very similar to Washburn's law (that assumes a cylindrical channel). Using (7.60), the flow velocity V is then

$$V = \sqrt{\frac{d\gamma \cos\theta q(\varepsilon)}{\eta}} \frac{1}{\sqrt{t}}. \tag{7.64}$$

The velocity of the flow decreases like $1/\sqrt{t}$. This decrease results from a balance between the constant Laplace driving pressure and the increasing flow resistance in the channel. Even though the liquid may progress very far, as the velocity becomes small, it may also be stopped by pinning on any inhomogeneity at the surface of the channel walls.

From a numerical point of view, the capillary pumping described above can be modeled with Evolver (figure 7.21).

Note that, according to (7.64), the fluid velocity is proportional to $\sqrt{dq(d/w)}$. By reducing the distance between the walls, the capillary force increases but the friction increases more rapidly.

7.3.4   The Dynamics of Capillary Pumping: General

An interesting approach to the dynamics of capillary pumping has been proposed by Reyssat *et al.* [16]. In their approach, the channel is not necessarily of constant cross-section but can be slightly conical, i.e. the cross-section increases with the distance, such as a wedge or a cone.

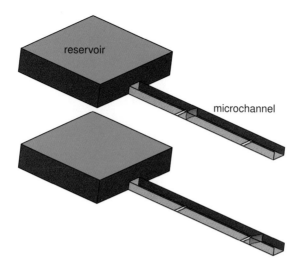

Figure 7.21 If the reservoir conditions stay constant (here the reservoir is sufficiently large to have a nearly constant pressure), the interface progresses in the channel under the action of the capillary forces.

Let us examine the case of an axisymmetric cone whose central axis is the z-axis. Darcy's law states that

$$\frac{\eta V}{K} = -\frac{\delta P}{\delta z}, \tag{7.65}$$

where K is the channel permeability. For a circular cross-section of radius R,

$$K = \frac{R^2}{8}. \tag{7.66}$$

And the mass conservation law is

$$Q(t) = \pi R^2 V, \tag{7.67}$$

where Q is the volume flow rate. On the other hand, the capillary pressure at the meniscus is

$$P = \frac{c\gamma}{R}, \tag{7.68}$$

where c depends on the local geometry and contact angle. Combining (7.65), (7.66), (7.67) and (7.68) results in

$$-\frac{\delta P}{\delta z} = \frac{8\eta Q(t)}{\pi R(z)^4}. \tag{7.69}$$

Integration of (7.69) between the channel entrance ($z = 0$) and the location of the meniscus $z = L(t)$ yields

$$Q(t) = \frac{\pi c\gamma}{8\eta R(L(t))\int_0^{L(t)} R(z)^{-4}dz}. \tag{7.70}$$

The velocity of the front is given by

$$\frac{\Delta L}{\Delta t} = \frac{Q(t)}{\pi R^2}.$$ (7.71)

Then

$$\frac{\Delta L}{\Delta t} = \frac{c\gamma}{8\eta R(L(t))^3 \int_0^L L(t)R(z)^{-4}dz}.$$ (7.72)

For a cylindrical channel $R = R_0$, and $c = 2\cos\theta$. Equation (7.72) can be easily integrated, yielding the Washburn law

$$L(t) = \left(\frac{\gamma\cos\theta R_0 t}{2\eta} \right)^{\frac{1}{2}}.$$ (7.73)

Equation (7.73) is very similar to equation (7.63). However, when the channel is a straight cone, Reyssat *et al.* have integrated (7.72) and found that velocity of the meniscus was not following a square root law.

7.3.5   Examples of Capillary Pumping

We have just seen that the presence of additional wetting walls adds to the efficiency of capillary pumping. Different designs have been investigated: Let us present here some designs by Zimmermann *et al.* [17] which are shown in figure 7.22. In all theses case, the presence of posts in (a) and (c), subdivisions in (b) and (e), and fins in (d) and (f) brings additional capillary forces.

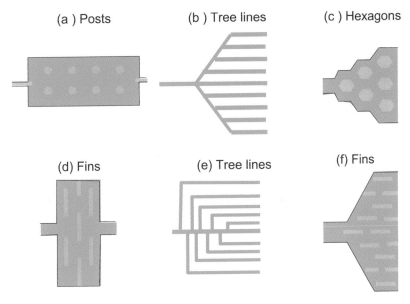

Figure 7.22 Different designs of capillary pumps according to [9]. The flow direction is from left to right.

As mentioned earlier, in biotechnology, capillary pumping is the preferred solution for lab-on-chips for point-of-care (POC). Point-of-care devices are highly portable, stand-alone devices that can be used to monitor biological targets in the human blood, such as DNA, proteins and cells. Figure 7.23 represents a typical sketch of such systems. Note the presence of fins to enhance the pumping.

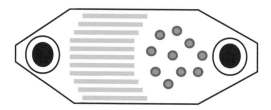

Figure 7.23 Sketch of a typical point of care device for blood testing: the parallel separation walls acts as a capillary pump.

Capillary pumping is also very useful in "open microfluidic systems"; this point will be detailed in the next chapter.

7.3.5.1   Spontaneous capillary flow in a composite channel

It is not always the case that the four walls of a microchannel have the same wetting properties. Microfabrication techniques often produce composite channels where the bottom parts (bottom and lateral walls) have a calibrated Young contact angle and the top (often glass, for visualization purposes) another contact angle. In such a case, what is the behavior of the fluid in the channel, and is a spontaneous capillary flow possible or not?

Take the example of a square channel with a very hydrophilic top (5°) obtained by atmospheric plasma deposition, and a moderately hydrophobic bottom (bottom and lateral walls) with a contact angle of 115°.

Let us first analyze the problem analytically. The capillary force is given by

$$F_{cap} = \gamma_{LG}\Sigma_i a_i \cos\theta_i = \gamma_{LG}[w\cos\theta_a + (2h+w)\cos\theta_r] \tag{7.74}$$

where $i = \{1,4\}$, w the channel width and h the channel height, and θ_a and θ_r the hydrophilic and hydrophobic contact angles. The meniscus does not move when the total capillary force is zero, i.e.

$$\frac{\cos\theta_r}{\cos\theta_a} = -\frac{1}{2\frac{h}{w}+1}. \tag{7.75}$$

This formula can be checked numerically using Evolver. Let take the case $h = 150\,\mu m$, $w = 250\,\mu m$. We find an equilibrium of the meniscus for $\theta_a = 5°$; $\theta_r \approx 118°$ (Fig. 7.24). On the other hand, the analytic formula is satisfied for $\theta_r \approx 117°$.

7.4   Capillary Valves

Besides pumping, the other function essential to controlling a microflow is valving. In this section, the principle of capillary valves is presented and some examples of such valves are shown.

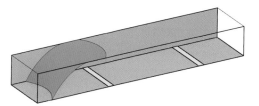

Figure 7.24 Meniscus at equilibrium in a channel of aspect ratio 3/5, $\theta_a = 5°$ and $\theta_r = 118°$

7.4.1   Principles of Capillary Valves

In recent years, fluidic microsytems have gained in complexity in order to integrate ever more potentialities, and valving has become essential for today's fluidic microsystems. Many types of valves exist, using different physical principles such as piezoelectric membrane actuation, flexible membrane deformation, magnetic plugs, and one shot sacrificial membranes [18,19]. However, capillary valves are the only ones to be "passive", i.e. they do not require another energy source than that of the pump that drives the flow in the system.

Tthe principle of capillary valves is very simple: it is based on the pinning of a fluid interface on geometrical edges. The valving pressure is the pressure needed to exit the canthotaxis mode. Most of the time the pinning of the incoming fluid is achieved on a sudden enlargement of the flow channel (figure 7.25). At the arrival of the liquid, a meniscus forms at the orifice by pinning of the interface on the edges. The bulging of the meniscus corresponds to a Laplace pressure given by

$$\Delta P = \gamma \kappa, \tag{7.76}$$

where κ is the curvature of the meniscus. Let us assume first that the surfaces are sufficiently hydrophobic that the meniscus stays pinned on the orifice edges. By following the spherical bulging of the interface, it can be shown there exists a maximum curvature. Indeed, the curvature varies continuously; at the liquid arrival, the meniscus has a small curvature (depending on the contact angle at the wall); the curvature increases to a maximum value and then decreases (figure 7.26). In the case of a circular channel of radius R, the maximum Laplace pressure is

$$\Delta P = 2\frac{\gamma}{R}. \tag{7.77}$$

If the pressure of the fluid is increased continuously, the interface will bulge out to the location of maximum curvature, and then suddenly the fluid will invade the larger channel. If the contact angle is not sufficient to pin the interface in the position of maximum curvature, the valving pressure will be lower [20]. For this reason, capillary valves are sometimes designed with an angle larger than 90°, as shown in figure 7.25.b.

7.4.2   Valving Efficiency and Shape of the Orifice

In this section the valving efficiency of a capillary valve is computed as a function of the shape and aspect ratio of the orifice.

For different shapes of orifice having the same hydraulic diameter, using Evolver, we have progressively incremented the liquid pressure until the rupture of the interface. The results are tabulated in table 7.1. The maximum pressure is obtained for the circular orifice and is the value given by equation (7.68). A square orifice of aspect ratio 3:1 is the least efficient valve.

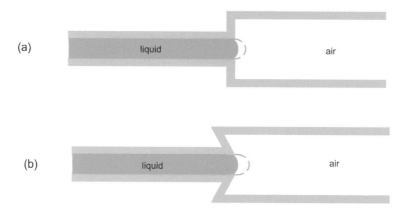

Figure 7.25 (a) Principle of a capillary valve: as long as the liquid pressure is smaller than the maximum Laplace pressure at the orifice the microflow is stalled.

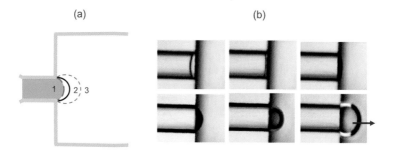

Figure 7.26 (a) A maximum curvature exists (meniscus 2) because then curvature is a continuously increasing function from meniscus 1 to meniscus 2, and continuously decreasing from meniscus 2 to meniscus 3. (b) View of a liquid front passing over an orifice.

The graph of the valving pressure is shown in figure 7.27. The maximum contact angle of the liquid surface with the solid wall is always larger than or equal to 90° (figure 7.28). In particular, rectangular openings have contact angles larger than 100°.

When designing a rectangular capillary valve, care should be taken about the canthotaxis possibility: in the case of a hydrophobic (lyophobic) contact angle of the order of, or less than, 100°, the pinning is lost before reaching the maximum pressure, and the efficiency of the valve is even lower than that predicted previously (in table 7.1 and figure 7.27).

7.4.3   Examples of Capillary Valves in Microsystems

7.4.3.1   Stop Valve

Practically, it is very difficult to achieve the type of valves shown in the preceding section, due to microfabrication constraints. In a sense the preceding section designs are ideal valves. Indeed, it is long and costly to form more than two etching depths during microfabrication. So capillary valves are either vertical slits going from the bottom to the top of the microchannel (only one etching depth), or steps (two etching depths), as schematized in figure 7.29.

Table 7.1 Values of the maximum pressure at the valve as a function of the geometry. a and b are the relevant dimensions of the gate, D_H is the hydraulic diameter, θ_{max} is the maximum contact angle, ΔP_{max} is the maximum pressure difference, and ΔP_* is the ratio to the ΔP_{max} for a circle geometry.

Geometry	a [μm]	b [μm]	a/b	D_H[μm]	θ_{max}[$^\circ$]	ΔP_{max}	ΔP_*
circle (theory)	50	-	-	100	90	2880	1
circle	50	-	-	100	90.5	2909	1.01
hexagon	57.7	-	-	100	102.5	2749	0.9545
square	100	100	1	100	107	2581	0.8962
rectangle	125	83.3	1.5	100	114.5	2552.4	0.8862
rectangle	150	75	2	100	115.05	2510	0.8715
rectangle	200	66.7	3	100	108.7	2470	0.8567
rectangle	250	62.5	4	100	105.7	2477	0.8601
rectangle	300	60.0	5	100	102.5	2504	0.8694
rectangle	350	58.3	6	100	101.3	2539	0.8816
rectangle	400	57.14	7	100	100	2570.3	0.8925

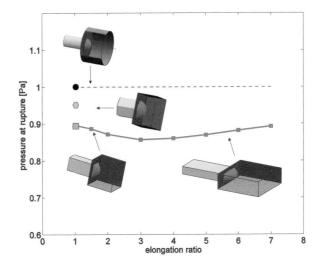

Figure 7.27 Rupture pressure as a function of the opening elongation: a minimum exists for a rectangular section of ratio 3 to 1 and a maximum for a circular orifice.

7.4.4   Delay Valves

Some "capillary valves" are designed not to stop the flow, but to create a delay or a slow down of the flow. Generally such an effect is obtained by an increase of the channel dimensions, exactly the opposite of the principle of figure 7.30. Consider a set of rectangular channels of dimensions w_1 and h_1 merging into a larger channel of dimensions w_2 and h_2. Assuming that the meniscus has approximately a bi-quadratic shape [21], the capillary pressure decreases from

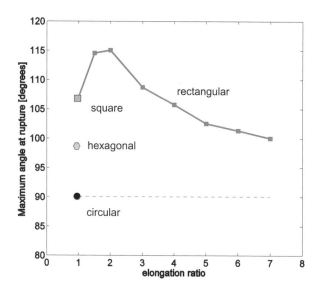

Figure 7.28 Maximum angle at the moment of rupture: rectangular capillary valves have a maximum angle larger than 100°.

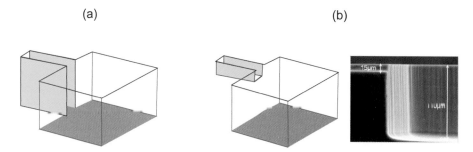

Figure 7.29 Sketch of typical capillary valves. (a) Vertical slit; (b) step.

the value

$$P_1 \approx \gamma \cos \theta \left(\frac{1}{w_1} + \frac{1}{h_1} \right) \tag{7.78}$$

to the value

$$P_2 \approx \gamma \cos \theta \left(\frac{1}{w_2} + \frac{1}{h_2} \right). \tag{7.79}$$

If the etching depths are the same ($h_1 = h_2$), then

$$\frac{P_2}{P_1} = \frac{w_1}{w_2}. \tag{7.80}$$

Relation (7.80) shows that the capillary pressure is lowered in the inverse ratio of the channel widths. It has been shown earlier that the flow velocity results from a balance between the

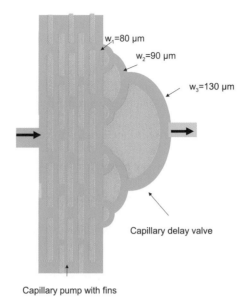

w₁=80 μm

w₂=90 μm

w₃=130 μm

Capillary delay valve

Capillary pump with fins

Figure 7.30 Example of design of a capillary delay valve at the flow exit of a capillary pump, from [20].

capillary force and the wall friction. Hence the flow velocity is reduced by the enlargement of the free cross-section. The device sketched in figure 7.30 shows such a "delay valve" [22].

7.5   Conclusions

Capillary forces are powerful at the geometrical scale of microsystems. They are sufficiently important to trigger capillary rise along vertical plates and fibers. Also, capillary forces can be use to achieve liquid pumping or valving. It is still more the case with the trend to further miniaturize components from microscale dimensions to nanoscale dimensions. Making use of capillary forces for pumping and valving has the advantage of avoiding the used of moving parts, which are not convenient at these scales.

In the following chapter we shall show that capillary forces can also be used when the liquid is not totally channeled between walls, a situation which is often called "open microfluidicsi".

7.6   References

[1] P.G. deGennes, F. Brochart-Wyart, D. Quèrè. *Drops, bubbles, pearls and waves.* Springer, 2005.

[2] H.M. Princen, "Capillary Phenomena in Assemblies of Parallel Cylinders. I. Capillary Rise between Two Cylinders," *J. Colloid and Interface Sci.* **30**, pp. 69-75, 1969.

[3] H.M. Princen, "Capillary Phenomena in Assemblies of Parallel Cylinders. II. Capillary Rise in Systems with More Than Two Cylinders," *J. Colloid and Interface Sci.* **30**, pp. 359-371, 1969.

[4] J. Bico, D. Quèrè, "Rise of liquids and bubbles in angular capillary tubes," *J. Colloid and Interface Science* **247**, pp. 162–166, 2002.

[5] O.J. Murphy and J. S. Wainright, "Interfacial Tension at Solid Electrode/Solution Interfaces: A Wilhelmy Plate Measurement Technique," *Langmuir* **5**, pp. 519–523, 1989.

[6] http://www.attension.com/surface-tension.aspx.

[7] http://www.kibron.com/company/science-technology/surface-tension-measurement-techniques/.

[8] L.H. Tang, Y. Tang, "Capillary rise in tubes with sharp grooves," *J. Phys. II France* **4**, p. 881, 1994.

[9] Y. Tsori, "Discontinuous liquid rise in capillaries with varying cross-sections," *Langmuir* **22**, pp. 8860–8863, 2006.

[10] T. Liu, K.F. Choi, Y. Li, "Capillary rise between cylinders," *Journal of Physics D: Applied Phsics* **40**, pp. 5006–5012, 2007.

[11] H. Bruus. *Theoretical microfluidics*. Oxford Master Series in Condensed Matter Physics, 2008.

[12] E.V. Washburn. "The Dynamics of Capillary Flow," *Phys. Rev.* **17** (3), pp. 273-283, 1921.

[13] M. Bahrami, M. M. Yovanovich, J. R. Culham, "Pressure drop of fully-developed, laminar flow in microchannels of arbitrary cross section," *Proceedings of ICMM 2005, 3rd International Conference on Microchannels and Minichannels, June 13-15, 2005*, Toronto, Ontario, Canada, 2005, pp. 1-12.

[14] J. Berthier, P. Silberzan. *Microfluidics for Biotechnology*. Second Edition, Artech House, 2010.

[15] J. Berthier, R. Renaudot, P. Dalle, G. Blanco-Gomez, F. Rivera, V. Agache, "COMSOL assistance for the determination of pressure drops in complex microfluidic channels," *Proceedings of the 2011 COMSOL European Conference,* Paris, 15-17 Nov. 2010, pp. 1-7.

[16] M. Reyssat, L. Courbin, E. Reyssat, H.A. Stone, "Imbibition in geometries with axial variations," *J. Fluid Mech.* **615**, pp. 335–344, 2008.

[17] Martin Zimmermann, Heinz Schmid, Patrick Hunziker and Emmanuel Delamarche, "Capillary pumps for autonomous capillary systems," *Lab Chip* **7** pp. 119-125, 2007.

[18] K.W. Ohl, C.H. Ahn. "A review of microvalves," *J. Micromech. Microeng.* **16**, pp. 13–39, 2006.

[19] M. Allain, J. Berthier, S. Basrour, P. Pouteau, "Electrically actuated sacrificial membranes for valving in microsystems," *Journal of Micromechanics and Microengineering,* **20**(3), pp. 35006-35012(7), March 2010,

[20] J. Berthier, F. Loe-Mie, V-M. Tran, S. Schoumacker, F. Mittler, G. Marchand and N. Sarrut, "On the pinning of interfaces on micropillar edges," *J. Colloid Interface Sci.* **338**(1), pp. 296–303, 2009.

[21] F.P. Man, C.H. Mastrangelo, M.A. Burns, D.T. Burke, "Microfabricated capillary-driven stop valve and sample injector." *Proceedings of the 11th Annual International Workshop on MEMS,* 25-29 Jan 1998, Heidelberg, Germany, pp. 45-50.

[22] M. Zimmermann, P. Hunziker, E. Delamarche, "Valves for autonomous capillary systems," *Microfluidics and Nanofluidics* **5**(3), pp. 395–402, 2008.

8

Open Microfluidics

8.1 Abstract

Historically, microfluidics was a mere size reduction of well-established laminar macrofluidics. Microflows were confined inside solid-wall channels and moved by pumps or syringes. Progressively new ways of doing microfluidics have emerged. Digital and droplet microfluidics are two techniques that have the advantage of confining targets such as biological objects of interest inside droplets of picoliter to nanoliter volumes. Recently, with the improving knowledge of capillary forces, it has been found that it is not always a necessity that microflows be confined between solid walls. Hydrophilic rails [1] or posts [2] could be sufficient to guide a microflow. The main advantage is to have direct access to the flowing liquid. The second advantage is the large liquid-gas surface area. This type of microfluidics is usually called "open microfluidics" or sometimes "virtual walls microfluidics", or HCMs for "hydrodynamically confined microflows." Open microfluidic systems are now common in nanotechnology, biotechnology, fuel cells, and space technology [3-5].

From a general point of view, open microfluidics is a general designation for liquids having

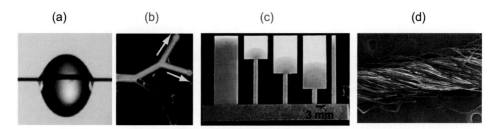

Figure 8.1 Four different types of open microfluidics: (a) droplet attached to a wire, reprinted with permission from [6], ©Elsevier 2004;(b) liquid progressing on rails and moved by EWOD, reprinted with permission from [1], c©ACS, 2005; (c) liquid-imbibing flat paper matrices, reprinted with permission from [7], ©Springer, 2010; (d) thread of polyester fibers creating a "porous" microchannel, reprinted with permission from [8], ©AIP, 2011, http://linkip.org/link/doi/10.1063/1.3567094.

one interface with air (fig. 8.1). There is a wide diversity in this family. It encompasses droplets pierced by wires [6], liquid bridges, and liquid spreading between rails and inside grooves [1]. Still more recently, new concepts have emerged, such as paper-based microfluidics – where liquid is absorbed by a microporous medium [7] – and thread-based microfluidics – where liquid is guided by a thread of twisted fibers [8].

In this chapter, we deal first with droplets and wires. In a second section the spontaneous capillary flow (SCF) of a liquid with partial contact with solid structures is presented. The case of fibers and imbibition is then investigated, leading to an introduction to paper-based and thread-based microfluidics.

8.2   Droplet Pierced by a Wire

In this section we focus on droplets pierced by a solid, hydrophilic or hydrophobic, cylindrical wire. The case of suspended droplets is investigated first, then the case of sessile droplets (droplet deposited on a solid substrate and pierced by a horizontal wire).

8.2.1   Suspended Droplet

We have seen in chapter 3 that the Bond number defines the relative importance of surface tension and gravitational forces. In the case of a wire, the situation is a little more complicated. The number of parameters is larger than for a simple sessile droplet. In this case there is the weight of the droplet, as before, but there is an additional parameter, which is the dimension of the wire radius. Hence, two non-dimensional numbers are usually defined: first the Goucher or the wire Bond number [9,10],

$$Go = \frac{r}{\kappa^{-1}} = \sqrt{\frac{\rho g r^2}{\gamma}}, \tag{8.1}$$

where $\kappa^{-1} = \sqrt{\frac{\gamma}{\rho g}}$ is the capillary length and r the radius of the wire (fig. 8.2), or equivalently

$$Bo = \frac{r^2}{\kappa^{-2}} = Go^2. \tag{8.2}$$

So far the volume of the droplet has not been taken into account. An additional scaling number – a length scale – is then used to take into account the droplet volume [8],

$$R = \frac{D'}{D}, \tag{8.3}$$

where D and D' are respectively the wire and droplet diameter.

Alternatively, Gilet *et al.* [11] have tried to group the two aspects by defining the Bond number as the ratio of the droplet weight to the capillary forces,

$$Bo - \frac{mg}{4\pi r \gamma} - \frac{\Omega}{4\pi r \kappa^{-2}}, \tag{8.4}$$

where Ω is the droplet volume. In conclusion, not only the intrinsic volume of the droplet has to be taken into account, but also the size of the wire, which conditions the magnitude of the capillary forces.

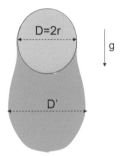

Figure 8.2 Sketch of a droplet suspended from a wire.

8.2.2 Small Droplet

First, we recall some theoretical aspects. Droplets on cylindrical fibers have been the subject of many investigations, in particular by Carroll [12] and McHale and Newton [13]. Carroll described the equilibrium shape of small droplets (smaller than the capillary length) deposited on thin fibers (thin enough for the droplet to surround the fiber). Their shape has been shown to be an unduloid. He calculated the length L, the liquid/air area A, the volume O and the pressure ΔP of the drop as a function of its maximal thickness h and of the fiber radius r (fig. 8.3).

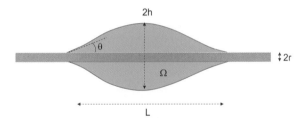

Figure 8.3 Sketch of a droplet on a horizontal wire (Goucher number smaller than 1). Note the unduloidal shape of the droplet.

If the drop size is smaller than the capillary length, its curvature is constant (when the gravitational effects are negligible) and proportional to its overpressure ΔP. Carroll [12] has demonstrated that if the liquid is perfectly wetting on the fiber then the overpressure is

$$\Delta P = \frac{2\gamma}{r+h}. \tag{8.5}$$

This formula is consistent with the geometry of a fiber: when the volume of the drop tends to zero (i.e. $h \to r$), ΔP is equal to γ/r, the Laplace pressure inside a cylinder of radius r. On the other hand, if the radius of the fiber tends to zero for a drop of given volume, we find $\Delta P = 2\gamma/h$, the overpressure in a sphere of radius h. Carroll's demonstration is detailed in an Appendix at the end of this chapter.

Although the droplet is not spherical, it is usual to define a characteristic size R_0 of a droplet of volume Ω by $\Omega = \frac{4}{3}\pi R_0^3$. Because the effect of the fiber on the droplet shape decreases when the ratio r/R_0 decreases – as the droplet shape becomes closer to a sphere – it is expected that the maximum height h varies as R_0 for large droplet volumes. Hence a log-log plot of the function $h(\Omega)$ should display an asymptote with a slope 1/3 when Ω increases. This asymptotic behavior is shown in figure 8.4, in which the numerical results have been compared to the experimental results of Lorenceau *et al.* [14].

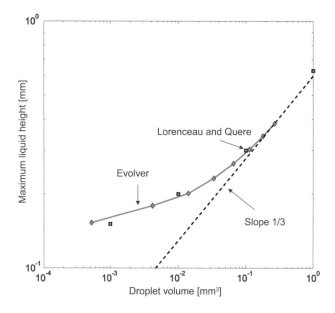

Figure 8.4 Calculated maximum height of a drop deposited on a fiber of radius $r = 150\mu m$, as a function of the volume of the drop. The oblique line indicates the slope 1/3.

Comparison between experiments and calculation are shown in figure 8.5.

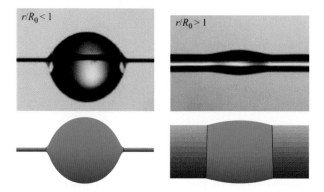

Figure 8.5 Droplet suspended on a horizontal wire: comparison between experimental view [14] and calculation. The liquid surface shows an unduloidal shape. Left: $r = 12\,\mu m$ and $R_0 = 200\,\mu m$; right: $r = 100\,\mu m$ and $R_0 = 80\,\mu m$. Reprinted with permission from [14], ©Elsevier 2004.

8.2.3   Effect of Gravity on Small Droplets

Even if the volume Ω of the liquid is small, and R_0 is of the order of the capillary length, gravitational forces affect the shape of the droplet when the contact angle is sufficiently large. In other words, the liquid does not envelop the wire at large Young contact angles. The relative importance of gravity and wetting properties are shown in figure 8.6. This point has already been observed by Carroll and McHale [12].

(a) $g = 0$ m/s^2, $\theta = 5°$ (b) $g = 9.81$ m/s^2, $\theta = 5°$

(c) $g = 0$ m/s^2, $\theta = 120°$ (d) $g = 9.81$ m/s^2, $\theta = 120°$

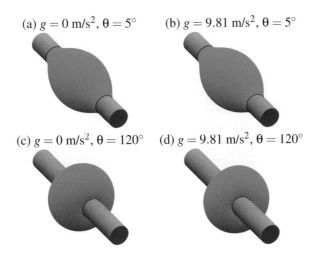

Figure 8.6 Water droplet pierced by a wire: (top) hydrophilic case; (bottom) slightly hydrophobic case (100° contact angle); (left) no gravity; (right) with gravity.

(a)   (b)

Figure 8.7 Evolver calculation of the different shapes of the droplet by changing the value of the contact angle: (a) barrel shape corresponding to a small contact angle (5°); (b) clamshell shape corresponding to a contact angle of 35° [15].

Note that the value of the contact angle deeply affects the position of the droplet on the wire. Depending on this value, the droplet can have a barrel shape or a clamshell shape. This property has been shown by Eral *et al.* by using electrowetting to modify the value of the contact angle (Fig.8.7) [15]. In their experiment Eral *et al.* have used an oil droplet immersed in water. When actuation is turned off, the contact angle is small and the drop has a "barrel" shape; when then actuation is turned on, the contact angle increases and the droplet adopts a "clamshell" shape. This kind of electrowetting effect is described later in this book, in chapter 10.

8.2.4   Large Droplet

In this subsection, the shape of larger droplets suspended from horizontal cylindrical wires is investigated (fig. 8.8). The droplet volume is large enough to be affected by gravitational forces, i.e. the Gilet-Bond number is larger than 1. The droplet moves below the fiber and, depending on its weight and the Young contact angle, stays attached to the wire or falls (fig. 8.9). As pointed out by Lorenceau *et al.* [6], there is a threshold value of the droplet volume Ω_M above which the droplet will fall. It is usual to define the threshold radius R_M by $\Omega_M = \frac{4}{3}\pi R_M^3$. If the radius of the wire is larger than the capillary length, i.e. $r > \kappa^{-1} = \sqrt{\gamma/\rho g}$, then the threshold radius is constant and given by $R_M \approx 1.6\kappa^{-1}$. If the radius of the wire is smaller than κ^{-1}, the threshold value is given by $R_M \approx 1.536 r^{1/3}\kappa^{-2/3}$.

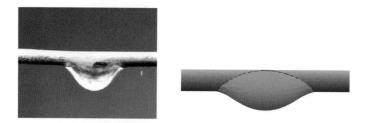

Figure 8.8 4 mm^3 water droplet suspended to a horizontal wire: comparison between experimental observation and Evolver calculation.

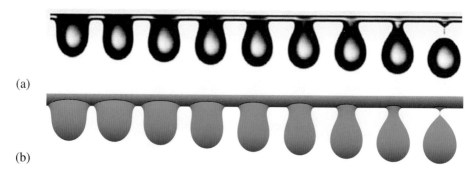

(a)

(b)

Figure 8.9 Large droplet suspended to a wire (hydrophobic case): (a) photographs of a detaching droplet (reprinted with permission from [6], ©Elsevier 2004); (b) modeling with Evolver, stages in one drop detaching. But recall that Evolver kinematics don't model physical dynamics.

8.2.5   Sessile Droplet Pierced by a Wire

Although one full chapter of this book (chapter 10) is dedicated to digital microfluidics, and particularly to electro-wetting on dielectric (EWOD), we analyze here the particular case of an "historical" open digital microfluidic device, where one of the two electrodes is a wire running parallel to the electrode (fig. 8.10). In such a case the droplets are pierced by the wire, which is used as the grounded electrode.

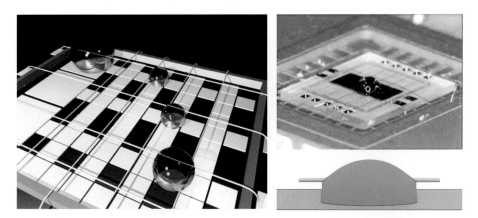

Figure 8.10 Photos and simulation of an electrowetting-based chip using wires as grounded electrodes (image CEA-Leti).

It is of interest to investigate the general case of a wire piercing a sessile droplet. A hydrophilic (wetting) wire elongates the droplet along its axis and, intuitively, it is expected that the droplet will move until it is symmetrical relative to the wire. Conversely, a hydrophobic (non-wetting) wire pinches the droplet, and, intuitively, it is expected that the droplet will move aside the wire. Figure 8.11 sketches the forces on the droplet and explains the difference between the hydrophilic and hydrophobic cases.

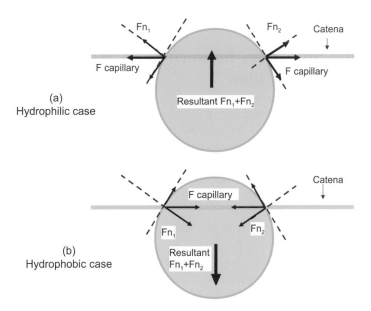

Figure 8.11 Sketch of the forces on the droplet: only the forces normal to the surface F_{n1} and F_{n2} contribute to the resultant.

This effect is pointed out by Evolver: Figure 8.12 shows the initial and equilibrium positions of a droplet in the two cases of a wetting and non-wetting wire.

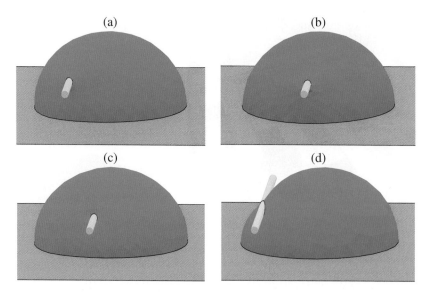

Figure 8.12 Sessile water droplet pierced by a wire: (a) hydrophilic case, initial position (not at equilibrium) ($\theta = 40°$); (b) the equilibrium position shows a symmetry on a vertical plane passing through the wire; (c) hydrophobic case ($\theta = 130°$), initial position (not at equilibrium); (d) the equilibrium position is reached when the drop moves aside from the wire.

8.3   Liquid Spreading Between Solid Structures – Spontaneous Capillary Flow

In this section, the capillary spreading at the contact of wetting walls is investigated. This phenomenon is usually called SCF for spontaneous capillary flow. We successively analyze the cases shown in figure 8.13: (a) two parallel rails, (b) and (c) U-grooves, (d) trapezoidal channels and (e) half-pipe. We derive first the conditions for spreading using an energy approach, and then, using an approach based on the forces acting on the liquid, we find a universal relation that encompasses all the cases.

8.3.1   Parallel Rails

In this section, the spreading of a wetting liquid between two parallel rails is investigated. Different biotechnological applications make use of this configuration. The approach was pioneered by Satoh *et al.* [1], who have demonstrated that liquid "fingers" can spread between two horizontal, parallel, solid "rails" (fig. 8.14). In their approach, the rails were switched from a hydrophobic state to a hydrophilic state by electrowetting actuation (see chapter 10), and a ramification of liquid "fingers" is progressively formed along a network of "rails," depending on the actuation at each bifurcation. A reconfigurable microfluidic network is then achieved.

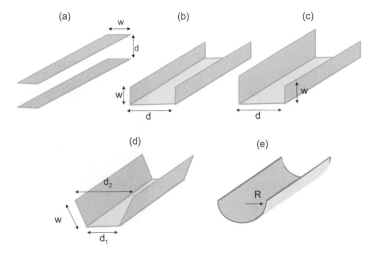

Figure 8.13 The different geometries of open channels.

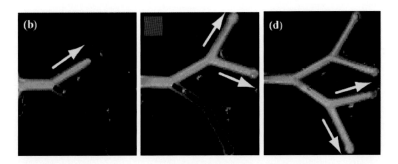

Figure 8.14 Open microflows moved by EWOD actuation, reprinted with permission from [1], ©ACS, 2005.

Another example is that of the device conceived and fabricated by Sung Hoon Lee *et al.*[16] to form microchambers to trap cells and study the different communication patterns between groups of cells (fig. 8.15). In their work, liquid fingers form under the action of capillary forces in a comb-like structure; once the fingers are achieved, a polymerization step is performed. Artificial walls are thus formed, separating microchambers. Different shapes of rails can be fabricated: straight fingers in a comb-like structure, and also spiral fingers.

8.3.2   Spontaneous Capillary Flow Between Parallel Rails

The capillary flow between two parallel rails is shown in figures 8.16 and 8.17, which have been obtained by an Evolver calculation.

Minimization of the Gibbs free energy $dG = -PdV + dE = -PdV + \gamma dA$ leads to

$$P = \frac{dE}{dV} = \left(\gamma_{SG} \frac{dA_{SG}}{dV} + \gamma_{SL} \frac{dA_{SL}}{dV} + \gamma_{LG} \frac{dA_{LG}}{dV} \right). \tag{8.6}$$

(a)   (b)

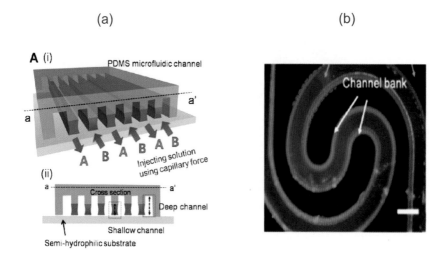

Figure 8.15 Liquid fingers invade the bridges between comb-like rails and a bottom plate [11]. Once polymerized, separated microchambers are formed. Reprinted with permission from [11], ©ACS, 2010

Considering that any change of A_{SL} is at the expense of A_{SG},

$$dA_{SG} = -dA_{SL}, \tag{8.7}$$

and using the Young equation,

$$\gamma \cos\theta = \gamma_{SG} - \gamma_{SL}, \tag{8.8}$$

equation (8.6) collapses to

$$P = \gamma_{LG}\left(\frac{dA_{LG}}{dV} - \cos\theta\frac{dA_{SL}}{dV}\right). \tag{8.9}$$

In this particular case, the increase of the solid-liquid contact surface A_{SL} for a progression of the front of length dx is

$$dA_{SL} = 2w\,dx, \tag{8.10}$$

where w is the width of the rail. On the other hand, the variation of the liquid-gas interface during the motion is

$$dA_{LG} = 2d\,dx. \tag{8.11}$$

Finally, the variation of the liquid volume is

$$dV = wd\,dx, \tag{8.12}$$

and (8.9) becomes

$$P = 2\gamma_{LG}\left(\frac{1}{d} - \frac{\cos\theta}{w}\right). \tag{8.13}$$

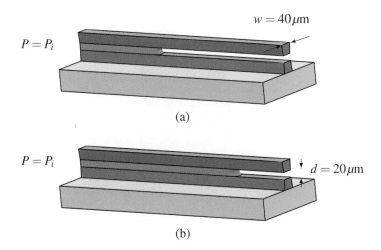

Figure 8.16 A liquid filament wetting the rails ($\theta = 30°$) may progress constantly, depending on the dimensions w and d and on the advancing contact angle, if the reservoir pressure P_i is kept constant.

As long as the inlet pressure is larger than P_c, the liquid continues its progression along the rails. Most of the time, the inlet pressure is the atmospheric pressure and the condition for a continuous flow is

$$P = 2\gamma_{LG}\left(\frac{1}{d} - \frac{\cos\theta}{w}\right) < 0. \tag{8.14}$$

This relation can be written as

$$d \leq w\cos\theta. \tag{8.15}$$

In the (w, d) coordinate system, the condition (8.15) is verified in a half plane (fig. 8.18). It is deduced from (8.15) that the vertical distance between the two rails cannot be too large compared to the width of the rails for obtaining the spreading of the liquid.

A comparison between the theoretical formula (8.15) and Evolver results is shown in figure 8.18. The results coincide for small dimensions, but there is a deviation of the two limits for large dimensions, because the theoretical approach does not take into account the curvature of the interface. This curvature effect increases with the geometrical dimensions.

8.3.3   Spontaneous Capillary Flow in U-grooves

Grooves etched in a solid substrate are very common in microsystems for biology. Let us investigate the conditions under which the liquid spreads. Figure 8.19 shows the shape of an advancing liquid filament in a rectangular groove for two different contact angles; the first one ($\theta = 35°$) does not respect the Concus-Finn conditions whereas the second ($\theta = 55°$) does.

An approach similar to that of the preceding section can be done. Let us start from (8.9),

$$P = 2\gamma_{LG}\left(\frac{1}{d} - \frac{\cos\theta}{w}\right). \tag{8.16}$$

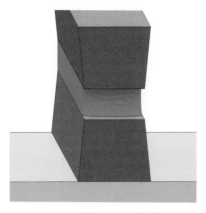

Figure 8.17 Close-up on the advancing front of the liquid finger.

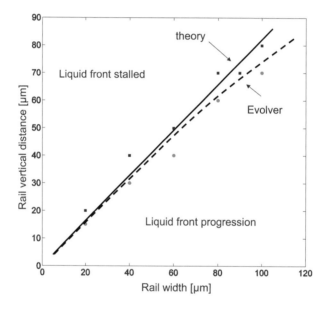

Figure 8.18 Comparison between the theoretical formula (8.15) and Evolver results: the green dots signify that the liquid front progresses according to Evolver and the red dots that the front is stalled. The continuous line is the theoretical limit (35°).

In this particular case, the increase of the solid-liquid contact surface A_{SL} for a progression of the front of a length dx is

$$dA_{SL} = (2w + d)\, dx, \tag{8.17}$$

where d is the width of the groove. On the other hand, the variation of the liquid-gas interface during the motion is

$$dA_{LG} = d\, dx. \tag{8.18}$$

Figure 8.19 A liquid filament wetting the groove may progress constantly, depending on the dimensions w and d and on the advancing contact angle. Here the liquid progresses in a horizontal groove under the action of capillary forces: (a) if the contact angle does not respect the Concus-Finn condition, liquid filaments precede the advancing liquid; (b) if the contact angle respects the Concus-Finn condition, the liquid front progresses uniformly.

Finally, the variation of the liquid volume is

$$dV = wd\,dx, \tag{8.19}$$

and (8.16) becomes

$$P = 2\gamma_{LG}\left(\frac{1}{w} - \cos\theta\frac{(2w+d)}{wd}\right) = \gamma_{LG}\left(\frac{1-\cos\theta}{w} - \frac{2\cos\theta}{d}\right). \tag{8.20}$$

For the liquid to spread, the capillary pressure has to be negative, hence

$$\frac{d}{w} \leq \frac{1-\cos\theta}{2\cos\theta}. \tag{8.21}$$

In the (w,d) coordinate system, the condition (8.21) is verified in the upper half plane (fig. 8.20). It is deduced from (8.21) that the width of the groove cannot be too large compared to the height of the walls for obtaining the spreading of the liquid.

A comparison between the theoretical formula (8.21) and Evolver results is shown in figure 8.20. The results coincide for small dimensions, but there is a deviation of the two limits for large dimensions, because the theoretical approach does not take into account the curvature of the interface. This curvature effect increases with the geometrical dimensions.

8.3.4   Spontaneous Capillary Flow in Asymmetric U-grooves – Spreading of Liquid Glue During Microfabrication

Glue is often used in micro-fabrication to seal the microdevice. The other alternative is the use of direct bonding, which requires restrictive conditions, such as very smooth, hydrophilic surfaces, resistance to heating (annealing) above 200°C, and small "void" chamber volumes in order that a large percentage of the cover surface is in contact with the substrate. In many cases, microsystems are sealed by a cover glued on top of the etched substrate. However, the sealing step can go wrong, because, depending on the conditions, liquid glue may infiltrate – under the action of capillary forces – the chip grooves and trenches and ruin the quality of the microfabrication (fig. 8.21). Glue is first liquid, with a viscosity of the order of 10 to 30 times that of water. It is later in the process solidified by UV polymerization. If glue has spread inside the chip, it becomes polymerized and cannot be removed. It is of interest to better understand under what conditions liquid can spread in an asymmetric groove like that shown in figure 8.21.

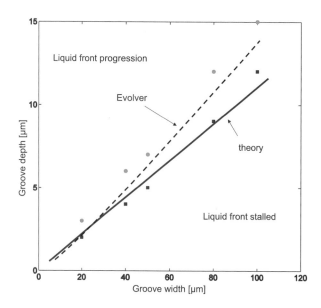

Figure 8.20 Comparison between the theoretical formula (8.21) and the results obtained with Evolver. The green dots signify that the liquid front progresses according to Evolver and the red dots that the front is stalled. The continuous line is the theoretical limit.

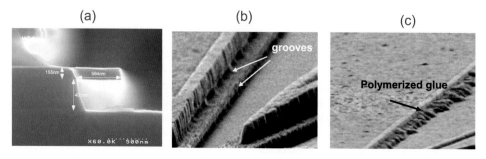

Figure 8.21 (a) View of a groove/trench in the microfabrication process of a NEMS (Nano-Electrical-Micro-System); (b) perspective view of the grooves that can be found in NEMS; (c) liquid glue has invaded the groove and has been polymerized (photo F. Baleras CEA-LETI).

This particular case is slightly different than that of the preceding section, because the liquid is not pinned on both sides of the groove. It can spread on one of the two walls limiting the groove (fig. 8.22). If the curvature of the free interface is neglected, the geometry can be sketched as in figure 8.23.

Let us again start from the relation

$$P = 2\gamma_{LG}\left(\frac{1}{d} - \frac{\cos\theta}{w}\right).$$

(8.22)

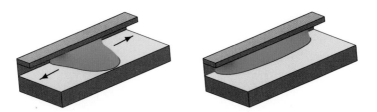

Figure 8.22 SCF of wetting liquid in an asymmetric groove (Evolver calculation).

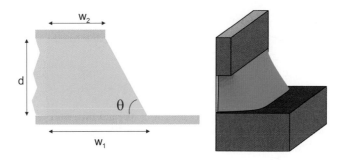

Figure 8.23 Left: schematic of a cross section of the liquid; right: the surface calculated with Evolver.

Using the notations of figure 8.23, we deduce

$$dA_{LG} = \sqrt{d^2 + (w_1 - w_2)^2}\, dx,$$
$$dA_{SL} = (d + w_1 + w_2)\, dx, \tag{8.23}$$
$$dV = d\frac{(w_1 + w_2)}{2}\, dx.$$

Note that w_2 is linked to w_1, d and θ by

$$w_1 = w_2 + \frac{d}{\tan\theta}. \tag{8.24}$$

Upon successive substitutions of (8.24) in (8.23) and (8.22), and using some algebra, the condition for liquid spreading is

$$\frac{d}{w_2} < \frac{2\sin(2\theta)}{1 - \sin(2\theta) - \cos(2\theta)}. \tag{8.25}$$

The relation (8.25) is plotted in figure 8.24 for the dimensions $w_2 = 584$ nm, $d = 401$ nm, corresponding to figure 8.22, and for four different values of θ (55°, 60°, 70° and 80°). For a given value of θ, the liquid spreads if the dimensions d and w_2 correspond to a point below the straight lines.

Let us now perform numerical simulations with Evolver. A result for $\theta = 54°$ and a ratio $d/w_2 = 0.4$ is shown in figure 8.25. With such conditions, the liquid spreads in the groove.

It is interesting to compare Evolver results and the simplified theory. Let us consider the theoretical wetting limit

$$\frac{d}{w_2} = \frac{2\sin(2\theta)}{1 - \sin(2\theta) - \cos(2\theta)} \tag{8.26}$$

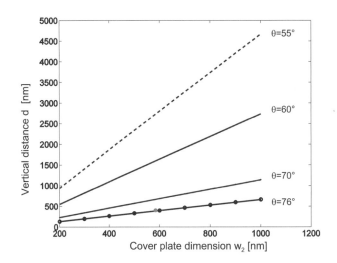

Figure 8.24 Plot of relation (8.25) for $\theta = 55°, 60°, 70°$ and $80°$, and $w_2 = 584$ nm, $d = 401$ nm. The green dot is the spreading observed in figure 8.22, suggesting that the liquid glue has a contact angle with the silicon substrate smaller than $76°$.

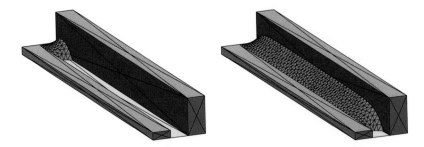

Figure 8.25 Evolver simulation of the capillary flow in an asymmetric groove. Contact angle $54°$.

and draw the plot of d/w_2 versus θ; we obtain the figure 8.26. According to Evolver, the wetting of the groove requires a contact angle about $2°$-$3°$ smaller than that predicted by the theory. This difference is attributed to the approximation of planar surfaces in the theory.

8.3.5   Spontaneous Capillary Flow in a Trapezoidal Channel

In this section, we analyze the case of trapezoidal channels. Let us consider a trapezoidal section as defined in figure 8.27. The same reasoning based on the energy formulation yields

$$
\begin{aligned}
dA_{LG} &= d_2\,dx, \\
dA_{SL} &= (2w + d_1)\,dx, \\
dV &= h\frac{(d_1 + d_2)}{2}\,dx,
\end{aligned}
\tag{8.27}
$$

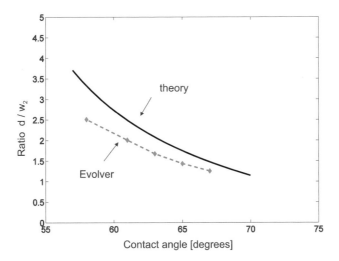

Figure 8.26 Comparison between Evolver results and theory for the wetting limit.

where h is the channel height. Upon substitution in (8.27), the condition for liquid spreading is

$$\frac{d_2}{(2w+d_1)} < \cos\theta. \tag{8.28}$$

Consider the particular case of $d_1 = w$ and $d_2 = 2d_1$. The preceding condition simplifies to $\cos\theta > 2/3$, which yields $\theta = 48.19°$. A calculation with Evolver shows that, in such a geometry, the liquid advances for $\theta = 47°$ and recedes for $\theta = 49°$ (fig. 8.27).

(a)   (b)

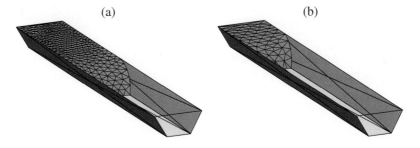

Figure 8.27 Evolver calculation showing (a) the liquid advancing for $\theta = 47°$ and (b) receding for $\theta = 49°$ ($d_1 = 40\ \mu m$, $w = 40\ \mu m$, and $d_2 = 80\ \mu m$).

8.3.6   Spontaneous Capillary Flow in a Half-pipe

Let us continue our investigations into capillary pumping in open microfluidic devices, and consider a half-pipe (fig. 8.28). The relation (8.16) becomes

$$P = \gamma_{LG}\left(\frac{2Rdx}{\pi\frac{R^2}{2}dx} - \frac{\cos\theta\pi Rdx}{\pi\frac{R^2}{2}dx}\right) = 2\frac{\gamma_{LG}}{R}\left(\frac{2}{\pi} - \cos\theta\right). \tag{8.29}$$

The condition for capillary pumping is then

$$\cos\theta > \frac{2}{\pi},$$ (8.30)

a condition that is satisfied for $\theta > \arccos(2/\pi) = 50.45°$.

(a)   (b)

Figure 8.28 Liquid in a half-pipe: (a) below the contact angle $\theta = 50.45°$ the front advances; (b) it recedes above this value.

8.3.7   A Universal Law for Capillary Pumping

In sections 8.3.2, 8.3.3, 8.3.5 and 8.3.6 we have seen three conditions for capillary pumping in different "open microfluidic" configurations. We show here that the two relations can be united under a unique condition under the assumption that the fluid reservoir is at zero pressure (large reservoir). Consider the ratio

$$p^* = \frac{p_F}{p_W},$$ (8.31)

where p_F is the free liquid perimeter and p_W the wetted perimeter. The general condition for spreading can be derived from (8.16), observing that the condition for a negative capillary pressure is

$$\frac{dA_{LG}}{dV} < \cos\theta \frac{dA_{SL}}{dV},$$ (8.32)

which can be rearranged as

$$\frac{dA_{LG}}{dA_{SL}} < \cos\theta.$$ (8.33)

Upon elimination of the dimension dx, one finds

$$\frac{p_F}{p_W} < \cos\theta.$$ (8.34)

In the case of the two parallel rails (section 8.3.3), the ratio p^* is equal to

$$p^* = \frac{p_F}{p_W} = \frac{d}{w}.$$ (8.35)

Upon introduction of the condition (8.35) in (8.15), one finds

$$p^* < \cos\theta. \tag{8.36}$$

Using (8.34), this inequality is equivalent to

$$\gamma p_F < \gamma p_W \cos\theta, \tag{8.37}$$

which means that for observing a spontaneous capillary flow, the capillary force $F_{cap} = \gamma p_W \cos\theta$ must be larger than the surface tension force $F_\gamma = \gamma p_F$:

$$F_{cap} = \gamma p_W \cos\theta > F_\gamma = \gamma p_F. \tag{8.38}$$

The surface tension force $F_\gamma = \gamma p_F$ is the force required to increase the free surface of the fluid. Remember the symbolic experiment shown in figure 8.29, where a force F is exerted on the rod to extend a liquid film. The work of this force is given by the relation

$$\delta W = F dw = 2\gamma L dx. \tag{8.39}$$

The coefficient 2 stems from the fact that there are two interfaces between the liquid and the air. In this particular case, $F_\gamma = 2\gamma L$. An analogy with our case yields the surface tension force: $F_\gamma = \gamma p_F$.

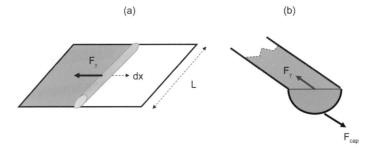

(a)   (b)

Figure 8.29 A tube placed on a rigid frame whose the left part is occupied by a soap film requires a force to be displaced towards the right; this force opposes the surface tension that tends to bring the tube to the left; (b) similar situation for a liquid in an open channel.

Hence relation (8.36) is universal in the case of the groove (section 8.3.3). Changing the notations to be consistent with the preceding case, the ratio p^* is equal to

$$p^* = \frac{d}{2w+d} = \frac{1}{2\frac{w}{d}+1}. \tag{8.40}$$

Upon introduction of the condition (8.21), we find for capillary pumping in a groove

$$p^* = \frac{p_F}{p_W} < \cos\theta, \tag{8.41}$$

which is the same condition as (8.27) for two parallel rails.

Consider now a channel having the shape of a semi-cylinder. It has been shown in section 8.3.5 that the condition for capillary pumping is $\cos(\theta) > 2/\pi$. On the other hand, the ratio p^* is

$$p^* = \frac{2R}{\pi R} = \frac{2}{\pi}. \tag{8.42}$$

Hence the condition for capillary pumping is

$$p^* = \frac{p_F}{p_W} < \cos\theta. \tag{8.43}$$

Again the same criterion applies. Note that for a completely closed channel $p_F = 0$, and the condition for liquid spreading is $\cos\theta > 0$ or $\theta < \pi/2$, which is the usual criterion.

Different open capillary pumping conditions are shown in the table 8.1.

Geometry	Condition for spreading	View
Closed channel	$0 < \cos\theta$	
Half pipe	$\frac{2}{\pi} < \cos\theta$	
Two rails	$\frac{d}{w} < \cos\theta$	
U-groove	$\frac{d}{w} < \frac{2\cos\theta}{1-\cos\theta}$	
Asymmetric U-groove	$\frac{d}{w_2} < \frac{2\sin(2\theta)}{1-\sin(2\theta)-\cos(2\theta)}$	
Trapezoidal open tube	$\frac{d_2}{2w+d_1} < \cos\theta$	
Triangular open tube	$\frac{d}{2w} < \cos\theta$	

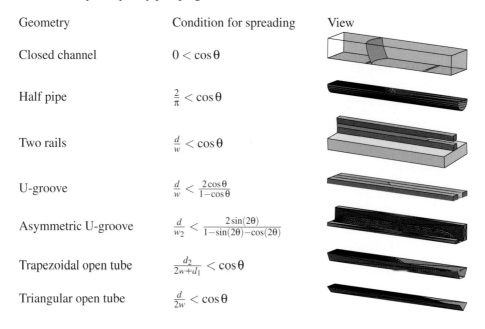

Table 8.1 Summary of the different conditions for liquid spreading under capillary action in different open microfluidic geometries.

8.3.8   Spontaneous Capillary Flows in Cracks

It is a well established fact – when manipulating liquid drops, or small amount of liquids – that these liquids are absorbed by the cracks, i.e. the cracks are easily wetted as soon as the liquid is slightly wetting, even if the contact angle is close to 90°.

If the crack is assumed, for simplicity, to have a triangular cross section, the condition for wetting is

$$\frac{p_F}{p_W} = \frac{d}{2w} = \sin\alpha < \cos\theta. \tag{8.44}$$

where 2α is the crack angle, d the width of the crack and w the length of the side of the crack. Relation (8.44) can be cast in the form

$$\theta < \frac{\pi}{2} - \alpha \tag{8.45}$$

which is the Concus-Finn limit. Thus, even poorly wetting fluids disappear by spontaneous capillary flow (SCF) into narrow cracks in the substrate (Fig. 8.30). It suffices that α be small enough.

Figure 8.30 (a) A sessile droplet placed over a breakdowned dielectric; (b) Evolver modeling of a droplet on a crack of the substrate; (c) Evolver modeling of a droplet located above two cracks in a solid substrate.

8.3.9   Spontaneous Capillary Flow Triggered by a V-groove

In the preceding section, the SCF in a V-groove has been investigated. It has been shown that a sharp V-groove was very favourable to SCF, even with relatively large contact angles 60 to 85°. This property, i.e. the triggering of SCF by Concus-Finn effect, can be used for the capillary filling of a microchamber, in the case where little or no spontaneous capillary filling would occur.

Let us consider the case of a U-slot used to perform blood analysis [35]. In this application, a liquid droplet of blood taken from the tip of the finger is analyzed. It first must fill a reaction chamber. In order to avoid any external pumping in this disposable device, capillary force is used: the blood is drained by capillarity from the tip of the device to the reaction chamber. A strong capillary effect must be actuated because the blood is very viscous and the cheap disposable material composing the device is a weakly hydrophilic plastic (PDMS).

In order to facilitate – and even make possible – the capillary filling, a V-groove is added at the bottom of the U-groove (fig. 8.31).

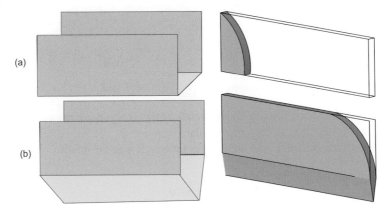

Figure 8.31 (a) U-groove; (b) U-groove modified to a V-groove. On the right are Evolver simulations showing how much better liquid penetrates a V-groove, with the same contact angles.

Let us investigate why the use of the V-groove triggers the capillary filling. A Concus-Finn effect – similar to that studied in the preceding section – creates a region of low pressure (below atmospheric pressure) that drives the SCF. In figure 8.31 a comparison between a simple U-groove and a U-groove modified to a V-groove performed with Evolver shows that the flow can stop in the first case, whereas it continues in the second case.

8.3.10   Anisotropic Superhydrophilicity

We have seen in the preceding section that SCF is increasingly used in "home test" devices for biotechnology. For example, home tests for blood analysis rely on the filling of a microchamber by capillarity. These tests must be affordable and disposable; that rules out using active systems to pump the liquid inside the device. Capillarity is then the obvious solution.

We have demonstrated that triangular V-grooves facilitate the spontaneous capillary filling of a microchamber. Now we focus on an outcome of these developments, which is the understanding of the use of the Concus-Finn limit for turning a hydrophilic surface into an anisotropic superhydrophilic surface.

Consider a rectangular U-groove, in the bottom of which parallel triangular V-grooves have been etched, as shown in figure 8.32. Let us assume that the width w is divided into n segments of width w/n and triangular V-grooves of depth $h = w/k$. The SCF condition can be written

$$\frac{w}{2d + 2n\sqrt{\left(\frac{w}{2n}\right)^2 + \left(\frac{w}{k}\right)^2}} < \cos\theta. \tag{8.46}$$

In terms of the aspect ratio w/d, relation (8.46) yields

$$\frac{w}{d} < \frac{2\cos\theta}{1 - \cos\theta\sqrt{1 + \left(\frac{2n}{k}\right)^2}}. \tag{8.47}$$

Using the V-groove half-angle α,

$$\tan\alpha = \frac{\frac{w}{2n}}{\frac{w}{k}} = \frac{k}{2n}. \tag{8.48}$$

Then (8.47) can be written

$$\frac{w}{d} < \frac{2\cos\theta}{1 - \cos\theta\sqrt{1 + \left(\frac{1}{\tan\alpha}\right)^2}} = \frac{2\cos\theta}{1 - \frac{\cos\theta}{\sin\alpha}}. \tag{8.49}$$

Inequality (8.49) shows that for sufficiently small angles α, even a large contact angle θ can lead to SCF. This is not the case for a rectangular U-groove.

Let us investigate the limit case where the ratio w/d can be infinite. From (8.49), this corresponds to

$$\cos\theta = \sin\alpha. \tag{8.50}$$

If (8.50) is satisfied, then there is a SCF even when $d = 0$. In fact, (8.50) is another form of the Concus-Finn limit

$$\theta - \frac{\pi}{2} - \alpha. \tag{8.51}$$

Hence we deduce the following result: if the triangular V-grooves etched into the bottom surface have a half-angle such that the Concus-Finn relation is satisfied, then the surface is superhydrophilic, and the liquid (aqueous) spreads completely in the direction of the grooves, even if no

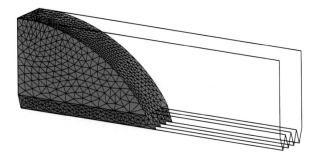

Figure 8.32 SCF in a rectangular U-groove with striations.

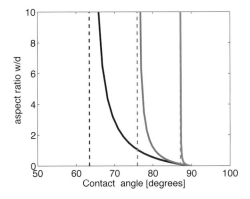

Figure 8.33 Relation between the aspect ratio and the contact angle below which a SCF occurs, for ($w = 200\,\mu$m, $n = 10$, and $k = 10, 5, 1$. The vertical dotted lines are the Concus-Finn angles corresponding to three groove angles, $2\alpha = 53.1°$, $28.1°$ and $5.7°$ respectively.

side walls are present. SCF is then reached whatever the aspect ratio of the channel. The preceding theoretical developments have been checked numerically with Evolver. Figure 8.32 shows SCF occurring in a channel etched with many V-grooves.

Relation (8.49) can be plotted in a $\{\theta, w/d\}$-axis coordinates system. In figure 8.33, the plots for $w = 200\,\mu$m, $n = 10$, and $k = 10, 5$ and 1 have been superposed. As mentioned above, the curves have an asymptote at the Concus-Finn limit. For a given Young angle, any value of the aspect ratio w/d smaller than that indicated by the corresponding curve will lead to SCF.

At the asymptote, SCF occurs even if $d = 0$ (no side walls). In such a case, an anisotropic superhydrophilic state is reached. A flat surface etched with micro V-grooves can be rendered superhydrophilic in the direction of the grooves, even for a large contact angle – as long as it is smaller than $90°$. Of course this result is theoretical and the quality of the dihedral at the bottom of the grooves is essential.

Figure 8.34, obtained with Evolver, illustrates the liquid spreading indefinitely.

In conclusion, triangular V-grooves facilitate SCF in an open micro-chamber. The shape of the V-groove must be chosen according to the Young contact angle. Even for contact angles of the order of $85°$, a SCF can be reached. The smaller the angle of the V-groove, the larger is the capillary force. If the Concus-Finn limit is reached, SCF occurs in all cases.

Extrapolation of the preceding conclusions shows that a surface can be rendered superhydrophilic (in one direction) by etching sharp V-grooves. This property may be of interest for

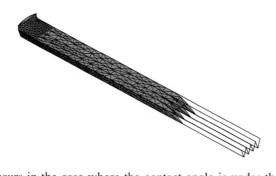

Figure 8.34 SCF occurs in the case where the contact angle is under the Concus-Finn limit, even if no side walls are present (contact angle 85°).

electrowetting, where the saturation limit is a drawback. In the case of microscopic sharp V-grooves, an electric potential rise will lower the contact angle below the value where complete spreading will occur.

8.3.11   Spontaneous Capillary Flow in Diverging U-grooves

The general relation (8.34) can be used to predict the length that a capillary flow will reach in a widening U-channel (fig. 8.35). At each cross-section of the channel, the condition for the flow to continue is

$$\frac{d}{w} < \frac{2\cos\theta}{1-\cos\theta}.$$
(8.52)

Because the width d increases as the front progresses, (8.52) will break down at some point and the flow will stop. This limit is given by

$$d_{stop} = w\frac{2\cos\theta}{1-\cos\theta}.$$
(8.53)

Figure 8.35 shows the distance reached by the capillary flow in two different widening channels. A comparison between the theoretical limit and the Evolver calculation is shown in figure 8.36. A very good agreement is reached, considering that the theory does not account for the shape of the free surface.

Experiments performed at the University of Wisconsin confirm the stopping of the flow in diverging U-grooves. PDMS diverging U-grooves have been fabricated using a SU8 mold. A colored ethanol solution has been pipetted in all the entry ports and it is checked that the flow stops when the liquid front reaches a threshold width of the channel (fig. 8.37).

8.3.12   Spontaneous Capillary Flow in Diverging-converging U-grooves

An interesting application of both the SCF universal law and the Laplace law is given by considering an X-shaped U-groove, i.e. a converging-diverging U-groove (Fig. 8.38). The liquid in the X-shaped channel is at equilibrium when the two Laplace pressures at the two ends are equal. From the preceding analysis, the capillary forces at both ends must balance each other. This statement implies that

$$\frac{d_{left}}{d_{left}+2w} = \frac{d_{right}}{d_{right}+2w},$$
(8.54)

(a)   (b)

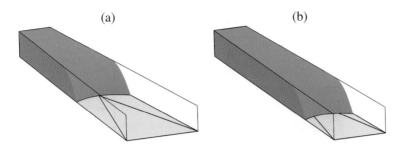

Figure 8.35 Maximum length reached by a capillary flow in diverging U-grooves, calculated with Evolver: (a) groove half-angle 3.34°; (b) groove half-angle 1.43° (contact angle 57°, channel depth 30 μm, inlet width 40 μm, outlet widths 110 and 80 μm).

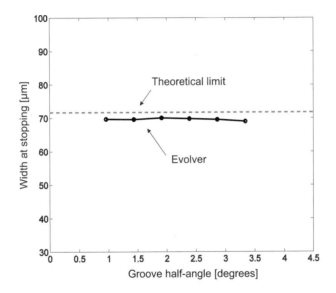

Figure 8.36 Width of the channel at the stopping front versus the stopping length (length of channel filled by the liquid): for a contact angle of 57° the theoretical limit is 71.8 μm; Evolver results indicate a value around 70 μm.

where w is the height of the walls and d the distance between the walls. Equation 8.54 yields

$$d_{left} = d_{right}. \tag{8.55}$$

The droplet is at equilibrium when the two ends have the same width, as shown in figure 8.38.

8.3.13   Capillary Flow Over a Hole

Can a capillary flow pass over a hole? This question is the focus of this section. Consider a U-groove of constant width d with a circular hole of diameter δ in the bottom plate (fig. 8.39).

The condition of capillary flow in the "plain" U-groove must be fulfilled, i.e. $d/(2w + d) < \cos\theta$. But this condition is not sufficient when the contact line reaches the hole. A more

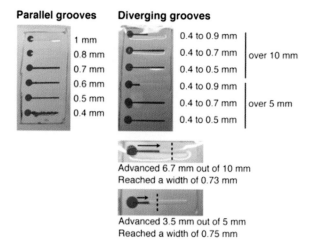

Figure 8.37 Experimental verification of the stopping of the flow in a diverging U-groove (©E. Berthier and B. Casavant, University of Wisconsin, by permission).

(a)   (b)

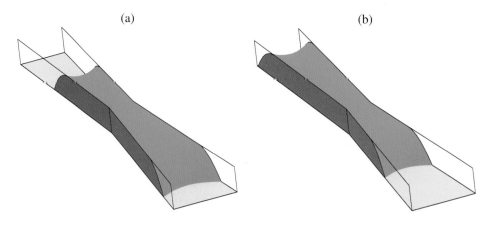

Figure 8.38 Liquid in a X-shaped U-groove, with wider end in the front: (a) Initial droplet; (b) equilibrium position.

restrictive condition is imposed by the presence of the hole. Let us assume first that the contact angle is above the Concus-Finn limit, i.e. for a 90° corner $\theta > 45°$. The condition for SCF is then

$$\frac{p_F}{p_w} = \frac{d+\delta}{2w+d-\delta} < \cos\theta. \tag{8.56}$$

The last relation can be rewritten in the form

$$\frac{\delta}{d} < \frac{\cos\theta(2\frac{w}{d}+1)-1}{1+\cos\theta}, \tag{8.57}$$

showing that the SCF condition depends only on three parameters w/d, δ/d and θ. The theoretical relation for SCF has been obtained by assuming a geometrically simplified shape of the

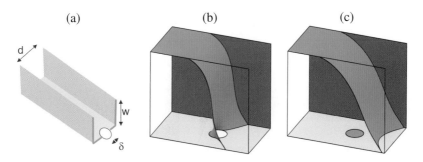

Figure 8.39 (a) Sketch of a U-groove with a hole. (b) The contact line moving past the hole. (c) After the contact line has fully passed the hole.

liquid surface. Its validity can be checked against Evolver calculations. In figure 8.40 the limit given by (8.57) has been plotted and compares well with the SCF limit produced by Evolver.

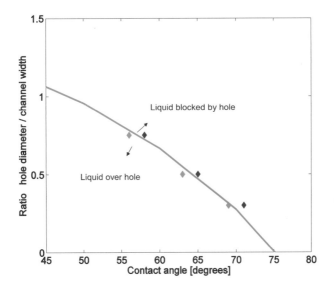

Figure 8.40 Theoretical and Evolver limit for SCF in a U-groove with a hole (here $w/d = 1.5$).

The situation however is much more complex when the contact angle is below the CF limit. In such a case, two filaments stretch in the two corners. These filament can eventually merge together downstream from the hole and may fill the hole, depending on the ratio w/d and the value of the contact angle (fig. 8.41).

8.3.14 Suspended Microfluidics

Recently, a new type of open microfluidics has appeared which is called "suspended microfluidics." This type of flow can be defined as a capillary flow that is partly suspended over gas

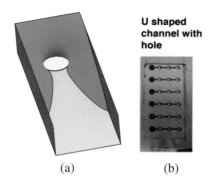

(a)   (b)

Figure 8.41 (a) Capillary flow going around a hole ($\theta = 40°$); (b) filaments merging downstream without filling the hole (©E. Berthier and B. Casavant, University of Wisconsin, by permission).

or vapor. One example is that of the rails described before in this chapter when these rails are turned vertically. The SCF still works under the same conditions – if the distance between the rails is not too large, and gravity is still negligible. Another example is the SCF in U-grooves pierced with holes.

Such a solution can be interesting to build new microfluidic devices. Let us take the example of the solution proposed by the University of Wisconsin [17]. In this case, a suspended liquid polymer SCF propagates between two rails, or rather between the parallel vertical boundaries of two horizontal plates (Fig.8.42); when the flow has filled the whole allowed domain, a polymerization step is made and a microporous membrane is formed. This solution is extremely attractive for cell-chips where diffusion of chemical species is controlled by membranes, or where migration of cells in microporous media is investigated.

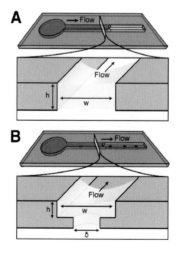

Figure 8.42 Two examples of suspended microfluidics. A: suspended SCF in a channel, B: suspended SCF over holes [17]. (©E. Berthier and B. Casavant, University of Wisconsin, by permission)

8.3.15   Application to Droplet Dispensing in EWOD/LDEP Systems

The general rule given by (8.34) explains the impossibility of achieving droplet dispensing from a reservoir in an open EWOD microsystem. EWOD microsystems will be presented in detail in Chapter 10. But this example is characteristic of the use of the general rule (8.34). Let us just mention here that in EWOD systems, the electric actuation of an electrode, located beneath a hydrophobic dielectric upon which a sessile aqueous droplet is deposited, reduces the apparent contact angle, i.e. makes the surface "apparently" hydrophilic.

A sketch of the dispensing device is shown in figure 8.43a. If dispensing were possible in such a geometry, a liquid "tongue" would advance along the electrode alley. Let dx be an infinitesimal progression of the liquid front along the electrode alley (width e). This progression corresponds to an approximate increase of the interfacial area of

$$dE \approx \gamma \pi \frac{e}{2} dx. \tag{8.58}$$

It has been assumed here that the side contact angle – limited by the side boundary of the electrode – is close to 90° (fig. 8.43b); it has also been assumed that the volume of the liquid "tongue" is very small compared to the remaining liquid volume in the reservoir. The surface tension force is then

$$F_{ST} \approx \gamma \pi \frac{e}{2}. \tag{8.59}$$

On the other hand, the capillary force is

$$F_{cap} = \gamma e \cos \theta_a, \tag{8.60}$$

where θ_a is the actuated contact angle. In other words, the free perimeter is $p_F = \pi e/2$ and the wetted perimeter $p_W = e$. Then using (8.34), there is motion if the ratio of the free to the wetted perimeter is less than $\cos \theta_a$. Obviously, this is impossible since

$$\frac{p_F}{p_W} = \frac{\pi}{2} > \cos \theta_a. \tag{8.61}$$

Evolver modeling of the dispensing is shown in figure 8.43.c, showing that it is not even possible to obtain the progression of a liquid "tongue." We shall see in chapter 10 that covered EWOD systems are used when dispensing is required [18,19], or that dielectrophoresis (LDEP systems) is used if one wants to keep an open geometry [20].

8.3.16   Restriction of the Theory in the Case of Rounded Corners

We have derived in the preceding sections the conditions for observing a spontaneous capillary flow (SCF) of a liquid in a groove with an ideally sharp corner. However, in the reality, ideally sharp corners do not exist and the quality of the corner depends on microfabrication. In this section we analyze the consequences of non-ideal corners on the spontaneous capillary flow by assuming that a small curvature radius is present in the corner. In the case where the wetting conditions are below the Concus-Finn limit, the analysis has been done by Kitron-Belinkov et al. [21].

Consider a 90° corner with a rounded shape as shown in figure 8.44. Upon applying the general rule for spontaneous capillary flow, we find

$$\frac{p_F}{p_W} = \frac{d}{2(w-R)+(d-2R)+\pi R} < \cos \theta_a. \tag{8.62}$$

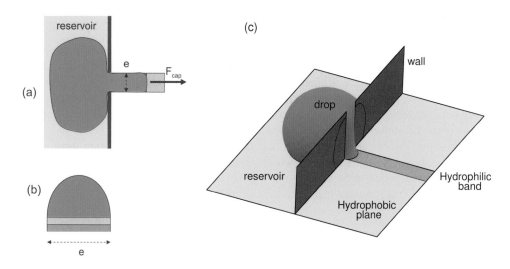

Figure 8.43 (a) Schematic of droplet dispensing device from a reservoir, view from top; (b) cross section of the tongue; (c) 3D Evolver modeling of the dispensing. Sketch courtesy E. Berthier and B. Casavant, University of Wisconsin.

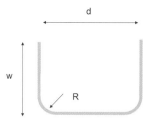

Figure 8.44 Sketch of a U-groove with rounded corners.

As the curvature radii are assumed to be small in the corners, $R \ll w$, or equivalently $R \ll d$. Then the preceding inequality can be written in the form

$$\frac{d}{2w+d} + \frac{dR(4-\pi)}{(2w+d)^2} < \cos\theta_a. \tag{8.63}$$

It is verified that when R goes to zero, the corrective term goes to zero and inequality (8.63) reduces to the inequality determining the SCF in a regular U-groove. The second (positive) term in the left hand side of (8.63) requires a smaller contact angle to obtain the SCF. A comparison between a perfect corner and a rounded corner is shown in figure 8.45 for the case of a dihedral. Clearly the CF condition is not sufficient to obtain a filament for a rounded corner. Remember that the CF effect is due to the flat interface profile in the direction parallel to the edge. This effect progressively vanishes with the curvature in the corner.

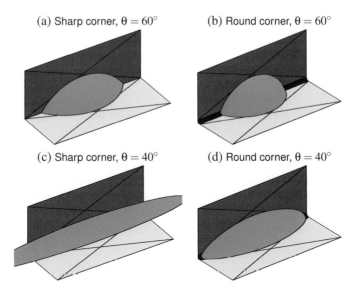

(a) Sharp corner, $\theta = 60°$ (b) Round corner, $\theta = 60°$

(c) Sharp corner, $\theta = 40°$ (d) Round corner, $\theta = 40°$

Figure 8.45 Comparison between a sharp corner and a rounded corner (R=20 μm): above the CF limit, the droplet stretches less; below the CF limit, for a contact angle of 40°, there is no filament in the rounded corner morphology.

8.4   Liquid Wetting Fibers

Another aspect of open microfluidics is that of liquid spreading between fibers. This domain is connected to the theory of imbibition. Applications can be found in microfluidics on paper and thread-based microfluidics, which are gaining momentum in biotechnology [7,8]. The advantages of "lab-on-paper" are simplicity, portability and robustness. Of course, not all the functions that conventional microfluidic systems offer are available on such systems, but they are well adapted to the fast-developing segment of home-testing. Figure 8.46 is a typical example of architecture of a lab-on-paper.

In this section, we first analyze the behavior of droplets between fibers, and then present a

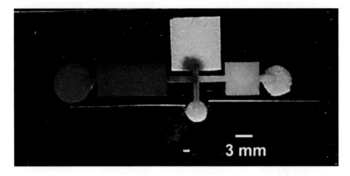

Figure 8.46 Example of lab-on-paper [7]: the different paper pads are dimensioned so that the three reagents arrive successively in a predefined order on the reaction pad. Reprinted with permission from [7]; ©Springer, 2010.

generalization for an assembly of fibers (tissue) with the Washburn and Darcy's laws.

8.4.1   Droplet Between Parallel Fibers

The spreading condition of liquid between two parallel rails fed by a large reservoir (assumed infinite) has already been investigated above in section 8.3.2. In this section, we analyze the behavior of a droplet between two parallel fibers. We show that the problem depends on the liquid contact angle with the fibers, the volume of liquid and the distance between the fibers.

Assume a droplet contacting two parallel, hydrophilic fibers as shown in figure 8.47. The capillary forces on the four contact lines with the fibers act to elongate the droplet, while the surface tension force acts oppositely to render the droplet more spherical. Depending on the volume and on the distance between the fibers, the liquid bulges out more or less.

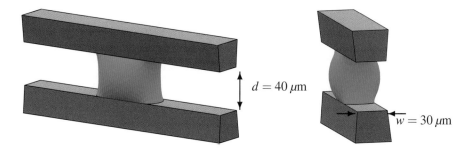

Figure 8.47 Liquid bridge between two parallel fibers ($\theta = 70i°$).

Assume now that the drop is evaporating or – equivalently – that the distance between the fibers is increased. As the volume of liquid decreases, the bulging out of liquid is progressively reduced, until the curvature changes sign and the droplet finally splits into two separate droplets attached to each fiber (fig. 8.48).

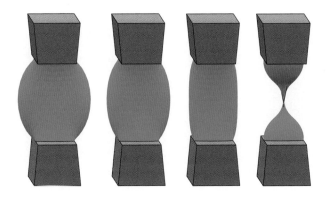

Figure 8.48 Shape of an evaporating droplet contacting two hydrophilic fibers.

8.4.2   Intersecting Fibers

The case of a droplet between two fibers making an angle is investigated here. We have already seen that a droplet placed between two plates making an angle or droplets on conical wires [14] are generally not in an equilibrium state. This behavior stems directly from the Laplace law, because the two ends of the droplets do not have equal curvature due to spatial evolution of the geometry.

In figure 8.49, the evolution of two water droplets between two hydrophilic fibers making an angle is shown. It is clearly observed that the curvature radius in the vertical plane is smaller on the small end of the droplet than on the other end. The Laplace pressure is then smaller (negative) in the small end of the droplet. The droplets find their equilibrium state once they have merged in the angle.

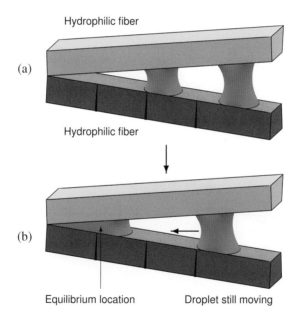

Figure 8.49 A water droplet between two hydrophilic fibers finds an equilibrium location at the junction.

8.4.3   Wicking in a Bundle of Fibers

What is the effect of a bundle of parallel fibers piercing a droplet? Will the droplet spread or not, and more specifically what is the value of contact angle that allows for spreading? The problem is difficult, because in this particular case the capillary forces on the triple lines that want to stretch the liquid are opposed by the free surface extension between the fibers, and also by the free surface of the source droplet. The first part – extension between the fibers – is easy to calculate. It suffices to start from equation (8.16)

$$P = \gamma_{LG} \left(\frac{dA_{LG}}{dV} - \cos\theta \frac{dA_{SL}}{dV} \right),$$
$$dA_{LG} = 4d\,dx, \tag{8.64}$$
$$dA_{SL} = 8a\,dx,$$
$$dV = (4ad + d^2)\,dx,$$

to find the condition

$$\frac{d}{2a} = \cos\theta. \tag{8.65}$$

However, the second part corresponding to the change of the free surface of the central droplet is difficult to calculate. Using Evolver, the limit contact angle for wicking can be derived for a given initial droplet volume (fig. 8.50).

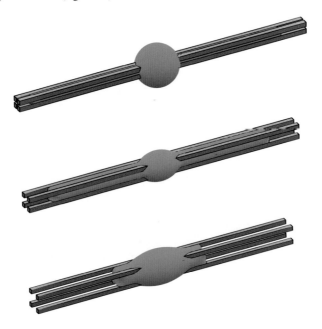

Figure 8.50 Droplet spreading between fibers: the cutoff contact angle depends on the spacing of the fibers. Top, 84°. Middle, 70°. Bottom, 50°.

The cutoff contact angle to obtain wicking has been plotted in figure 8.51 as a function of the reduced distance d/a. On the same figure, the contribution of the increase of free surface between the fibers ($d/2a = \cos\theta$) has been plotted.

In the preceding analysis, the dynamic effect is not taken into account and the contact angle is the static contact angle. In reality, an advancing contact angle should be taken into account. It has been shown [22] that in an imbibition process the dynamic contact angle θ_a is linked to the static contact angle θ_s by the relation

$$\cos\theta_a = \cos\theta_s - aCa^b, \tag{8.66}$$

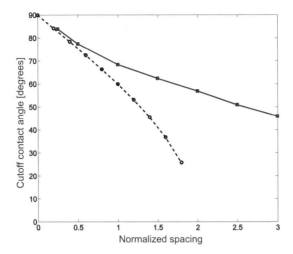

Figure 8.51 Imbibition contact angle as a function of the geometry: when the normalized spacing goes to 0, the fibers are getting very close together and the contact angle goes to 90°.

where Ca is the capillary number $Ca = \eta V / \gamma$, a and b are constant, and b is of the order of 0.4. For water, $\eta = 10^{-3}$ Pa.s and the surface tension with air is 72×10^{-3} N/m. For an approximate velocity of 1 mm/s, the capillary number is approximately 1.4×10^{-5}. Hence the correction for the advancing contact angle is quite small, and within the error margin for the value of the critical contact angle for imbibition.

8.4.4 Total Impregnation – Imbibition

The cases analyzed in the preceding section lead us to introduce shortly the laws governing imbibition in disordered matrices. The first law – called the Washburn law – concerns the flow during the wet-out process of a microporous medium. The second law – called the Darcy law – describes the flow in the fully wetted medium.

8.4.5 Washburn's Law

Let us recall the Lucas-Wasburn theory about impregnation in a bundle of parallel fibers [23,24]. The porous medium is assumed to consist of a bundle of parallel capillary tubes of the same size and the model uses a one-dimensional flow as its basis since the flow can just move along the axis of the assumed capillary tubes. For such an idealized porous medium, the Hagen-Poiseuille law [25,26] for laminar flow through pipes can be employed to find the equation of motion for the porous-medium flow. The Hagen-Poiseuille law states that the volumetric flow rate is proportional to the pressure drop along the tube length:

$$\Delta P = \frac{8\eta Q L}{R_c^2}, \tag{8.67}$$

where Q is the volume flow-rate, R_c is the tube radius, η is the fluid viscosity, L is the length of the wetted tube, and ΔP is the net driving pressure (pressure drop across the length L). If the only driving pressure is the capillary pressure (for example if the tube is horizontal), then

$$\Delta P = \frac{2\gamma\cos\theta}{R_c}. \tag{8.68}$$

Eliminating ΔP between (8.67) and (8.68) yields

$$LQ = \frac{\pi\gamma\cos\theta}{4\eta}R_c^3. \tag{8.69}$$

Note that the volume flow-rate in a tube can be expressed in terms of the rate of change of the wetted length of the tube as

$$Q = \pi R_c^2 \frac{dL}{dt}. \tag{8.70}$$

Upon combining (8.68), (8.69) and (8.70) and further integration, the expression of the length of the wetted region as a function of time is found to be

$$L = \sqrt{\frac{\gamma\cos\theta R_c}{2\eta}t}. \tag{8.71}$$

This equation is commonly known as the Washburn equation [24-26] and is used quite frequently in wicking applications [4]. Upon substitution of (8.71) in (8.70), the flow rate of liquid is given by

$$Q = \pi\sqrt{\frac{\gamma\cos\theta R_c^3}{8\eta}}\frac{1}{\sqrt{t}}. \tag{8.72}$$

Hence the flow rate decreases as $1/\sqrt{t}$. Note that the Lucas-Washburn model can be extended to porous media in general by taking into account the tortuosity of the flow path [27],

$$L = \sqrt{\frac{\gamma\cos\theta R_p}{2\tau^2\eta}t}, \tag{8.73}$$

where R_p is the average radius of the pores, and the tortuosity τ is the square of the ratio between the length L_e of the shortest path between two points through the liquid, and the straight line distance L between these two points:

$$\tau = \left(\frac{L_e}{L}\right)^2. \tag{8.74}$$

Note that the porous medium has been assumed to be rigid, i.e. does not deform when the liquid penetrates. In rigid porous media, the sizes of the pores remain constant during the wicking process. There exist in the literature models derived from the Washburn law that take into account the deformation of a porous medium [28].

8.4.6 Fully Wetted Flow – Darcy's Law

Continuous flow in a wetted porous or microporous medium can be described by Darcy's law [29]. This applies also to flows in paper channels. Darcy's law links the average flow velocity $\vec{U}$ to the pressure gradient:

$$\vec{U} = -\frac{K}{\eta}\nabla P, \tag{8.75}$$

where η is the fluid viscosity and K the medium permeability (unit m^2). In the case of an anisotropic medium, for example if the fibers in the paper matrix are not randomly disposed, the permeability is a symmetric matrix, and the Darcy law becomes [30]

$$\vec{U} = -\frac{[K]}{\eta}\nabla P. \qquad (8.76)$$

Darcy's law, with the corresponding boundary conditions, produces the hydrodynamics in the paper matrix. The motion of a chemical or biochemical species in the wetted paper matrix can be followed by solving the convection-diffusion equation for the species concentration:

$$\frac{\partial c}{\partial t} + \vec{U} \cdot \nabla c = D\delta c, \qquad (8.77)$$

where c is the species volume concentration, and D its diffusion coefficient (units m^2/s). In a Cartesian coordinate system, equation (8.77) can be written in the form

$$\frac{\partial c}{\partial t} + u\frac{\partial c}{\partial x} + v\frac{\partial c}{\partial y} + w\frac{\partial c}{\partial z} = D\left(\frac{\partial^2 c}{\partial x2} + \frac{\partial^2 c}{\partial y^2} + \frac{\partial^2 c}{\partial z^2}\right). \qquad (8.78)$$

Note that in a flat paper matrix the flow is nearly two-dimensional and the z-component can be dropped:

$$\frac{\partial c}{\partial t} + u\frac{\partial c}{\partial x} + v\frac{\partial c}{\partial y} = D\left(\frac{\partial^2 c}{\partial x^2} + \frac{\partial^2 c}{\partial y^2}\right). \qquad (8.79)$$

Generally the diffusing species concentration is small and does not modify the porous matrix, so that the two equations (8.76) and (8.79) are decoupled and can be solved sequentially. Figures 8.52 and 8.53 show the results of a calculation with the COMSOL numerical program [31] and a comparison with experimental results [7].

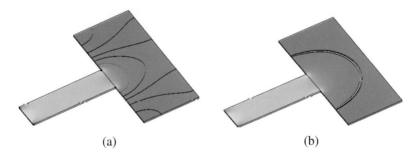

(a)   (b)

Figure 8.52 (a) Flow in a paper matrix calculated with COMSOL using the "Darcy" option; (b) calculation of the concentration kinetics of a diffusing species.

8.4.7 Paper-based Microfluidics

In biotechnology, the development of "labs-on-paper" – sometimes called μPAD for microfluidic paper-based analytical devices – has been motivated by the need for inexpensive biological tests [32,33]. As home-tests and point-of-care tests are quickly developing, there is a

Concentration isosurfaces at t = 2 s

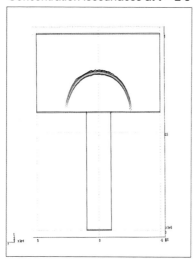

Figure 8.53 Comparison between modeling results (left) and experimental observation (right) for convection-diffusion in a flat paper matrix. Reprinted with permission from [7], ©Springer, 2010.

strong incentive to reduce the costs while maintaining sufficient test sensitivity. The other advantages of "labs-on-paper," or μPADs, are compactness and portability. In parallel, using the same concepts of porous medium flow, the development of smart bandages is under way. These smart bandages are aimed at monitoring the progression of body fluids in the bandage and at the same time at analyzing the presence of pathogen bacteria [34].

The technique of μPADs relies on liquid imbibition and progression in a laminated porous matrix. The wet-out process of paper strips of depends on shape: sudden expansions slow the progression of the fluid, and sudden contractions have the opposite effect. Hence, in "lab-on-paper" devices, sudden changes of section can be used to control the arrival of the fluids. An example of a paper network with three legs shows the successive arrival of reagents in the central detection regions (fig. 8.46). Successive delivery of reagents can thus be achieved using in paper networks.

8.4.8   Thread-based Microfluidics

An interesting approach, yet in its infancy, has been proposed by Ballerini *et al.* at Monash University [8]. They have shown that polyester threads can be used to replace solid wall channels. In a sense, thread-based microfluidics is a mix of the open microfluidic concepts and paper-based microfluidics (fig. 8.54 and 8.55).

Hydrophilic threads were obtained by twisting together multiple polyester fibers. Fluid penetration was found to increase with the number of twisted fibers. A qualitative model for the motion of liquid along such threads can be derived from the Poiseuille and Laplace laws, leading to the Washburn equation if the threads are in a horizontal plane. If the thread is vertical, the Washburn equation must take into account gravity. The equation is derived by combining

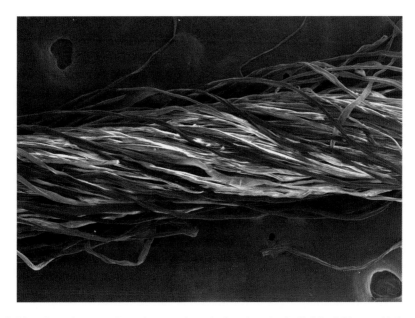

Figure 8.54 A SEM image of a polyester thread, showing the individual fibers which comprise the overall structure. Capillary channels are formed by the gaps between fibers, enabling the thread to conduct flow. Reprinted with permission from [8], ©AIP, 2011, http://linkip.org/link/doi/10.1063/1.3567094.

the Hagen-Poiseuille law

$$\frac{dh}{dt} = \frac{r^2 \Delta P}{8\eta h},$$

(8.80)

and the Laplace pressure with a gravity term

$$\Delta P = \frac{2\gamma \cos\theta}{r} - \rho g h.$$

(8.81)

Substitution of (8.81) in (8.80) leads to

$$\frac{dh}{dt} = \frac{r^2}{8\eta h} \left(\frac{2\gamma \cos\theta}{r} - \rho g h \right),$$

(8.82)

which constitutes the kinetic equation for a vertical thread.

8.5 Conclusions

Open microfluidics relies heavily on spontaneous capillary flow (SCF) to move and actuate the fluids. Once the conditions for SCF are satisfied, pumping is achieved. Electrowetting techniques go well with open microfluidics because they can be used for valving: when the electrode is actuated the flow passes, and conversely the flow is stopped when the electrode is off. A detailed study of electrowetting is done in chapter 10.

Open microfluidic technique is a vast domain, which is increasingly used in biotechnology and optofluidics. Used in microsystems, it has the advantages of offering access to the fluid,

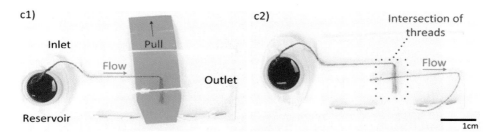

Figure 8.55 Folding style switches can be modified to be pull-tab actuated: (c1) the pull-tab switch with inlet connected to an ink reservoir; (c2) the pull-tab switch with tab removed, allowing flow across the device. Reprinted with permission from [8] ©AIP, 2011, http://linkip. org/link/doi/10.1063/1.3567094.

easier visualization and good monitoring. Another advantage is avoiding the often tricky operation of sealing the system cover plate. For these reasons, many new designs have been proposed, each one dedicated to a specific task.

8.6   References

[1] Wataru Satoh, Hiroko Hosono, Hiroaki Suzuki "On-chip microfluidic transport and mixing using electrowetting and incorporation of sensing function," *Anal. Chem.* **77**, pp. 6857–6863, 2005.

[2] E. Kreit, M. Dhindsa, S. Yang, M. Hagedon, K. Zhou, I. Papautsky, J. Heikenfeld, "Laplace barriers for electrowetting thresholding and virtual fluid confinement," *Langmuir* **26**(23), pp. 18550-18556, 2010.

[3] H.A. Stone, A.D. Stroock, A. Ajdari, "Microfluidics towards a lab-on-a-chip," *Ann. Rev. Fluid Mech.* **36**, pp. 381-411, 2004.

[4] B. Zhao, J.S. Moore, D.J. Beebe, "Surface-directed liquid flows inside microchannels," *Science* **281**, 1023-1026, 2001.

[5] P. Concus, R. Finn, M. Weislogel, "Measurement of critical contact angle in a microgravity space experiment," *Exp. Fluids* **28**, 197–205, 2000.

[6] E. Lorenceau, C. Clanet, D. Quèrè, "Capturing drops with a thin fiber," *Journal of Colloid and Interface Science* **279**, pp. 192–197, 2004.

[7] E. Fu, S. A. Ramsey, P. Kauffman, B. Lutz, P. Yager, "Transport in two-dimensional paper networks," *Microfluidics and Nanofluidics Journal* **10** , pp. 29-35, 2011.

[8] D.R. Ballerini, XU LI, Wei Shen, "Flow control concepts for thread-based microfluidic devices," *Biomicrofluidics* **5**, 014105, 2011.

[9] J. Tallmadge, R. Labine, B. Wood, "Films adhering to large wires upon withdrawal from liquid baths," *I & EC Fundamentals* **4**(4), p. 403, 1965.

[10] L.S. Hung, S.C. Yao, "Experimental investigation of the impaction of water droplets on cylindrical objects," *International Journal of Multiphase Flow* **25** pp. 1545-1559, 1999.

[11] T. Gilet, D. Terwagne, and N. Vandewalle, "Digital microfluidics on a wire," *Appl. Phys. Lett.* **95**, 014106, 2009.

[12] B.J. Carroll, "The equilibrium of liquid drops on smooth and rough circular cylinders," *J. Colloids Interface Sci.* **97**(1), pp. 195–200, 1984.

[13] G. McHale, M.I. Newton, "Global geometry and the equilibrium shapes of liquid drops on fibers," *Colloids Surf. A* **206**(103), pp. 79–86, 2002.

[14] E. Lorenceau, D. Quèrè, "Drops on a conical wire," *J. Fluid Mech.* **510**, pp. 29–45, 2004.

[15] H.B. Eral, R.Ruiter, J.Ruiter, J.M. Oh, C. Semprebon, M. Brinkmann, F. Mugele, "Drops on functional fibers: from barrels to clamshells and back," *Soft Matter*, 7, 5138–5143, 2011.

[16] Sung Hoon Lee, A.J. Heinz, Sunghwan Shin, Yong-Gyun Jung, Sung-Eun Choi, Wook Park, Jung-Hye Roe, Sunghoon Kwon, "Capillary based patterning of cellular communities in laterally open channels," *Anal. Chem.* **82** (7), pp. 2900-2906, 2010.

[17] B.P. Casavant, E. Berthier, J. Berthier, Lauren Bischel, K. Brakke, N. Keller and D.J. Beebe, "Suspended Microfluidics," (submitted).

[18] J. Berthier, Ph. Clementz, O. Raccurt, D. Jary, P. Claustre, C. Peponnet, Y. Fouillet, "An analytical model for the prediction of microdrop extraction and splitting in digital microfluidics systems," *Proceedings of the 2005 Nanotech conference,* Anaheim, USA, 8-12 May 2005, pp. 664-667.

[19] J. Berthier, Ph. Clementz, O. Raccurt, D. Jary, P. Claustre, C. Peponnet, Y. Fouillet, "Mechanical behavior of micro-drops in EWOD systems: drop extraction, division, motion and constraining," *Proceedings of the 2005 Nanotech conference,* Anaheim, USA. 8-12 May 2005, pp. 688-691.

[20] R. Ahmed and T.B. Jones, "Optimized liquid DEP droplet dispensing," *J. Micromech. Microeng.* **17**, pp. 1052–1058, 2007.

[21] M. Kitron-Belinkov, A. Marmur, T. Trabold and G. Vyas Dadheech, "Groovy Drops: Effect of Groove Curvature on Spontaneous Capillary Flow," *Langmuir 23*, pp. 8406–8410, 2007.

[22] P.G. de Gennes, "Wetting: statics and dynamics," *Reviews of Modern Physics* **57**(3), pp. 827–863, 1985.

[23] R. Lucas, "Rate of Capillary Ascension of Liquids," *Kollid Z.* **23**, p. 15, 1918.

[24] E.V. Washburn, "The Dynamics of Capillary Flow," *Phys. Rev.* **17**, p. 273, 1921.

[25] H. Bruus. *Theoretical microfluidics*. Oxford University Press, 2008.

[26] J. Berthier, P. Silberzan. *Microfluidics for Biotechnology.* Second edition, Artech House, 2010.

[27] M. Reyssat, L. Courbin, E. Reyssat, H.A. Stone, "Imbibition in geometries with axial variations," *J. Fluid Mech.* **615**, pp. 335–344, 2008.

[28] R. Masoodi, K.M. Pillai, "Darcy's law-based model for wicking in paper-like swelling porous media," *AIChE Journal* **56**(9), p. 2257, 2010.

[29] R.G. Larson, "Derivation of generalized Darcy equations for creeping flow in porous media," *Ind. Eng. Chem. Fundam.* **20**, pp. 132–137, 1981.

[30] D. Benavente, P. Lock, M. Á. García Del Cura and S. Ordonez, "Predicting the Capillary Imbibition of Porous Rocks from Microstructure," *Transport in Porous Media* **49**(1), pp. 59–76, 2002.

[31] COMSOL Multiphysics, http://www.comsol.com.

[32] W.A. Zhao, A. van den Berg, "Lab on paper," *Lab Chip* **8**(12), pp. 1988–1991, 2008.

[33] A.W. Martinez, S.T. Phillips, "Diagnostics for the developing world: microfluidic paper-based analytical devices," *Anal. Chem.* **82**(1), pp. 3-10, 2010.

[34] P. Singer, "Smart bandage detects bacteria with silicon sensor," *Semiconductor International* **24**, p. 34, 2001.

[35] A. Williamsson, S. Wahlqvist, S. Nilsson, J. Lilja, L. Jansson, B. Nilsson, patents EP 0 821 784, US005674457A, 1997.

8.7   Appendix: Calculation of the Laplace Pressure for a Droplet on a Horizontal Cylindrical Wire

In this Appendix, we detail how (8.5) was derived by Carroll *et al.* At equilibrium, the free energy of a drop is at minimum. The energy G that must be minimized can be written in the form

$$G = \gamma A - \lambda \Omega, \tag{8.83}$$

where A is the liquid/air interface area, γ is the liquid/air surface tension, Ω is the volume of the drop and λ is a Lagrange multiplier related to the condition of a constant volume. Denoting as z the distance from the fiber axis Ox, the free energy G is equal to

$$G = \pi \int \left[\gamma \left(2z\sqrt{1 + \dot{z}^2} \right) - \lambda(z^2 - r2) - \lambda \Omega \right] dx, \tag{8.84}$$

where $\dot{z} = dz/dx$, and r is the fiber radius. Let us denote by $f(z, \dot{z})$ the integrand in (8.84); then (8.84) can be rewritten in the form

$$G = \pi \int f(z, \dot{z}) \, dx. \tag{8.85}$$

As we are searching for the extremum of G, the function f satisfies the Euler-Lagrange relation

$$-\frac{d}{dx} \frac{\partial f}{\partial \dot{z}} + \frac{\partial f}{\partial z} = 0. \tag{8.86}$$

Replacing f by its expression in (8.86) yields

$$\gamma \left[\frac{-\ddot{z}}{(1 + \dot{z}^2)^{3/2}} + \frac{1}{z(1 + \dot{z}^2)^{1/2}} \right] = \lambda. \tag{8.87}$$

This equation is the Laplace equation, since the term in the brackets is the curvature in an axisymmetric coordinate system. Then

$$\lambda = \Delta P, \tag{8.88}$$

and the Lagrange multiplier λ is the pressure inside the droplet. Moreover, (8.87) can be further integrated to

$$-\Delta P \frac{z^2}{2} + \gamma \frac{z}{(1 + \dot{z}^2)^{1/2}} = K, \tag{8.89}$$

where K is a constant. Using the boundary conditions $\dot{z} = 0$ when $z = r$ (assuming that the liquid wets the solid), we deduce

$$K = -\Delta P \frac{r^2}{2} + \gamma r. \tag{8.90}$$

A second substitution with the condition $\dot{z} = 0$ and $z = h$ eventually yields

$$\Delta P = \frac{2\gamma}{r + h} = 0. \tag{8.91}$$

9

Droplets, Particles and Interfaces

9.1 Abstract

Solid particles, bubbles, droplets or gelled droplets are commonly found in microfluidic systems. Very often, these objects interact together or contact existing interfaces.

The different combinations of contact are numerous. In this chapter we restrict ourselves to the study of the behavior of liquid droplets in contact with other droplets and interfaces, and on solid spheres at interfaces. In the first part, we focus on the question of engulfment of a liquid droplet by another droplet or in another liquid phase across an interface. On one hand, engulfment of a droplet immersed in a carrier liquid into another liquid across the interface is crucial in encapsulation applications [1,2]; on the other hand, engulfment of a droplet in another droplet is of importance in chemical micro-reactors using reactions performed inside droplets [3-6].

In a second section, we focus on solid spheres and their behavior in contact with droplets and interfaces (fig. 9.1). In particular, the crossing of an interface by a solid sphere is presented. There are many occurrences of particles at an interface in the domain of microfluidic systems; for example, the understanding of interaction of solid spheres and interfaces is essential in self-assembly applications [7].

9.2 Neumann's Construction for Liquid Droplets

Young's law is obtained by projection of the surface tension forces at the triple line on the plane defined by the solid surface. In reality there is also a normal constraint in the solid at its surface to balance the normal component of the surface tensions [8]. But, in the case of solid substrates, there is no visible deformation (there is a nanoscopic effect, though, which was shown by de Gennes [9, 10]). However, a droplet deposited on a deformable surface, such as a micro-cantilever, or a viscoelastic substrate or a liquid surface makes clearly apparent the 3D balance of the surface tension forces [8,9] (fig. 9.2). This 3D balance is called Neumann's construction. Neumann's construction has already been presented in Chapter 1. Let us recall that Neumann's construction is the three-dimensional version of Young's law, or, more precisely, that Young's law is the two-dimensional projection of Neumann's construction:

$$\vec{\gamma}_{L1L2} + \vec{\gamma}_{L1G} + \vec{\gamma}_{L2G} = 0. \tag{9.1}$$

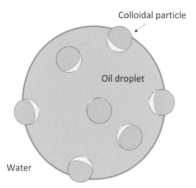

Figure 9.1 Example of PMMA (polymethyl methacrylate) particles stuck on the surface of an oil droplet (image inspired by the Manoharan Lab at Harvard).

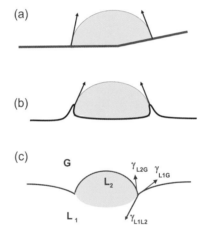

Figure 9.2 Neumann's construction: (a) a sessile droplet bends a sufficiently sensitive canti-lever; (b) a sessile droplet deforms the surface of a viscoelastic medium (gel, for example); (c) a droplet deposited on the surface of an immiscible liquid. G stands for gas, L for liquid.

9.3   The Difference Between Liquid Droplets and Rigid Spheres at an Interface

Even if the Neumann's force balance always applies at a triple line, there is a considerable difference between a deformable liquid droplet and a non-deformable body at an interface. In the liqiud droplet case, the vertical profile of the droplet has a singularity at the intersection of the triple line, i.e. the vertical profile is a continuous but not differentiable function at the intersection with the triple line (fig. 9.3a). In this case the Neumann law has to be used to characterize the contact. On the other hand, a rigid body – for example a rigid sphere – behaves as a solid planar surface and the Young law can be used to define the contact at the triple line (fig. 9.3b).

In the case of a rigid body, the question is: what is the effect of the non-zero resultant perpendicular to the surface? In the case of a spherical rigid body placed at the surface of a

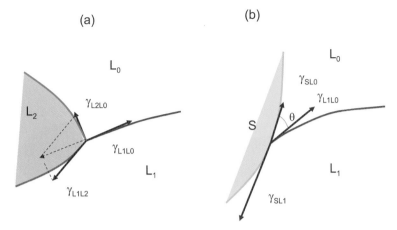

Figure 9.3 (a) Triple contact line between three liquids; (b) triple contact line between a solid and two liquids.

liquid with a symmetric contact line, the answer is straightforward: the normal resultants along the triple line cancel out by symmetry (fig. 9.4).

$$\int \left(\vec{\gamma}_{L0L1} + \vec{\gamma}_{SL0} + \vec{\gamma}_{SL1} \right) \cdot \vec{n} = 0. \tag{9.2}$$

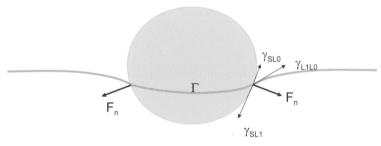

Figure 9.4 In the case of a sphere, the normal resultants cancel out by symmetry.

In the case of a particle of nonspherical shape, the particle will rotate until it finds an equilibrium position where the resultants of the normal components of the surface tensions cancel out. An interesting investigation of the equilibrium position of orthorhombic particles at an interface has been performed by Morris and colleagues [11]. They have numerically shown that the shape of the particle, especially its aspect ratio, modifies the equilibrium position of floatation (fig. 9.5).

9.4 Liquid Droplet Deposited at a Liquid Surface

In this section we investigate the behavior of a droplet deposited on a horizontal liquid-liquid or liquid-gas interface.

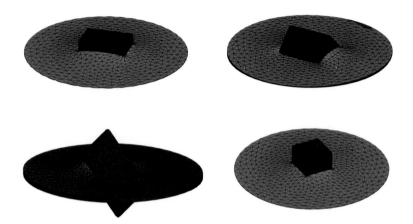

Figure 9.5 The position of a solid orthorhombic particle at the surface of a liquid depends on the aspect ratio of the particle (reprinted with permission from [11], ©Elsevier, 2011).

9.4.1   Introduction

The relative position of the droplet at the interface is governed by the relative importance of the three surface tensions. For simplicity, we denote with the indices D the droplet, A the air – or the liquid above the interface – and L the liquid below the interface.

If the droplet has sufficiently small dimensions or is buoyancy neutral, the gravitational effect is negligible. This behavior corresponds to the case where the Bond number is much smaller than 1,

$$Bo = \frac{\Delta \rho g R^2}{\gamma},$$ (9.3)

where $\Delta \rho$ is the density difference between the droplet and the liquid, and R is the equivalent radius defined by

$$Volume = \frac{4}{3} \pi R^3,$$ (9.4)

Consider first a water droplet initially placed on top of an oil bath (fig. 9.6a). If the Bond number is very small, the gravity effect is negligible and the water droplet finds an equilibrium position at the interface where most of the droplet volume lies below the interface. Moreover, the oil surface remains nearly flat. A detail analysis of the forces at the triple line is shown in figure 9.7.

In the case of a larger droplet, for which the gravity (buoyancy) force is not negligible, the interface is deformed, and the droplet falls into the oil bath if its relative weight is sufficient (fig. 9.6b).

Figure 9.8 shows some typical sketches of very small floating droplets obtained by numerical simulation with the Surface Evolver for an equivalent radius $R = 200 \ \mu m$.

The first case (fig. 9.8a and b) corresponds to the spreading of an oil droplet on the water surface: $\gamma_{DL} = \gamma_{WO} = 45$ mN/m, $\gamma_{LA} = \gamma_{WA} = 72$ mN/m, $\gamma_{DA} = \gamma_{OA} = 30$ mN/m. In this case $\gamma_{LA} > \gamma_{DA} \simeq \gamma_{DL}$.

The second case (fig. 9.8c and d) is an image of a pre-gelled droplet of alginate at an oil-water interface. In such a case, the alginate-water surface tension is small compared to the two

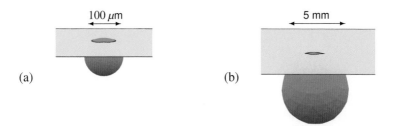

Figure 9.6 Water (W) droplet deposited on oil (O): (a) a small water droplet stays at the surface, which remains nearly flat; (b) a large water droplet falls through the interface under the action of gravity. $\gamma_{DL} = \gamma_{WO} = 45$ mN/m, $\gamma_{LA} = \gamma_{WA} = 72$ mN/m, $\gamma_{DA} = \gamma_{OA} = 30$ mN/m.

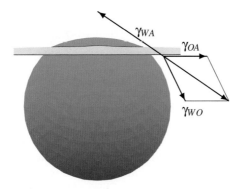

Figure 9.7 Detail of a gravity-free water droplet at the oil interface: Neumann's construction at the triple line. The parallelogram shows the sum of the oil-air and water-oil tensions adding to produce the opposite of the water-air tension.

other surface tensions. The third case (fig. 9.8.e) corresponds to the case of a small surface tension between the liquid and the air in comparison with the two other surface tensions. This is for instance the case of a small droplet of mercury. The fourth case (fig. 9.8.f) is an example of a very high drop-liquid surface tension.

The values of the three surface tensions determine the position of the droplet at the interface. Let us now investigate the conditions under which a droplet can penetrate into a liquid across the interface, i.e. be engulfed in the other phase.

9.4.2 Liquid Droplet Crossing an Interface

We distinguish two cases: a first case where gravity is neglected – in such a case the interface is not necessarily horizontal, it can have any orientation – and a second case where gravity is taken into account – in such a case the interface is horizontal.

9.4.2.1 Spreading and Engulfment

In the preceding section, we have seen that the relative values of the three surface tensions determine the position of the droplet. Now let us ask the following questions: under which

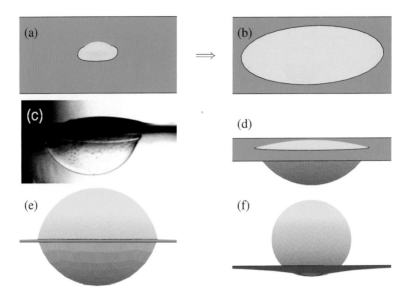

Figure 9.8 Different configurations for a droplet at an interface: (a) a droplet of oil initially deposited at the surface of water with the sum of the oil/air and oil/water surface tensions less than the oil/air tension spreads endlessly at the surface of the water (b); (c) and (d): a droplet can adopt a non-symmetrical shape depending on the values of the three surface tensions, here the drop-liquid surface tension is small compared to the two others, corresponding to the case of a pre-gelled alginate droplet (with a core still liquid) at an oil/water interface; (e) sketch of a droplet when the surface tension between the liquid and the air is small compared to the two others; (f) a droplet can "float" on an interface, nearly like a hydrophobic droplet on a plane, if the surface tension between droplet and underlying liquid is very large compared to the two others.

condition a droplet completely spreads, and under which condition a droplet is totally engulfed in a liquid?

We have seen that, when the liquid-air surface tension is large compared to the two other surface tensions, the droplets have a tendency to spread. Assume now a liquid-air surface tension larger than the sum of the two others:

$$\gamma_{LA} > \gamma_{DA} + \gamma_{DL}. \qquad (9.5)$$

In such a case, there cannot be a Neumann's balance of the three surface tensions (fig. 9.9) and the droplet completely spreads at the surface of the liquid.

A similar argument can be used for engulfment. Let us assume that the droplet-liquid surface tension is larger than the sum of the two other surface tensions:

$$\gamma_{LA} > \gamma_{DA} + \gamma_{DL}. \qquad (9.6)$$

In such a case, the Neumann's balance of the surface tension forces at the interface cannot be met (fig. 9.10) and the droplet is engulfed in the liquid.

On the other hand, if

$$\gamma_{DL} > \gamma_{DA} + \gamma_{LA}. \qquad (9.7)$$

the droplet will not penetrate in the liquid, as sketched in figure 9.11.

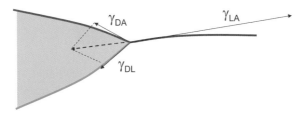

Figure 9.9 Sketch of an impossible balance of the surface tension forces leading to spreading.

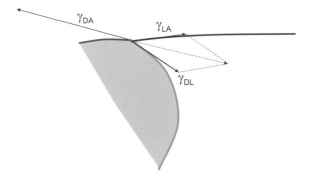

Figure 9.10 Sketch of an imbalanced surface tension forces leading to droplet engulfment in the liquid phase.

9.4.2.2   Engulfment with Gravity

In the case of a horizontal interface, and a sufficiently heavy droplet, engulfment may occur even if condition (9.6) is not exactly satisfied. Let us consider a locally flat interface and a droplet with sufficient weight and/or sufficient capillary forces for crossing the interface: $R = 5$ mm, $\Delta\rho = 150$ kg/m^3, $\gamma_{DL} = 45$ mN/m, $\gamma_{LA} = 30$ mN/m, $\gamma_{DA} = 72$ mN/m. With such values of the surface tension, the condition (9.6) is not exactly met, but engulfment occurs as shown in figure 9.12. Note that the shape of the droplet is considerably modified by the crossing of the interface. This behavior is closely linked to the Neumann's construction – which predicts that the triple line is a singularity line.

The deformation of a droplet crossing an interface can be experimentally observed. Let us

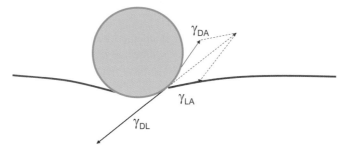

Figure 9.11 Sketch of a droplet that cannot penetrate in the liquid.

Figure 9.12 Progressive descent of a droplet into an immiscible liquid: the motion is slow and governed by capillary and gravity forces. Note the progressive change of the directions of the different surface tension forces.

take the example of an alginate droplet gently dropped on an oil layer (fig. 9.13). The droplet adopts a "pear" shape during the crossing of the interface between oil and air; this shape can be easily observed after polymerization (gelling).

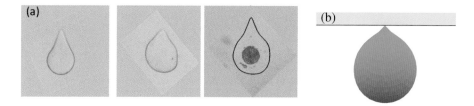

Figure 9.13 (a) Alginate droplet after having crossed an oil layer (this shape has been later frozen by polymerization in an aqueous layer of calcium located below); the last image on the right shows a pancreatic cell encapsulated in the alginate droplet; (b) modeling of a liquid aqueous droplet traversing an air/oil interface (photo Prisca Dalle, CEA-Leti).

9.5   Immiscible Droplets in Contact and Engulfment

In multi-phase microfluidic systems – like those used in biotechnology – droplets composed of different, immiscible liquids can travel together in microchannels. Water (or aqueous solution), organic (oil), gas (air), ionic liquids, etc, are examples of immiscible liquids. In microchannels, these droplets may accidentally or on purpose come into contact. The droplets may stay in contact or one can engulf the other. The behavior of droplets contacting each other is analyzed in this section.

9.5.1   Introduction

First, let us analyze the case of droplets staying in contact. It is again Neumann's construction that governs the mutual position of the droplets as shown in figure 9.14.

A promising way to perform biological and chemical reactions in micro-volumes is to use plugs transported by an immiscible carrier fluid [3-5]. The principle utilizes a T-junction (or a flow-focusing device, FFD) to form droplets in a continuous carrier phase. Figure 9.15 shows the principle of such reactions. Different reagents transported by independent plugs are successively mixed with a solution containing a chemical or biochemical species. A condition for the proper functioning of such plug flow reactions is that the plugs do not coalesce. Coalescence would bring contamination between the liquids. Coalescence may occur when the plugs are not moving at the same velocity, due to differences of viscosity and surface tensions. In order

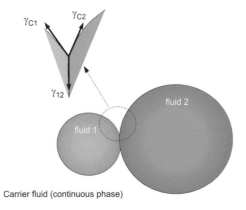

Figure 9.14 Droplets in contact, embedded in a carrier fluid.

to keep the plugs separated, Chen *et al.* [6] use spacer plugs constituted of a third immiscible liquid. Let us place ourselves in the case of slow moving liquids (small Weber and capillary numbers), so that surface tension forces dominate inertia and viscous forces, and assume that the carrier fluid is the only fluid wetting the solid walls.

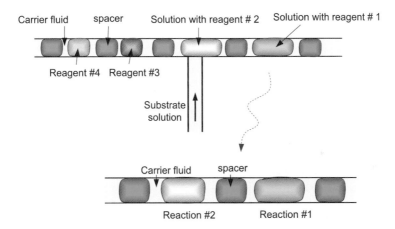

Figure 9.15 Principle of three-phase flow reactions: spacer plugs of immiscible liquid prevent coalescence of droplets.

9.5.2   Physical Analysis

A condition for the efficiency of spacer plugs is that the liquid plugs do not engulf each other. Figure 9.16 shows the satisfactory arrangement of the plugs (a), and engulfment (b). The condition for stability of plugs in contact is given by the balance of the surface tension forces at the triple line (Neumann's construction)

$$\vec{\gamma}_{c1} + \vec{\gamma}_{12} + \vec{\gamma}_{c2} = \vec{0}. \tag{9.8}$$

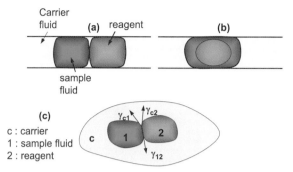

Figure 9.16 (a) Sketch of two droplets with no engulfment; (b) total engulfment of reagent by the sample fluid; (c) schematic of the contact with the surface tensions on the contact line (Neumann's construction).

Neumann's relation can be satisfied only if the magnitude of every force is smaller than the sum of the magnitudes of the other two forces. This statement can be easily verified by remarking that if the magnitude of a force is larger than the sum of the magnitudes of the two others, equilibrium cannot be reached (fig. 9.17).

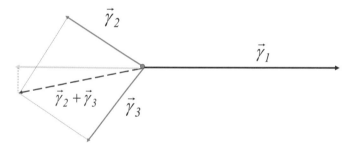

Figure 9.17 Assuming that γ_1 is larger than $\gamma_2 + \gamma_3$, the resultant of the forces projected on the direction of γ_1 cannot equilibrate.

Hence, it can be shown that the conditions for non-engulfment are

$$\gamma_{c1} < \gamma_{c2} + \gamma_{12}, \tag{9.9}$$

$$\gamma_{c2} < \gamma_{c1} + \gamma_{12}. \tag{9.10}$$

If the first condition (9.9) is not met, a droplet of liquid 1 is engulfed inside liquid 2. Conversely, if the second condition (9.10) is not met then liquid 2 is engulfed by capsule 1 (in the case where capsule 1 is liquid and not a solid). Gas bubbles usually satisfy both relations. However, gas bubbles can induce pressure fluctuations in the system. In order to avoid this drawback, Chen and colleagues [6] have found organic spacer liquids (SID, DTFS) adapted to fluorinated carrier fluids and aqueous reagent plugs.

On the other hand, if

$$\gamma_{12} > \gamma_{c1} + \gamma_{c2}, \tag{9.11}$$

then two droplets in contact separate from each other. Relations (9.9-9.11) will be illustrated in the following section.

9.5.3   Numerical Approach

Droplet engulfment is dominated by capillary forces; it can be modeled using the Surface Evolver. Figure 9.18 shows two characteristic cases: the first one is that of partial engulfment when relations (9.9,9.10) are satisfied. In this particular case, $\gamma_{C1} = 50$ mN/m, $\gamma_{C2} = 30$ mN/m, $\gamma_{12} = 26$ mN/m, and the droplet of liquid 1 is partially engulfed by liquid 2. The second case is total engulfment.

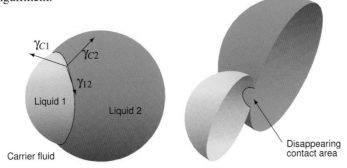

Figure 9.18 Left: Partial engulfment of sphere of liquid 1 by droplet of liquid 2 when conditions (9.9,9.10) are met. Right: Progressing separation of the two droplets in the case where $\gamma_{12} = \gamma_{c1} + \gamma_{c2}$.

9.5.4   Total Engulfment

Total engulfment occurs when either one of inequalities (9.9) or (9.10) is not met. This is the case shown in figure 9.19, where $\gamma_{c1} = 58$ mN/m; $\gamma_{c2} = 30$ mN/m and $\gamma_{12} = 26$ mN/m. In the particular case where the two droplets are composed of the same liquid (but with surfactants at the interface that delay the contact), relation (9.9) or (9.10) can never be satisfied, and coalescence will occur as soon as the thin film of carrier fluid initially separating the two droplets is drained out.

Figure 9.19 Total engulfment of droplet 1 by droplet 2 ($\gamma_{c1} = 58$ mN/m; $\gamma_{c2} = 30$ mN/m and $\gamma_{12} = 26$ mN/m).

So far we have only presented capillary effects for engulfment and coalescence phenomena, because our goal was to determine if engulfment is occurring or not. Note that dynamic aspects are also present: for example, when a droplet or a capsule approaches an interface, a thin interstitial liquid film remains for some time, and interfacial contact occurs only when the film starts to drain out [12]. Figure 9.20 shows such an interstitial thin oil film between an alginate capsule and the water interface. Other dynamical aspects, like the effect of inertia, fluid viscosities, etc. have been widely studied [13].

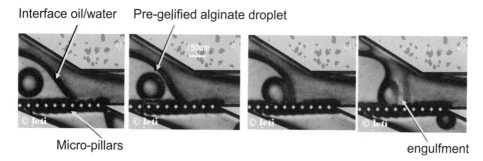

Figure 9.20 Image of an alginate droplet crossing an oil-water interface; here the problem combines capillary and inertia forces (photo P. Dalle).

9.6   Non-deformable (Rigid) Sphere at an Interface

Solid or gelled spheres are frequently used in biotechnology. In this section, we examine the problem of a solid spherical capsule at the interface between two liquids, or between a liquid and a gas. Such a situation frequently occurs in biotechnology. For example, this is the case in encapsulation applications where gelled droplets have to be transferred from an organic phase to an aqueous phase, as shown in figure 9.20. It also occurs in material sciences, for example in capillary force assembly applications (CFA) [7,14].

As pointed out at the beginning of this chapter, the behavior of rigid spheres at an interface is not the same as that of liquid droplets, and the question we want to address is the following: under what conditions the capsule crosses the interface or stays pinned on the interface? For simplicity, we will treat only the case of a spherical capsule.

9.6.1   Introduction

Solid spherical – or nearly spherical – particles are frequently used in biotechnological microsystems. For example, functionalized polystyrene spheres, often having a paramagnetic core, are used to search and bind to biological targets – such as DNA – in suspension in a carrier fluid [15]. Classically, they disperse in the bulk by Brownian motion, bind to the targets, and then magnetic properties are used to concentrate the hybridized spheres. Functionalized gold spheres may be used as active surfaces for the hybridization of DNA or proteins with on-line SPR (Surface Plasmon Resonance) detection [16]. In encapsulation applications, gelled capsules containing cells or any biological target or chemical compound are formed in flow-focusing devices. Quantum dots are semi-conductor capsules of approximately 10 nm diameter that emit a characteristic wavelength upon excitation [17,18]. Due to quantum confinement, the emitted color depends on the size of the quantum dot. Quantum dots are useful markers for cell biology.

In two-phase microflows, all these particles may encounter interfaces. What is their behavior at an interface? Will they stay attached to the interface, or pass across the interface under only the action of capillarity, or is an additional force needed for this passage? These questions are dealt with in this section.

9.6.2   Capillary Problem – No Body-force

In this section, it is assumed that capillary forces are the only forces acting on the spherical particle. The spheres are supposed small enough or buoyancy neutral so that gravity can be neglected.

9.6.2.1   Energy Approach

Let us calculate the surface energy of the system. We might think of the two possible configurations for the sphere at the interface shown in figure 9.21. In the absence of an external body force, the position of minimum energy is when the fluid1/fluid2 surface remains flat. The total interfacial energy of the capsule relative to the capsule-free surface is then given by the relation

$$E = E_{S1} + E_{S2} - E_{12}. \tag{9.12}$$

The third term on the right hand side of (9.12) corresponds to the exclusion of the interface S_{12} cut out by the capsule. Let us develop (9.12) further:

$$E = \gamma_{S1} S_{S1} + \gamma_{S2} S_{S2} - \gamma_{12} S_{12}, \tag{9.13}$$

where the γ's are the surface tensions and S's the surface areas. Using spherical cap surface expressions where h is the height of the liquid surface above the sphere center, we find

$$E = \gamma_{S1} [2\pi R(R - h)] + \gamma_{S2} [2\pi R(R + h)] - \gamma_{12} \pi (R^2 - h^2). \tag{9.14}$$

The sphere places itself in a minimum energy configuration, corresponding to

$$\frac{dE}{dh} = 0 = -\gamma_{S1} R + \gamma_{S2} R + \gamma_{12} h. \tag{9.15}$$

In the case where the sphere is at equilibrium at the interface, the equilibrium position is given by

$$h = \frac{R(\gamma_{S1} - \gamma_{S2})}{\gamma_{12}}. \tag{9.16}$$

9.6.2.2   Force Approach – Young's Equation

Young's law can be substituted in 9.16, and we obtain

$$h = \frac{R(\gamma_{S1} - \gamma_{S2})}{\gamma_{12}} = \frac{R(\gamma_{12} \cos \theta)}{\gamma_{12}} = R \cos \theta. \tag{9.17}$$

This relation could have been directly derived, using the sketch of figure 9.22.

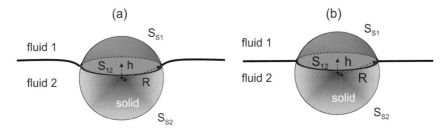

Figure 9.21 Sketches of a solid sphere at the interface between two fluids: in absence of gravitational energy, the minimum energy configuration is the figure at the right, because the free surface is reduced.

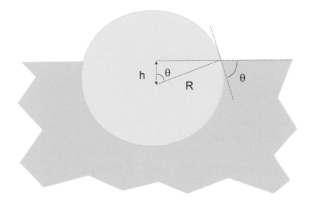

Figure 9.22 Sketch of a neutrally buoyant rigid sphere at a liquid interface.

9.6.2.3   Discussion

If $\gamma_{S1} < \gamma_{S2}$, then h is negative and the capsule center is located inside liquid 2. Is the engulfment total? From (9.17), the condition for total engulfment in liquid 2 is

$$h = \frac{R(\gamma_{S1} - \gamma_{S2})}{\gamma_{12}} < -R. \tag{9.18}$$

The condition for total engulfment in phase 1 is then

$$\gamma_{S1} + \gamma_{12} < \gamma_{S2}. \tag{9.19}$$

Conversely, if h is positive and the capsule centre is located inside liquid 1, then the condition for total engulfment in liquid 1 is given by

$$h = \frac{R(\gamma_{S1} - \gamma_{S2})}{\gamma_{12}} > R. \tag{9.20}$$

This condition is best written in the form

$$\gamma_{S2} + \gamma_{12} < \gamma_{S1}. \tag{9.21}$$

An easy way to remember the results of (9.19) and (9.21) is to keep in mind that a sufficiently large surface tension γ_{S1} or γ_{S2} pulls in on the liquid-liquid interface and triple line until the sphere is totally engulfed in the other liquid. Also note that if $\gamma_{S1} = \gamma_{S2}$, then $h = 0$ and half of the capsule is located in phase 1, and the other half in phase 2. The capsule will stay at the interface if

$$\gamma_{S2} + \gamma_{12} > \gamma_{S1} > \gamma_{S2} - \gamma_{12}, \tag{9.22}$$

or identically

$$\gamma_{S1} + \gamma_{12} > \gamma_{S2} > \gamma_{S1} - \gamma_{12}. \tag{9.23}$$

9.6.2.4   Numerical Approach

In this section an approach using the Surface Evolver is performed. First we verify that the sphere is centered on the interface if $\gamma_{S1} = \gamma_{S2}$ (fig. 9.23).

Figure 9.23 The sphere is centered on the interface if $\gamma_{S1} = \gamma_{S2}$.

Second, we examine the case of encapsulation of alginates, which has been mentioned in the introductory section. In that case, the two fluids are respectively mineral oil (letter O) and water (letter W), and the sphere is a capsule of gelled alginate (letter A). In this case, $\gamma_{S1} = \gamma_{AW} \approx 2$ mN/m, $\gamma_{S21} = \gamma_{AO} \approx 26$ mN/m and $\gamma_{12} = \gamma_{WO} \approx 26$ mN/m. From (9.17), the distance h is equal to $-(12/13)R$ and the sphere should be principally located in the water phase. The numerical model confirms the analysis, showing a very small surface area of the capsule in contact with the oil (fig. 9.24).

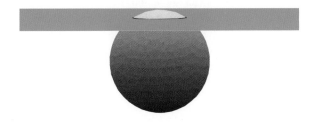

Figure 9.24 An alginate droplet is preferentially located in the aqueous phase.

9.6.3   Body-force: Gravity

In this section we consider a flat horizontal interface between water and air, for example, and spherical particles of sufficient size to have non-negligible buoyancy. The gravity forces are then acting on the particles together with the capillary forces. In this case the interface cannot remain flat. Four different situations can be considered: a "hydrophobic" sphere that wants to stay out of the water, and a "hydrophilic" sphere that wants to engulf in the water. Moreover, the sphere density can be larger or smaller than that of water, i.e. $\Delta\rho > 0$ or $\Delta\rho < 0$.

Let us first examine the energy of the system: we can restart from equation 9.14 modified to take into account the buoyancy forces:

$$E = \Delta\rho g h + \gamma_{S1}[2\pi R(R-h)] + \gamma_{S2}[2\pi R(R+h)] - \gamma_{12}\pi(R^2 - h^2). \tag{9.24}$$

Again taking the derivative in respect to h,

$$\frac{\delta E}{\delta h} = \Delta\rho g - 2\pi R\gamma_{S1} + 2\pi R\gamma_{S2} + 2\pi h\gamma_{12}. \tag{9.25}$$

The equilibrium position is given by $\frac{\delta E}{\delta h} = 0$, leading to

$$\frac{\Delta\rho g}{2\pi} = R\gamma_{S1} - R\gamma_{S2} - h\gamma_{12}. \tag{9.26}$$

Using the Young relation, we obtain

$$h = R\cos\theta - \frac{\Delta\rho g}{2\pi\gamma_{12}}. \tag{9.27}$$

This relation shows that the sphere floats "up" when $\Delta\rho < 0$ and "down" when $\Delta\rho > 0$. By floating "up" and "down" we mean that the water/air interface is deformed upwards or downwards in the vicinity of the droplet. For a hydrophilic sphere, the two possible positions are shown in figure 9.25.

The case of a hydrophobic sphere is shown in figure 9.26. A similar conclusion can be obtained numerically using Evolver (fig. 9.27).

Following D.D. Joseph [19], one can calculate the resultant of the vertical forces exerted on the sphere. First observe that the two components of the capillary force are

$$F_{c,v} = \gamma_{WA}2\pi(R\sin\alpha\sin(\alpha-(\pi-\theta)) = -\gamma_{WA}2\pi(R\sin\alpha)\sin(\alpha+\theta), \tag{9.28}$$
$$F_{c,h} = \gamma_{WA}2\pi(R\sin\alpha)\cos(\alpha-(\pi-\theta)) = -\gamma_{WA}2\pi(R\sin\alpha)\cos(\alpha+\theta), \tag{9.29}$$

where R is the sphere radius, θ the contact angle and α the angle between the vertical axis and the contact line. In both cases, $\frac{3\pi}{2} > \alpha + \theta > \pi$, but in the hydrophobic case $\alpha < \frac{\pi}{2}$, whereas in the hydrophilic case $\alpha > \frac{\pi}{2}$. Hence, in the hydrophobic case, the two components of the capillary force are positive. In the hydrophilic case, the vertical component is negative and the horizontal component positive. However, the sphere can be at equilibrium, even if the vertical capillary force is downwards. Indeed the vertical mechanical equilibrium can be written

$$F_{c,v} + F_P + G = 0, \tag{9.30}$$

where G is the weight of the particle, $G = -\frac{4}{3}\pi R^3\rho g$, and F_p is the vertical resultant of the pressure around the sphere (Archimedes' resultant),

$$F_P = \rho_W g v_W + \rho_A g(V - V_W) - (\rho_W - \rho_A)ghA, \tag{9.31}$$

where ρ_W and ρ_A are densities of the liquid and the air, respectively; h is the depression generated by the sphere; h is positive in the hydrophobic case and negative in the hydrophilic case. The sphere volume is

$$V = \frac{4}{3}\pi R^3 \tag{9.32}$$

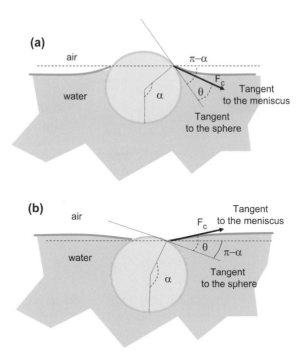

Figure 9.25 A hydrophilic sphere can float up or down depending on its buoyancy: (a) sphere less dense than water; (b) sphere denser than water.

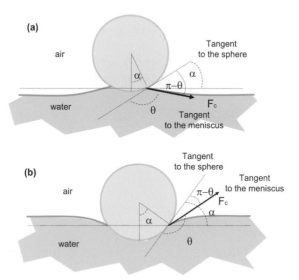

Figure 9.26 A hydrophobic sphere can float up or down depending on its buoyancy: (a) sphere less dense than water; (b) sphere denser than water.

and the volume of the spherical cap V_W is

$$V_W = \frac{1}{3}\pi R^3(2 - 3\cos\alpha + \cos^3\alpha).$$ (9.33)

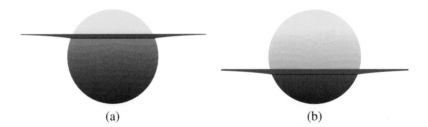

(a)   (b)

Figure 9.27 Equilibrium positions for a droplet of density 9000 kg/m3 at the surface of water: (a) hydrophilic case; (b) hydrophobic case. Droplet diameter 600 μm.

The area A delimited by the triple line is

$$A = \pi(R\sin\alpha)^2. \qquad (9.34)$$

The first two terms at the right hand side of (9.31) are in agreement with Archimedes' principle, while the last term accounts for the meniscus effect. Substitution of (9.32), (9.33) and (9.34) in (9.31) produces the expression of the resultant of the pressure on the sphere. Note that F_p is always positive (upwards), G is always negative (downwards) and F_c can be positive (hydrophobic case) or negative (hydrophilic case).

Let us now examine the case where equilibrium cannot be reached because of a too great weight of the sphere, and compare the penetration of two similar spheres (same size, same weight) with two different coatings, making one hydrophilic, the other one hydrophobic. At the same level down into the water, the pressure forces are equal in the two cases (same h); on the other hand, the capillary force facilitates the penetration in the water for a hydrophilic sphere whereas it opposes the penetration for a hydrophobic sphere. Hence, the force balance justifies the intuitive thought that a hydrophilic sphere penetrates more easily into the water (fig. 9.28).

(a)   (b)

Figure 9.28 The same sphere sinks through a water/air interface when it is hydrophilically coated (a), whereas it floats when hydrophobically coated (b).

On the other hand, a solid sphere of a given density crosses an interface or not depending on its size. Indeed, the gravitational force is proportional to the cube of the sphere radius ($G \approx R^3$); on the other hand, the crossing of the interface triggers the deformation of the interface on a surface approximately equal to the half sphere; hence the resisting force is the surface tension force on a half sphere ($F_\gamma \approx R^2$) [9]. Thus, there is a threshold radius above which the sphere penetrates the liquid. This behavior can be checked numerically (fig. 9.29). Note that the threshold also depends on the hydrophilic or hydrophobic characteristic of the sphere surface, as was shown before.

Figure 9.29 Comparison of the position of a rigid hydrophobic sphere at an interface according to the particle dimension; in this particular case the sphere density is 1900 kg/m³.

9.6.4   The Cases of Nearly Rigid Spheres

Let us note that liquid droplets can be "nearly rigid", as for example gelled polymer spheres and mercury droplets.

The first case corresponds to the encapsulation technique in biotechnology where liquid alginate droplets – containing biological objects like cells – are polymerized or gelled in order to obtain transportable and stable capsules [1,8,20-23]. On one hand, the polymerized capsule behaves like a semi-solid gel and is not easily deformable; on the other hand, the surface tension between the gelled polymer and water is very low. Figure 9.30 shows the resulting configuration.

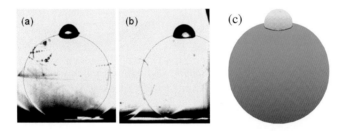

Figure 9.30 Oil droplet on a pre-gelled, still liquid alginate droplet in an aqueous phase with calcium ions. In (a) and (b) the contour of the alginate droplet has been outlined for an easier visualization; (c) represents the result of an Evolver calculation with $\gamma_{S1} = \gamma_{AW} \approx 2$ mN/m, $\gamma_{S2} = \gamma_{AO} \approx 26$ mN/m and $\gamma_{12} = \gamma_{WO} \approx 26$ mN/m (photo Prisca Dalle, CEA-Leti).

The second case is that of mercury. Mercury is a heavy liquid metal ($\rho = 13000$ kg/m³), with a very high value of the surface tension, with water as well as air ($\gamma_{MW} = 485$ mN/m and $\gamma_{MA} = 415$ mN/m). Hence a mercury droplet at the surface of water behaves like a rigid sphere. Figure 9.31 compares the behavior of a droplet of mercury at the water-air interface with that of trichloroethylene ($\rho = 1470$ kg/m3, $\gamma_{TrW} = 34$ mN/m and $\gamma_{TrA} = 28$ mN /m). Both liquids are denser than water; however the mercury droplet sinks into the water while the trichloroethylene spreads on top of the water.

9.6.5   Rigid Spheres Attached to a Meniscus

Rigid spheres of small size often stay attached to the interface. They may cross the interface if other forces are applied, such as magnetic or electric forces, or if the capillary conditions are extremely favorable. In the absence of additional forces, small size particles often stay at the interface. However, they are not immobile on the interface; often a lateral motion is observed on the surface until the spheres have reached their equilibrium position. For example, in the

Figure 9.31 Comparison of the behavior of mercury and trichloroethylene droplets at the surface of water.

case of multiple spheres, self-assembly is often observed (fig. 9.32) [19,24,25]. In the case of a curved interface (meniscus), spheres either move to the middle of the meniscus or to its side, depending on the conditions. Such behaviors are analyzed in this section.

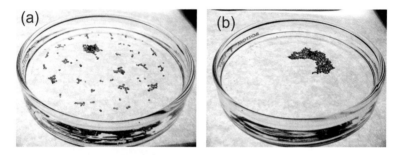

Figure 9.32 Free motions leading to self-assembly of floating particles (sands in .01 aqueous polyox solution). Reprinted with permission from [19], ©Springer, 2008.

9.6.5.1   Spheres Self-assembly Due to the Local Curvature of the Interface

Consider a locally flat interface between water and air, on which rigid spheres are floating. If the size of the spheres is not too small, then their weight is not negligible and the interface is deformed by the presence of the spheres. In the case of "hydrophobic" spheres, i.e. principally located outside the liquid, the liquid surface is locally curved downwards (fig. 9.33). Conversely, in the hydrophilic case, an uplift of the interface may occur, as sketched in figure 9.34.

The deformation of the interface induces a tilt of the outer spheres. But the tilt itself is not enough to induce an imbalance of the horizontal resultant of the capillary forces. Figure 9.2 shows the net force on the sphere, and that does not change if the whole configuration is tilted. It is gravity pulling the spheres downhill that gathers them into the hollow. A simulation with the Evolver confirms this behavior (fig. 9.35). The spheres create a depression in the interface in which the particles converge together.

Likewise, buoyant hydrophilic spheres can create a bulge, and outer buoyant spheres can float up the bulge sides to gather at the top.

9.6.5.2   Spheres Displacement on a Curved Meniscus

In general, menisci are curved. But a curved meniscus as such need not generate any lateral force on an imbedded sphere. In the absence of gravity, a sphere imbedded in a spherical

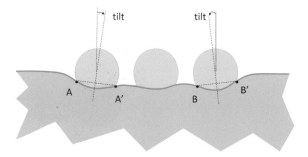

Figure 9.33 The weight of some hydrophobic spheres creates a hollow region in the interface, resulting in the spheres located on the side sliding inward.

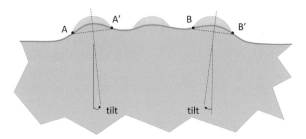

Figure 9.34 The presence of the light hydrophilic spheres creates a uplifted region in the interface, resulting in the tilt of the meniscus. Outer spheres float upwards, bringing the spheres together.

meniscus does not distort the meniscus, and its energy is the same whereever it is located. Thus there are no lateral forces. It is gravity that can pull denser spheres to the bottom of a curved meniscus, as in figure 9.36. Likewise, a buoyant sphere will float upwards towards the side.

If the curvature of the meniscus is not uniform, then there can be forces towards or away from regions of high or low curvature, but the details of that will not be discussed here.

Bead behavior encountering a meniscus is experimentally observed. Let us take the example of a special FFD (flow-focusing device) where "moustaches" have been added to the discontinuous phase channel in order to maintain a stable meniscus (fig. 9.37). In this particular case, the discontinuous phase is an alginate aqueous solution and the continuous phase is mineral oil. In absence of beads in the aqueous flow, the meniscus remains perfectly stable. Let us investigate now what happens when a polystyrene bead is transported by the aqueous phase: if the polystyrene bead transported by the discontinuous phase arrives at the meniscus in a corner close to the wall, it will be progressively displaced to the middle of the meniscus, exactly as in figure 9.36. The bulging out of the meniscus – with the sphere at its center – is then sufficient for a droplet to detach, carrying away the polystyrene bead in its motion. By the analysis above, static forces do not explain this, so it must be due to the dynamical effects of the flowing oil entraining the bead.

9.6.6   Droplet Attached to a Solid Sphere

A reciprocal picture of a solid sphere at a liquid interface is that of a droplet on a solid sphere. This case has been studied by Eral and colleagues [26], using electrowetting actuation to vary

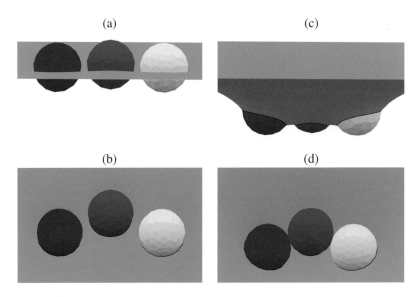

Figure 9.35 Three hydrophobic spheres floating at the water surface (a and b) move until they come to contact (c and d) as shown by an Evolver calculation. (a),(c) side view; (b),(d) top view.

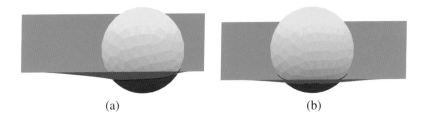

Figure 9.36 A dense sphere floating on a concave meniscus falls downhill to the center, here from (a) to (b).

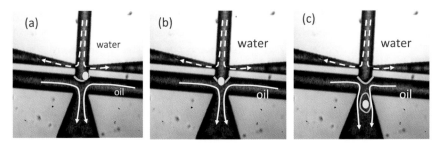

Figure 9.37 A polystyrene droplet that impacts a stabilized interface on its contact line with the solid (a) moves to the middle of the channel (b) before being ejected by the flow-focusing device in a droplet (photo courtesy Jayesh Wagh, CEA-Leti).

the contact angle. We just mention here the results of the simulation of such a situation done with Evolver. As in the case of a droplet on a wire studied in the preceding chapter, the electrowetting effect is equivalent to a change of the contact angle, or to a change of the solid-liquid surface tension. Indeed Lippmann's law states that the solid-liquid surface tension decreases with the electric field intensity. This approach is the easiest for numerical simulation; the results for two different values of the solid-liquid surface tension are shown in figure 9.38. As discussed earlier in the chapter, the balance of the three surface tensions and the liquid weight determines the equilibrium position.

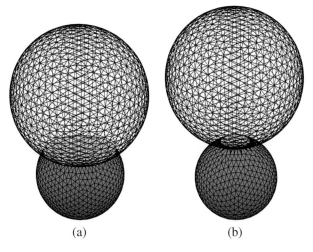

(a)   (b)

Figure 9.38 Liquid droplet attached to a solid sphere: (a) $\gamma_{SL} = 0.002$ mN/m, (b) $\gamma_{SL} = 0.042$ mN/m.

9.6.7   Body Force: Magnetic Force

Magnetic micro-beads are currently used in biotechnology, as carriers of biologic species, such as DNA or cells. Figure 9.39 illustrates a biologic protocol developed in combination with EWOD (Electro Wetting On Dielectric, see chapter 10). Labeled micro-magnetic beads are first used to bind to biologic targets, then to carry and trap the targets in a magnetic aggregate; finally, EWOD is used to extract the remaining liquid, leaving a concentrate aggregate of magnetic beads and targets.

In this section, we focus on single magnetic beads at an interface, and examine the conditions for bead pinning on the interface and conversely bead piercing the interface.

9.6.7.1   Physical Analysis

The interfacial energy E_{cap} is given by (9.13). Usually super-paramagnetic beads are used [15], and the magnetic force depends on the applied magnetic field $\vec{H}$ according to

$$\vec{F}_m = \frac{1}{2}\mu_0 V_p (\chi_p - \chi_f)\nabla H^2, \tag{9.35}$$

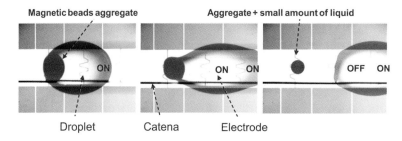

Figure 9.39 Use of magnetic beads aggregate to extract minute amounts of liquid from a droplet. Left: labelled magnetic beads are aggregated using a magnet. Middle: droplet is displaced by electrowetting forces while the aggregate is maintained immobilized by magnetic forces. Right: separation of a minute amount of liquid containing the aggregate from the rest of the droplet.

where χ_p is the magnetic susceptibility of the particle, χ_f that of the carrier fluid and V_p the bead volume [7,26,27]. The associated magnetic energy E_m is

$$E_m = -\frac{1}{2}\mu_0 V_p(\chi_p - \chi_f)H^2. \tag{9.36}$$

We are interested here only in a magnetic force perpendicular to the interface. A parallel component of the force results in a drift of the bead on the interface. Besides, micro-magnetic beads currently used in biotechnology are smaller than 1 μm, and we neglect the change of the value of the magnetic field between a bead centered at $h = -R$ and $h = R$. Hence, a bead is trapped at the interface if there is a value of h such that

$$\frac{\delta(E_m + E_{cap})}{\delta h} = 0. \tag{9.37}$$

Upon substitution of (9.14), we find

$$h = -\frac{F_m}{2\pi\gamma_{12}} + \frac{R(\gamma_{S1} - \gamma_{S2})}{\gamma_{12}}. \tag{9.38}$$

The condition for a bead to be pinned on the interface is $-R < h < R$. The magnetic force is then comprised in the interval

$$2\pi R(\gamma_{S1} - \gamma_{S2} + \gamma_{12}) > F_m > 2\pi R(\gamma_{S1} - \gamma_{S2} - \gamma_{12}). \tag{9.39}$$

Using Young's law, the preceding relation becomes

$$2\pi R\gamma_{12}(1 + \cos\theta) > F_m > 2\pi R\gamma_{12}(\cos\theta - 1). \tag{9.40}$$

9.6.7.2   Numerical Approach

In many applications the magnetic beads have little inertia at the interface, and it is reasonable to consider only the capillary and magnetic forces acting on the beads. Due to the bead's very small size, the magnetic field may be assumed constant on all the bead volume and the resultant of the magnetic force is applied at the mass center of the bead. Figure 9.40 shows the crossing of an interface by a magnetic sphere calculated with Evolver.

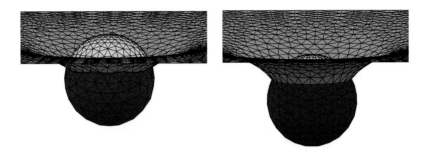

Figure 9.40 (left) Spherical magnetic bead submitted to a downward magnetic force. (right) Above a critical value (right), the bead pierces the interface and penetrates into the other liquid.

9.7   Droplet Evaporation and Capillary Assembly

Interfaces between an aqueous phase and air are very common; in such cases, evaporation is likely to occur. Hence, it is important to address the problem of evaporation, especially that of droplets. This topic constitutes the first part of this section. Also, when beads are present, their behavior is considerably affected by the receding motion of the interface, or the contact line. In case there are a large number of beads, the receding motion of the interface of an evaporating liquid is even used to re-arrange bead deposition on solid surfaces. This technique is called self-alignment or capillary alignment, and will be developed in the second part of this section.

9.7.1   Introduction

Capillary forces can be utilized to act not only on a single object but also collectively on a population of objects with the aim of coating or forming a new, larger object with specific physical and chemical properties. This technique is usually called capillary force assembly, or CFA. For example nanoscale gold colloidal dimers can be assembled by a CFA technique [7], colloidal optical waveguides have been built with the same technique [28], and coatings of complicated geometries have been achieved [29,30] (fig. 9.41).

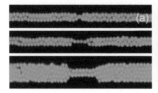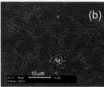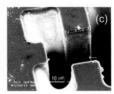

Figure 9.41 (a) Optical waveguide assembled by capillary force (reprinted with permission from [28], ©AIP, 2011); (b) coating of a surface with monodisperse polystyrene beads; (c) nearly uniform coating of a microfabricated channel with monodisperse beads (reprinted with permission from [51]; ©Elsevier 2004).

Different techniques of CFA exist but one of the most popular is the droplet evaporation technique. It originated with Deegan's observations of deposition of colloidal rings during the evaporation of droplets [31,32]. The idea stemming from Deegan's work is to use evaporation as a tool for assembling colloidal particles. In this technique, a contact line recedes during a controlled evaporation of a liquid droplet (fig. 9.42). For example, a receding contact line pulls back the colloidal particles in suspension in the droplet into traps or geometrical features such as square holes or grooves [7].

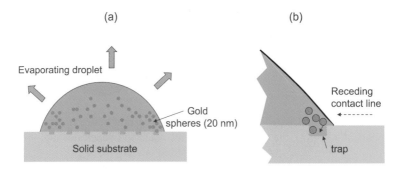

Figure 9.42 (a) Sketch of an evaporating droplet with colloidal particles in suspension; (b) the receding contact line pulls back the colloids into traps etched in the substrate.

In the following section the evaporation of sessile drops is first investigated. Then we show some examples of capillary assembly techniques using a receding triple line.

9.7.1.1   Evaporation of Sessile Droplets

One would expect that the evaporation process of liquid droplets was investigated many years ago. Indeed, Wittaker, Morse, and Langmuir set the basis for a theory in the 1910's; however, it is just recently that a more complete understanding of the evaporation process of micro-drops has been completed.

In this section we investigate evaporation of sessile droplets. We assume that there is no Marangoni type convection inside the droplet – Marangoni convection increases the evaporation rate [33] – and that the surrounding gas is at rest. Liquid evaporation is strongly associated with the diffusion of the vapor away from the liquid/gas interface (fig. 9.43). Note that if the gas is saturated with vapor (dew point for water), evaporation is stopped: this property has led to the notion of sacrificial droplets and the definition of the evaporation number [34]. Sacrificial droplets are used to saturate the closed environment and prevent a noticeable evaporation of the droplets of interest (fig. 9.44).

9.7.1.2   Experimental Observations

It has been observed experimentally [35-39] that wetting and non-wetting droplets do not evaporate in the same way.

Most of the time, non-wetting droplets keep a constant contact angle during evaporation. The case of textured surfaces constitutes a counter-example [35]; but if the solid surface is sufficiently smooth, this general rule applies.

The case of a wetting droplet is quite different: it has been observed that the contact radius remains constant during a long part of the evaporation process, except at the very end. During most of the evaporation process, the contact angle decreases gradually. This might seem strange since there is no motion of the contact line. The explanation of this behavior is not clear at the present time. It is linked to a "sort of pinning" of the triple line: the contact angle decreases to a receding value which is linked to the surface of the solid surface. Asperities and surface roughness, nanoparticles, nanostructuration, and like surface properties may modify the value of the

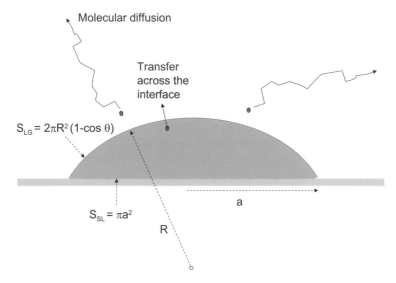

$S_{LG} = 2\pi R^2 (1-\cos\theta)$

$S_{SL} = \pi a^2$

Figure 9.43 The two mechanisms occuring during evaporation (assuming that there are no convective motions): Evaporation is limited by diffusion in the gas phase rather than by the transfer rate across the interface.

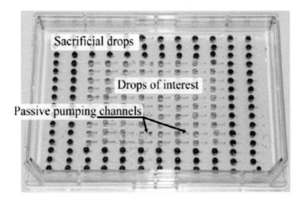

Figure 9.44 Principle of sacrificial droplets: Evaporation from sacrificial droplets maintains the water vapor pressure in a closed box and prevents noticeable evaporation of droplets of interest. Reprinted with permission from [34], ©RSC, 2008, http://dx.doi.org/10.1039/B717423C.

receding angle. However, it seems that the drying behavior of a wetting sessile droplet occurs even with extremely smooth solid surfaces and pure liquids. Figure 9.45 shows an experimental observation of the phenomenon and figure 9.46 schematizes this behavior.

The different shapes of a water droplet during evaporation can be calculated with Evolver; they are shown in figure 9.47.

In the first case (wetting substrate) the mass rate of liquid loss is proportional to the height of the drop, whereas in the second case (non-wetting substrate), the mass rate is proportional to the square of the contact radius. In the following section, we derive expressions for the contact angle change with time in the wetting case, and for the contact radius change with time in the

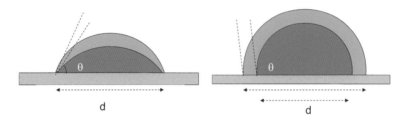

Figure 9.45 Evaporation of a water droplet on silanized silicon (top) and on silicon oxide (bottom): recall that the silanization process makes the surface hydrophobic, whereas the silicon oxide is hydrophilic. In the first case (top) the contact angle remains constant; in the second case (bottom) the contact radius remains constant [36].

Figure 9.46 The two schemes for sessile droplet evaporation according to the wettability of the surface [36], hydrophilic on the left and hydrophobic on the right.

non-wetting case.

9.7.1.3   Theoretical Model

Consider a complete spherical droplet. Following the approach of Birdi and colleagues [40], the evaporation rate is derived by using Fick's law:

$$\frac{dm}{dt} = \rho\frac{dV}{dt} = -D\int \nabla c \cdot dS = -D\int \frac{\delta c}{\delta n}dS, \qquad (9.41)$$

where m is the mass of the liquid, V its volume, ρ the density of the liquid, D is the diffusion coefficient of the vapor and c its concentration. By approximating the concentration gradient by

$$\frac{\delta c}{\delta n} \approx \frac{c_0 - c_\infty}{R} \qquad (9.42)$$

where R is the droplet radius, c_0 and c_{infty} are respectively the vapor concentration at the interface and at infinity, the evaporation rate then becomes

$$\frac{dm}{dt} = \rho\frac{dV}{dt} = 4\pi RD(c_0 - c_\infty), \qquad (9.43)$$

showing that the rate of mass evaporation is proportional to the radius of the spherical drop.

The case of a sessile droplet can be treated with the same reasoning. Equation (9.41) still holds, but the integration is made on the surface of the spherical cap. The difference with the preceding is that the assumption of a radially constant gradient is not strictly verified, but only approximately [38]. The evaporation flux is radially uniform except close to the triple line, as shown in figure 9.48. This effect is caused by a Marangoni convection that creates an axisymmetric convective torus. As we have seen before, the wetting and non-wetting cases must be treated differently.

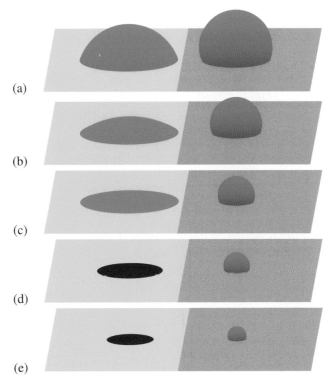

(a)

(b)

(c)

(d)

(e)

Figure 9.47 Shapes of evaporating water droplet on hydrophilic and hydrophobic surfaces using Evolver. In the case of a hydrophilic contact (left side), the contact line is first pinned (a,b,c) until a receding value of the contact angle is reached. During this phase the contact radius is constant. Then both the contact radius and the contact angle decrease (d and e). In the case of a hydrophobic contact (right side), the contact angle is constant during all the evaporation process. The static contact angles for the calculation with Evolver are respectively $\theta_1 = 70°$ and $\theta_2 = 110°$. Note that the kinetics of evaporation is not respected and the two droplets do not evaporate at the same speed.

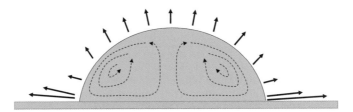

Figure 9.48 Evaporation flux along the surface droplet according to Hu [62].

9.7.1.3.1   Wetting (Lyophilic) Case: $\theta < 90°$ An expression for the spherical cap surface is

$$S_{LG} = 2\pi R^2 (1 - \cos\theta) = 2\pi Rh, \tag{9.44}$$

where h is the height of the sessile drop, R its radius and θ the contact angle. The rate of reduction of volume due to evaporation can be approximated by

$$\frac{dV}{dt} = -\frac{D(c_0 - c_\infty)}{\rho} \int \frac{dS}{R} = -\frac{D(c_0 - c_\infty)}{\rho} \int \frac{2\pi R}{R} dh = -\frac{2\pi D(c_0 - c_\infty)}{\rho} h. \qquad (9.45)$$

Equation (9.45) can be written in the form

$$\frac{dV}{dt} = -\lambda h \qquad (9.46)$$

with

$$\lambda = -\frac{2\pi D(c_0 - c_\infty)}{\rho}. \qquad (9.47)$$

Thus the evaporation rate is proportional to the height of the droplet. Using the expression of the droplet volume as a function of the contact angle,

$$V = \frac{1}{3}\pi R^3 (1 - 3\cos\theta + \cos^3\theta) = \frac{1}{3}\pi r^3 \frac{(1 - 3\cos\theta + \cos^3\theta)}{\sin^3\theta}, \qquad (9.48)$$

where r is the (constant) contact radius, we derive

$$\frac{dV}{dt} = \frac{\delta V}{\delta\theta}\frac{d\theta}{dt} = \frac{1}{3}\pi r^3 \frac{d}{d\theta}\left[\frac{(1 - 3\cos\theta + \cos^3\theta)}{\sin^3\theta}\right]\frac{d\theta}{dt} = \pi r^3 \frac{(1 - \cos\theta)^2}{\sin^4\theta}\frac{d\theta}{dt}. \qquad (9.49)$$

After substitution in (9.45), using the relation $h = r\tan(\frac{\theta}{2})$ and the trigonometric relation $\tan(\frac{\theta}{2}) = \frac{(1 - \cos\theta)}{\sin\theta}$, we obtain the following differential equation:

$$\frac{d\theta}{dt} = -\frac{\lambda \sin^3\theta}{\pi r^2 (1 - \cos\theta)}. \qquad (9.50)$$

The variables in equation (9.50) can be separated to obtain

$$\frac{(1 - \cos\theta)}{\sin^3\theta} d\theta = -\frac{\lambda}{\pi r^2} dt. \qquad (9.51)$$

Note that λ is constant. Integration of this differential equation yields

$$F(\theta) = \ln(\tan\frac{\theta}{2}) + \frac{(1 - \cos\theta)}{\sin^2\theta} = -\frac{\lambda}{\pi r^2}(t - t_0), \qquad (9.52)$$

where t_0 is a constant of integration. Equation (9.52) is implicit; however it has been shown that for a wetting angle in the interval $30° < \theta < 90°$ the function $F(\theta)$ is nearly linear:

$$F(\theta) = -1.592 + 1.632\theta. \qquad (9.53)$$

Then we find a linear decrease of the contact angle with time during evaporation,

$$\theta = \theta_0 - \frac{\lambda}{\pi r^2}(t - t_0). \qquad (9.54)$$

This law fits the experimental data extremely well in the range $30° - 90°$, as shown in figure 9.49.

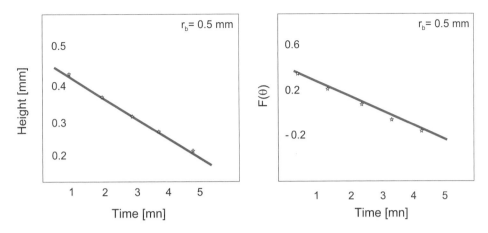

Figure 9.49 Left: time variation of the droplet height; right: time variation of the function $F(\theta)$. Droplet radius of 0.5 mm. Data from [36].

9.7.1.3.2   Non-wetting (Lyophobic) Case: $\theta > 90°$

This case has been studied by McHale and colleagues [39]. It can be shown that the equation

$$\frac{dV}{dt} = -\lambda h \tag{9.55}$$

still holds. But from this point on, the treatment of the equation changes: this time the contact radius is not kept constant, but the contact angle remains approximately constant (fig. 9.50) Then, keeping θ constant,

$$\frac{dV}{dt} = \frac{\delta V}{\delta r}\frac{dr}{dt} = -\lambda h. \tag{9.56}$$

Again we obtain a differential equation where the variables are separated,

$$r\frac{dr}{dt} = -\frac{\lambda \sin^2 \theta}{\pi(1 - \cos\theta)(2 + \cos\theta} \tag{9.57}$$

which can be solved to give

$$r^2 = r_0^2 - \frac{2\lambda \sin^2 \theta}{\pi(1 - \cos\theta)(2 + \cos\theta)}t. \tag{9.58}$$

Equation (9.58) shows that the square of the contact radius is proportional to the time. This is confirmed by McHale's experimental results (fig. 9.51).

9.7.1.4   Discussion

Equation (9.45) shows that the variation rate of the droplet volume is related to the diffusion coefficient of vapor in the surrounding gas, through the coefficient λ. This can be used as a method for estimating the diffusion coefficient D. From the curves of figure 9.51, the diffusion coefficient can be estimated as $2.32 \times 10^{-5} m^2/s$, whereas the *CRC Handbook of Chemistry and Physics* indicates a value of 2.56×10^{-5} m^2/s for the same conditions of temperature and pressure.

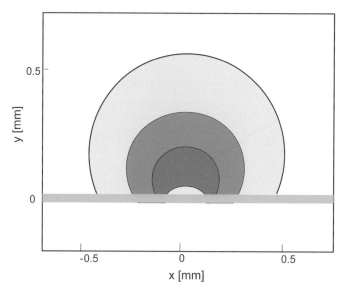

Figure 9.50 Vertical profile of an evaporating droplet; the case of a lyophobic surface and an initial contact radius of 0.48 mm. The contact angle remains approximately constant except at the end of the evaporation process. Reprinted with permission from [39]; ©1998; American Chemical Society.

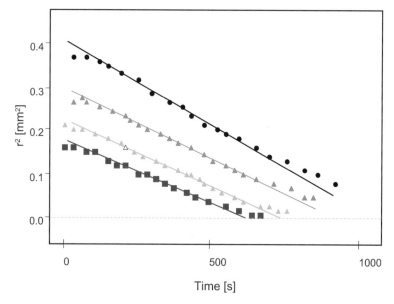

Figure 9.51 Time variation of the contact radius r for water droplets on Teflon (hydrophobic case), showing the proportionality of r^2 with time except at the very end of the evaporation process. Reprinted with permission from [39]; ©1998; American Chemical Society.

We have seen in the preceding paragraphs that the evaporation behavior of wetting and non-wetting droplets is quite different. There is no clear explanation for this difference at this time.

According to McHale and colleagues [39], the difference could be attributed to the fact that, in the non-wetting case, there is a ring of saturated vapor trapped near the contact line, maintaining constant conditions and a constant Young angle during the evaporation process. But this observation does not explain the constant contact surface in the wetting case. As we have seen earlier, in the wetting case, the surface of the substrate and/or the impurities in the liquid that gather in the sharp corner near the triple line may explain the pinning of this contact line. On the other hand, the notion of "receding angle" implies that the surface is not completely smooth. However, as we have noticed above, this behavior occurs even with extremely smooth solid surfaces and pure liquids. Another explanation of the pinning of the contact line could be the precursor film that would pin the contact line, preventing it from retreating. Let us recall that, for a wetting sessile droplet, a precursor film of nanometric height stretches beyond the apparent boundaries of the droplet. Such precursor films have been observed for different wetting situations, as shown in figure 9.52. Note the extreme thinness of the film in the photographs (of the order of a nanometer, or less).

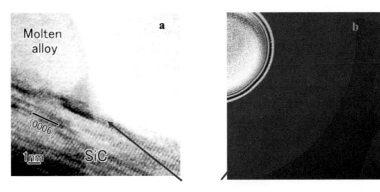

Figure 9.52 (a) Precursor film of spreading molten alloy; (b) scan of a liquid crystal precursor film; the precursor film spreads on a distance of 1 mm; (a) reprinted with permission from [41], ©Elsevier 2002; (b) reprinted with permission from [42] (Sarfus Image: Courtesy Nanolane).

A molecular description of this film is shown in figure 9.53. The precursor film can be explained by thermodynamic considerations; because a jump between the chemical potential of the gas and of the solid is not physical, liquid molecules intercalate between the gas and the solid [43]. Molecules of liquids progressively spread under the action of the "disjoining pressure" caused by the van der Waals interactions between the liquid and solid molecules [44].

9.7.2   Evaporation Rings

In this section, we investigate the first step towards capillary assembly: it is the case of colloidal suspensions that leave evaporation rings during evaporation. Most of the time, micro-drops are laden with particles, macromolecules, or polymers. A typical feature of evaporation of these droplets is evaporation rings. Ring formation is well documented in the literature; it has been studied by numerous authors such as Deegan [31], Deegan *et al.* [32], Zheng *et al.* [45], Popov *et al.* [46], and Yunker *et al.* [47].

During the evaporation of a wetting liquid droplet, the liquid at the edge of the droplet is "pinned" to the underlying surface by the deposited particles. This pinning prevents the droplet

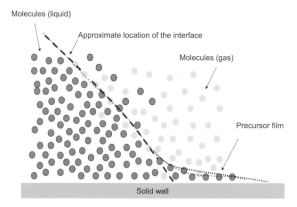

Figure 9.53 Interface with precursor film.

from shrinking. It implies that the footprint of the droplet remains constant. The evaporated liquid at the edge is replenished by liquid from the bulk of the droplet. This means that there is a flow of liquid moving towards the edge of the droplet. If the liquid droplet contains particles, as in a droplet of coffee, these particles will be transported outwards by the flow and deposited near the contact line (fig. 9.54). Finally, when all the liquid has evaporated, the colloidal particles will form a ring of stain (figure 9.55).

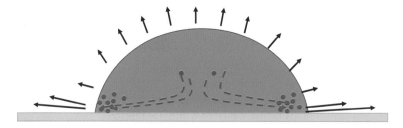

Figure 9.54 Sketch of an evaporating droplet: the two segments near the contact line represent the ring where colloidal particles are deposited. The evaporation rate is stronger near the contact line, provoking a radial convective motion from the center towards the contact line. Here it is supposed that there is no additional motion due to Marangoni convection [47].

However, the particle deposition is conditioned by the shape of the particles: spherical particles leave a deposition ring whereas cylindrical particles do not, except if surfactants are used. The convective pattern is complicated and the transport of the particles depends on their shape and aggregation potential, as demonstrated by Yunker and colleagues [47].

In digital microfluidics these rings are often a drawback. They modify locally the surface energy, and it becomes difficult to displace new droplets over these evaporation rings. Careful – and sometimes difficult – washing of the surface is needed to restore the quality of the surface.

9.7.3 Evaporation Stains

Evaporation rings at the droplet periphery are not a ubiquitous feature of droplet evaporation. Rings form only if the particles or macromolecules are sufficiently large to sediment or adhere

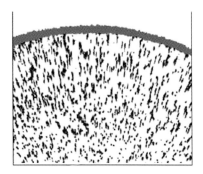

Figure 9.55 Evaporation rings left by an evaporating droplet containing microparticles. Reprinted with permission from [31]; ©APS 2000.

to the solid surface in the corner near the triple line. If the particles or macromolecules are very small, they are agitated by the convective motion and will not sediment near the contact line. As a result, the center part of the droplet – the last region to dry – has more solute deposition than the border (fig. 9.56). Chen *et al.* [48] have experimentally shown the difference between a droplet evaporating on a smooth horizontal plane and in a micro-well. Due to pinning on the edges of the micro-well, the last volume of liquid to evaporate is a ring adjacent to the rim of the well.

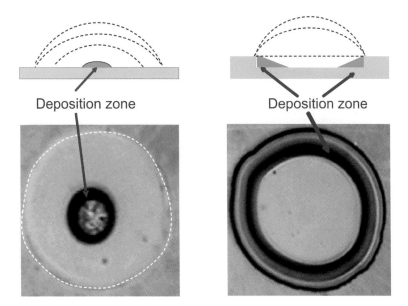

Figure 9.56 Top: schematic of a drying droplet; the droplet sitting on a smooth horizontal plane reduces progressively to a small central region, whereas the droplet in a micro-well reduces to a ring adjacent to the rim of the well. Bottom: Experimental view of the deposit. Reprinted with permission from [48]; ©Elsevier 2006.

9.7.4   The Use of Evaporation and Capillary Assembly

In the preceding section, we have seen that a receding triple line can leave a deposit at some preferential locations on a substrate. This observation has been the start of the development of capillary assembly techniques, where capillary forces are used to move, align or arrange in a given pattern, microscopic objects, for example gold nanospheres or DNA. In this section, we present some interesting uses of this technique.

9.7.4.1   Drop Evaporation as a Tool for DNA Stretching

Evaporation is a relatively slow process compared to molecular motion, and the receding contact line is geometrically (locally) linear at the scale of a group of molecules. Thus, particles and molecules have time to rearrange during the receding of the triple line. This feature has been used for specific applications in modern biotechnology, biology, and nanoassembly.

In genomics – the study of the sequences, functions, and interactions of genes – DNA stretching is of great importance. Usually, DNA strands bunch up and stretching them is a necessary step before any observation of the DNA chain, and a very precise immobilization is needed. This process is called molecular combing. One of the first methods – and still a very commonly used one – to comb DNA was proposed by Bensimon and colleagues [49,50]. This method takes advantage of liquid droplet evaporation (fig. 9.57). A DNA strand attached by one end to a glass surface stretches progressively as the gas/liquid interface retreats during evaporation. When all the liquid has vanished, the DNA strand forms a linear segment on the glass plate.

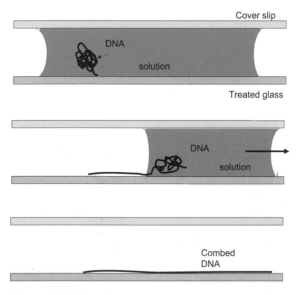

Figure 9.57 Combing DNA using droplet evaporation.

9.7.4.2   Drop Evaporation as a Tool for Nanoassembly

Evaporation can be used for depositing beads for coating the internal surfaces of microfluidic systems. Such coatings may be used to modify the chemical composition of the walls, or to hide a boundary of wettability between two zones of the microsystem, as is the case in composite microsystems where different materials – like plastic and glass or silicon – are used.

If the volatile liquid is laden with the appropriate particles, the evaporating liquid deposits the particles in an ordered pattern along the solid walls of the microsystem. Vengallatore and colleagues [51] have shown how to coat the interior silicon walls of a microsystem with a layer of spherical polystyrene or silica beads by evaporating the carrier fluid (fig. 9.58). An annealing treatment – heating of the coating for a brief moment – may be performed to stabilize the coating.

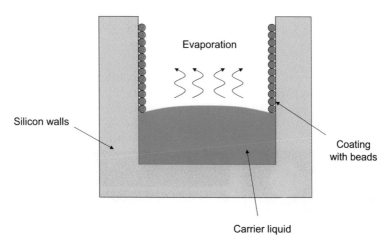

Figure 9.58 Schematic illustration of convective colloidal self-assembly: evaporation of the colloidal suspension leads to particle assembly on the internal surfaces.

The result is shown in figure 9.58 above. Particles predominantly assemble in hexagonal close-packed configurations, and, if sufficient precautions are taken, they can uniformly coat the internal walls of the microsystem.

9.7.4.3 Capillary Assembly of Gold Nano-spheres

Assembly of metal nanospheres (of the order of 10 nm) is particularly interesting.

Surface plasmon resonance (SPR) is a useful technique for measuring the adsorption of nanoparticles – or for following the hybridization of targets – on a solid substrate, typically gold and silver [16]. Surface plasmons are surface electromagnetic waves that propagate in a direction parallel to the metal surface. Since the wave is on the boundary of the metal and the external medium (usually air or water), these oscillations are very sensitive to any change of this boundary, such as adsorption of molecules, or hybridization of DNA sequences, to the metal surface.

An interesting aspect of SPR to detect hybridization of DNA strands is to use gold nano-spheres as a substrate for hybridization. In order to enhance SPR, the nanospheres have to be located at an optimal distance from each other; hence a precise spatial arrangement of the nanospheres must be achieved to obtain efficient detection. This arrangement can be done by capillary forces produced by a fast receding interface caused by evaporation enhanced by controlled air suction [7,52] (fig. 9.59). The nanospheres concentrate at the triple line under the joint action of the receding contact line and the internal convection inside the evaporating liquid.

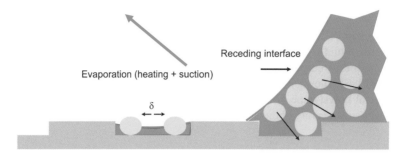

Figure 9.59 Principle of capillary assembly of gold nanospheres in traps [7].

Another interest involving the assembly of gold nanospheres is the formation of optical waveguides by assembling these spheres in micro-trenches or grooves [28]. Many different patterns can be achieved; figure 9.60 shows the two examples mentioned above.

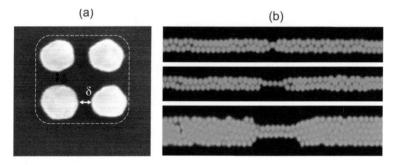

Figure 9.60 (a) Gold nanospheres in a square trap (reprinted with permission from [52], ©AIP, 2010; (b) gold nanospheres aligned in grooves to form waveguides. Reprinted with permission from [28], ©AIP,2011.

9.7.4.4   Conclusion

Deposition of macromolecules or particles using droplet evaporation is of great interest. In most cases, the evaporation is just sufficiently slow to give time for particles to self-assemble at the solid surface, leaving a precisely monitored coating. Often, annealing is performed to stabilize the coating. In the literature, there have been many recent publications on this topic. Let us cite here the work of Bormashenko and colleagues [53], who used different polymer concentrations in an evaporating droplet of solvent to obtain patterned coating, and the publication of Schnall-Levin et al. [54], who showed the patterns made by the rearrangement of particles during evaporation.

9.8   Conclusion

In the first part of this chapter we have analyzed the behavior of liquid droplets and that of rigid spheres at an interface. As expected, it has been shown that the behavior of liquid droplets at an interface is different than that of rigid beads. The conditions for engulfment of droplets or rigid spheres have been enounced.

In the second part, the mechanisms and effects of evaporation have been presented, leading to the concept of self-assembly and capillary assembly. These concepts are now extensively used in materials science to produce planar arrangements of particles at a solid surface that constitute ad hoc coatings.

9.9   References

[1] S. Matosevic and B. M. Paegel, "Stepwise Synthesis of Giant Unilamellar Vesicles on a Microfluidic Assembly Line," *J. Am. Chem. Soc.* **133** pp. 2798–2800, 2011.

[2] J. Berthier, P. Dalle, F. Rivera, R. Renaudot, S. Morales, P-Y. Benhamou, P. Caillat, "The Physics of Pressure Powered Micro-Flow Focusing Device for the Encapsulation of Live Cells," *Proceedings of the 2011 NSTI Nanotech Conference, June 13-16*, Boston, MA, USA, 2011, pp. 458-461.

[3] H. Song, J. D. Tice, R. F. Ismagilov, "A microfluidic system for controlling reaction networks in time," *Angew. Chem.* **42**, pp. 767–771, 2003.

[4] J.D. Tice, H. Song, A. D. Lyon, R. F. Ismagilov, "Formation of droplets and mixing in multiphase microfluidics at low values of the Reynolds and the Capillary numbers," *Langmuir* **19**, pp. 9127–9133, 2003.

[5] M.R. Bringer, C. J. Gerdts, H. Song, J. D. Tice and R. F. Ismagilov, "Microfluidic systems for chemical kinetics that rely on chaotic mixing in droplets," *Phil. Trans. R. Soc. Lond. A* **362**, pp. 1087–1104, 2004.

[6] D.L. Chen, Liang Li, S. Reyes, D.N. Adamson, and R.F. Ismagilov, "Using three-phase flow of immiscible liquids to prevent coalescence of droplets in microfluidic channels: criteria to identify the third liquid and validation with protein crystallization," *Langmuir* **23**, pp. 2255–2260, 2007.

[7] O. Lecarme, T. Pinedo-Rivera, K. Berton, J. Berthier, D. Peyrade, "Plasmonic coupling in nondipolar gold collidal dimers," *Applied Physics Letters* **98**, 083122, 2011.

[8] J. Berthier. *Microdrops and digital microfluidics.* William Andrew Publishing, 2008.

[9] B Roman and J Bico, "Elasto-capillarity: deforming an elastic structure with a liquid droplet," Topical review, *J. Phys.: Condens. Matter* **22** 493101, 2010.

[10] P-G de Gennes, F. Brochart-Wyart, D. Quèrè. *Capillary and wetting phenomena: drops, bubbles, pearls, waves.* Springer, 2002.

[11] G. Morris, S.J. Neethling, J.J. Cilliers, "An investigation of the stable orientations of orthorhombic particles in a thin film and their effect on its critical failure," *J. Coll. Int. Sci.*, advance article, 2011.

[12] A.N. Zdravkov, G.W.M. Peters, H.E.H. Meijer, "Film drainage between two captive drops: PEO-water in silicon oil," *J. Coll. Int. Sci.* **266**, pp. 195–201, 2003.

[13] T. Gilet, K. Mulleners, J.P. Lecmte, N. Vandewalle, S. Dorbolo, "Critical parameters for the partial coalescence of a droplet," *Phys. Rev. E.*, **75**, 036303, 2007.

[14] S. Vengallatore, Y. Peles, L.R. Arana, S.M. Spearing, "Self-assembly of micro- and nano-particles on internal micromachined silicon surfaces," *Sensors and Actuators A* **113**, pp. 124–131, 2004.

[15] J. Berthier, P. Silberzan. *Microfluidics for Biotechnology.* Second Edition, 2010.

[16] H. Raether. *Surface plasmons on smooth and rough surfaces and on gratings.* Springer Verlag, Berlin, 1988.

[17] D.J. Norris, "Measurement and assignment of size dependent optical spectrum in cadmium selenide (CdSe) quantum dots," PhD thesis, MIT, 1995.

[18] M. Dahan, S. Lévi, C. Luccardini, P. Rostaing, B. Riveau, A. Triller, "Diffusion dynamics of glycine receptors revealed by single-quantum dot tracking," *Science* **302**(5644), pp. 442–445, 2003.

[19] D.D. Joseph. *Interrogations of Direct Numerical Simulation of Solid-Liquid Flows.* eFluids book, http://www.efluids.com/efluids/books/efluids_books.htm.

[20] J. Atencia, D.J. Beebe, "Controlled microfluidic interfaces," *Nature* **437**, pp. 648–654, 2005.

[21] A. Katsen-Globa, P.F., F. Ehrhart, M. M. Weber, H. Zimmermann, "Alginate encapsulation improves viability and integrity of cryopreserved pancreatic islets and multicellular spheroids: combined fluorescence, scanning and block-face scanning electron microscopy," *Microsc. Microanal.* **13** (suppl.3), 2007.

[22] D. Lim, A.M Sun, "Microencapsulated islets as bioartificial endocrin pancreas," *Science* **210** pp. 908–910, 1980.

[23] V.L. Workman, S.B.D., P. Kille, D.D. Palmer, "Microfluidic chip-based synthesis of alginate microspheres for encapsulation of immortalized human cells," *Biomicrofluidics* **1**, 014105, 2007.

[24] P. Singh, D.D. Joseph, "Fluid dynamics of floating particles," *J. Fluid Mech.* **530**, pp. 31–80, 2005.

[25] Weixing Song, Yang Yang, Helmuth Moehwald, Junbai Li, "Two dimensional polyelectrolyte hollow spheres array at a liquid-air interface," *Soft Matter.*, **7**, pp. 357-362, 2011.

[26] J.A. Oberteuffer, "Magnetic separation: a review of principles, devices, and applications," *IEEE Trans. On Magnetics* **MAG-10**(2), pp. 223–238, June 1974.

[27] Aharoni, A., "Traction force on paramagnetic particles in magnetic separators," *IEEE Trans. On Magnetics* **MAG-12**(3), pp. 234–235, May 1976.

[28] O. Lecarme, T. Pinedo-Rivera, L. Arbez, T. Honegger, K. Berton, D. Peyrade, "Colloidal optical waveguides with integrated local light sources built by capillary force assembly," *J. Vac. Sci. Technol. B* **28**(5), p. C6011, 2010.

[29] S. Vengallatore, Y. Peles, L.R. Arana, S.M. Spearing, "Self-assembly of micro- and nano-particles on internal micromachined silicon surfaces," *Sensors and Actuators A*, **113**, pp. 124–131, 2004.

[30] E. Bormashenko, R. Pogreb, A. Musin, O. Stanevsky, Y. Bormashenko, G. Whyman, O. Gendelman, Z. Barkay, "Self-assembly in evaporated polymer solutions: Influence of the solution concentration," *Journal of Colloid and Interface Science* **297**, pp. 534–540, 2006.

[31] R.D. Deegan, "Pattern formation in drying drops," *Physical Review E* **61**(1), pp. 475–485, 2000.

[32] R.D. Deegan, O. Bakajin, T.F. Dupont, G. Huber, S.R. Nagel, T.A Witten, "Capillary flow as the cause of ring stains from dried liquid drops," *Nature* **389**, pp. 827–829, 1997.

[33] J. J. Hegseth, N. Rashidnia and A. Chai, "Natural convection in droplet evaporation," *Phys. Rev. E* **54**(2), pp. 1646–1644, 1996.

[34] E. Berthier, J. Warrick, H. Yu, D.J. Beebe, "Managing evaporation for more robust microscale assays. Part 1. Volume loss in high throughput assays," *Lab Chip* **8**(6), pp. 852–859, 2008.

[35] S.A. Kulinich, M. Farzaneh, "Effect of contact angle hysteresis on water droplet evaporation from super-hydrophobic surfaces," *Applied Surface Science* **255**, pp. 4056–4060, 2009.

[36] R.A. Meric, H.Y. Erbil, "Evaporation of sessile drops on solid surfaces: pseudospherical cap geometry," *Langmuir* **14**, pp. 1915-1920, 1998.

[37] S.M. Rowan, M.I. Newton, G. McHale, "Evaporation of micro-droplets and the wetting of solid surfaces," *J. Phys. Chem.* **99**, pp. 13268–13271, 1995.

[38] Hua Hu, R.G. Larson, "Evaporation of a sessile droplet on a substrate," *J. Phys. Chem. B* **106**, pp. 1334–1344, 2002.

[39] G. McHale, S.M. Rowan, M.I. Newton, M.K. Banerjee, "Evaporation and the wetting of a low-energy solid surface," *J. Phys. Chem. B.* **102**, pp. 1964–1967, 1998.

[40] K.S. Birdi, D.T. Vu, A. Winter, "A study of the evaporation rate of small water drops

placed on a solid surface," *J. Phys. Chem.* **93**, pp. 3702–3703, 1989.

[41] C. Iwamoto, S. Tanaka, "Atomic morphology and chemical reactions of the reactive wetting front," *Acta Materiala* **50**, p. 749, 2002.

[42] Nanolane: http://www.nano-lane.com/wetting-afm.php.

[43] L.M. Pismen, B.Y. Rubinstein, I. Bazhlekov, "Spreading of a wetting film under the action of van der Waals forces," *Physics of Fluids* **12**(3), 2000.

[44] P.G. deGennes, "Wetting: statistics and dynamics," *Rev. Mod. Phys.* **57**, p. 827, 1985.

[45] Rui Zheng, Y. O. Popov, T. A. Witten, "Deposit growth in the wetting of an angular region with uniform evaporation," *Physical Review E* **72**, 046303, 2005.

[46] Y. O. Popov, "Evaporative deposition patterns: Spatial dimensions of the deposit," *Phys. Rev. E* **71**, 036313, 2005.

[47] P. J. Yunker, T. Still, M. A. Lohr, A. G. Yodh, "Suppression of the coffee-ring effect by shape-dependent capillary interactions," *Nature* **476**, pp. 308–311, 18 August 2011.

[48] Chin-Tai Cheng, Fan-Gang Tseng, Ching-Chang Chieng, "Evaporation evolution of volatile liquid droplets in nanoliter wells," *Sensors and Actuators A* **130-131**, pp. 12–19, 2006.

[49] J.F. Allemand, D. Bensimon, L. Jullien, A. Bensimon, V. Croquette, "pH-dependent specific binding and combing of DNA," *Biophysical Journal* **73**, pp. 2064–2070, 1997.

[50] D. Bensimon, A. J. Simon, V. Croquette, A. Bensimon, "Stretching DNA with a receding meniscus: experiments and models," *Phys. Rev. Lett.* **74**, pp. 4754–4757, 1995.

[51] S. Vengallatore, Y. Peles, L.R. Arana, S.M. Spearing, "Self-assembly of micro- and nano-particles on internal micromachined silicon surfaces," *Sensors and Actuators A* **113**, pp. 124–131, 2004.

[52] M. J. Gordon, D. Peyrade, "Separation of colloidal nanoparticles using capillary immersion forces," *Appl. Phys. Lett.* **89**, 053112, 2006.

[53] E. Bormashenko, R. Pogreb, A. Musin, O. Stanevsky, Y. Bormashenko, G. Whyman, O. Gendelman, Z. Barkay, "Self-assembly in evaporated polymer solutions: Influence of the solution concentration," *Journal of Colloid and Interface Science* **297**, pp. 534–540, 2006.

[54] M. Schnall-Levin, E. Lauga, M. Brenner, "Self-assembly of spherical particles on an evaporating sessile droplet," *Langmuir* **22**, pp. 4547–4551, 2006.

10

Digital Microfluidics

10.1 Abstract

From its beginning, digital microfluidics has been developed for biotechnological applications, with the aim to manipulate extremely small volumes of sample liquid. Volumes smaller than 60 nl can be manipulated in such microsystems and used to transport biologic targets. In parallel, digital microfluidics has seen many applications in the field of optofluidics, such as tunable lenses and electronic display screens.

Digital microfluidics consists of moving, merging, separating, and mixing droplets on a locally planar surface. The actuation force can be of two types: electric or acoustic. In the first case, the droplets are moved on a substrate paved with electrodes; this technique is called electrowetting. In the second case, droplets are actuated by acoustic surface waves (SAW) directionally guided. Although some interesting applications have been developed using acoustic methods [1-3], we will only present here electrowetting, and its practical form called "electrowetting on dielectric" (EWOD). The theory of electrowetting is presented first, then the different mechanisms for droplet manipulation are analyzed, and finally some applications in the fields of biotechnology and optofluidics are given.

10.2 Electrowetting and EWOD

In this section, we present the Berge-Lippmann-Young law, which is the basis for the practical developments of digital microfluidics, and we investigate the potentialities of EWOD by analyzing its physical limits.

10.2.1 Berge-Lippmann-Young Equation (BLY)

The principle of electrowetting derives from the Berge-Lippmann-Young (BLY) equation [4,5]. The first historical relation derived by Lippmann stated that the apparent contact surface tension of an electrically conducting liquid in contact with a solid is modified by an electric field. This apparent change is attributed to the double layer that forms on the substrate surface. Using

Gibbs' interfacial thermodynamics [6],

$$\frac{\gamma_{SL}^{eff}}{dV} = -q_{SL},$$ (10.1)

where V is the electric potential, γ_{SL}^{eff} the effective or apparent surface tension and q_{SL} the field-induced surface charge density in counter-ions. More recently, Berge has shown that the Lippmann relation results in a change of the apparent contact angle. The Berge-Lippmann-Young (BLY) equation describes the change of contact angle with the applied voltage,

$$\cos\theta = \cos\theta_0 + \frac{C}{2\gamma_{LG}}V^2,$$ (10.2)

where θ, θ_0 are the actuated and Young contact angles, C the capacitance of the substrate, γ_{LG} the liquid-air surface tension, and V the electric potential. According to equation (10.2), the contact angle decreases when an electric field is applied (fig. 10.1).

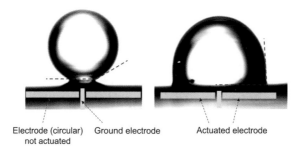

Electrode (circular)   Ground electrode   Actuated electrode
not actuated

Figure 10.1 Effect of an electric field on a conducting droplet: the contact angle θ decreases with the applied electric potential.

In fact, the BLY equation recovers only a part of the physics of electrowetting. In reality, the interface is locally deformed by the electric forces associated to the Maxwell tensor [7]. An electromechanical approach has been developed by Jones et al. [8,9], Kang [10], and recently reviewed by Zeng and Korsmeyer [11]. In the case of liquids, the body force density due to an electric field is given by the Korteweg-Helmholtz relation

$$\vec{f_k} = \sigma_f \vec{E} - \frac{\varepsilon_0}{2}E^2\nabla\varepsilon_f + \nabla\left[\frac{\varepsilon_0}{2}E^2\frac{\delta\varepsilon_f}{\delta\rho}\rho\right],$$ (10.3)

where ρ is the mass density of the liquid, ε_f the permittivity of the liquid, and σ_f the electric charge density. The first term on the right side of (10.3) corresponds to the electrostatic force; the second term to the dielectrophoretic force. The last term describes electrostriction and can be neglected here as the mass density of the liquid remains constant. The force acting on a volume element δV is obtained by integrating equation (10.3). It can be shown that it is equivalent to integrating the momentum flux density, i.e., the Maxwell stress tensor

$$T_{ik} = \varepsilon_0\varepsilon(E_iE_k - \frac{1}{2}\delta_{ik}E^2),$$ (10.4)

along the boundaries of the volume ΔV. In (10.4), the notation E^2 corresponds to $|\vec{E}|^2$ and δ_{ik} is the Kronecker delta function: $\delta_{ik} = 0$ if $i \neq k$ and $\delta_{ii} = 1$ if $i = k$; and $i,k = x,y,z$. Surface integration of (10.4) is much easier than volume integration of (10.3). The net force acting on the liquid volume element is

$$F_i = \int_\Omega T_{ik} n_k dA,$$

(10.5)

where the Einstein summation convention – summation on the repeated indices – has been used. At the surface of a perfectly conducting liquid, on the gas side, the electric field is perpendicular to the surface (figure 10.2), and related to the surface density of electric charge σ_s by Gauss' law:

$$\sigma_s = \varepsilon_0 \vec{E} \cdot \vec{n},$$

(10.6)

where $\vec{n}$ is the outward unit normal vector. Note that the electric field vanishes in the conducting liquid. If we consider the x, y, z axis system such that the x-axis is aligned with $\vec{n}$, then the electric field is $\vec{E} = (E_x, 0, 0)$ in the gas domain and $\vec{E} = (0, 0, 0)$ in the liquid domain. In the gas domain, the Maxwell tensor is

$$[T] = \varepsilon \left\{ \begin{array}{ccc} \frac{1}{2}\varepsilon_0 E_x^2 & 0 & 0 \\ 0 & -\frac{1}{2}\varepsilon_0 E_x^2 & 0 \\ 0 & 0 & -\frac{1}{2}\varepsilon_0 E_x^2 \end{array} \right\},$$

(10.7)

and vanishes in the liquid domain,

$$[T] = [0].$$

(10.8)

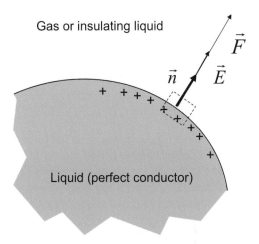

Figure 10.2 Electric force acting at the interface of a conducting liquid.

We can now integrate (10.5); the cross terms xy, yz, and zx are all zero, the forces in the y (respectively z) direction cancel out, and we find that the only non-vanishing contribution is a force directed along the outward normal $\vec{n}$:

$$\frac{\vec{F}}{\delta A} = P_e \vec{n} = \frac{1}{2}\varepsilon_0 E^2 \vec{n} = \frac{\sigma_s}{2}\vec{E},$$

(10.9)

where δA is an elementary surface area of the interface. In (10.9), P_e is the electrostatic pressure defined by $P_e = \frac{\varepsilon_0}{2}E^2$. The electrostatic pressure P_e acts on the liquid surface, and brings a negative contribution to the total pressure within the liquid. The liquid interface is distorted by the electric forces acting on it. The distortion depends on the distribution of charges σ_f at the surface. In the case of electrowetting, the electric charges at the liquid-gas interface are located close to the triple contact line, as sketched in figure 10.3, within a distance equal to the dielectric thickness d.

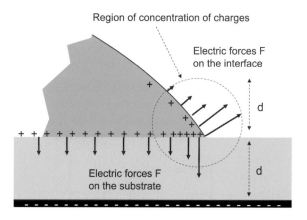

Figure 10.3 Schematic of the electric charge distribution in the vicinity of the triple contact line. Electric charges are located at the solid-liquid interface and at the liquid-gas interface, approximately within a distance d from the solid surface. In reality, this is a very simplified view since the liquid interface is distorted by the electric forces very close to the triple contact line.

The problem of the surface distortion has been solved by Kang [10] and Vallet *et al.* [12] using the Schwarz-Christoffel conformal mapping, assuming that the shape of the interface at the contact of the solid substrate is a wedge [13]. The electric potential ϕ satisfies the Laplace equation $\nabla^2\phi = 0$ and consequently is a harmonic function. The theory of analytic and harmonic functions states that there exists a conformal mapping (the Schwarz-Christoffel mapping) that transforms the fields $\vec{E}$ and ϕ for a half plane into the same fields for a wedge (figure 10.4).

This mapping shows that the field and charges concentrate at the tip of the wedge. In fact, as for any harmonic function near a geometric singularity, the electric field is singular at a wedge (see for example Thamida and Chang [14]), but the Maxwell pressure is integrable and produces a finite Maxwell force at the contact line after (10.5) is integrated:

$$F_{horizontal} = \frac{\varepsilon_0\varepsilon_d V^2}{2d}, \tag{10.10}$$

$$F_{vertical} = \frac{\varepsilon_0\varepsilon_d V^2}{2d}\frac{1}{\tan\theta}. \tag{10.11}$$

Note that the electric forces distribution on the liquid-gas interface due to the Maxwell stress is limited to a very small region close to the triple line. Note also that using (10.10) and making a balance of the forces – electric plus capillary – in the horizontal direction on the triple line results in the BLY equation.

The electromechanical approach shows that the BLY equation comes directly from the Maxwell stress on the liquid-gas interface near the contact line. In fact, since the liquid has a much

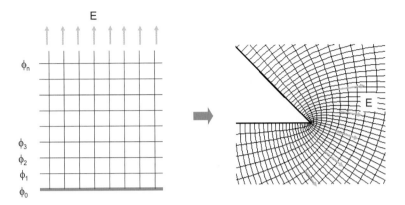

Figure 10.4 Principle of the Schwarz-Christoffel conformal mapping. The two fields $\vec{E}$ and ϕ are transformed from a simple half plane geometry (with an evident solution) to the geometry of a wedge. Note the concentration of the electric field around the tip of the wedge.

higher permittivity (and conductivity) than the gas, it is the wedge shape of the gas phase that is responsible for distortion of the contact angle, as is clearly shown by the Schwarz-Christoffel transformation. A consequence of equation (10.11) is that there is a vertical force on the liquid interface near the triple line. This vertical force increases quickly when the contact angle decreases. This could be an explanation for the phenomenon of contact angle saturation (the contact angle cannot go to zero), usually accompanied by ejection of nanodrops.

The wedge theory presented above is only a simplistic approximation of the reality. In fact, extremely close to the solid surface, the shape of the interface is complex. It has been shown that the real contact angle is still the Young contact angle, but within the width of the electrical double layer, the interface is deformed by electric forces, and the apparent contact angle is the BLY contact angle, as shown in figure 10.5 [8,9].

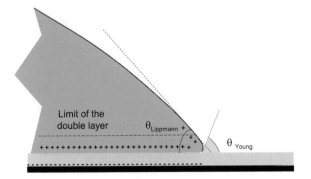

Figure 10.5 Close up view of the three-phase contact at the wall.

10.2.2   Electrowetting Force

In chapters 1 and 6 we have seen that the capillary line force density on a triple line is

$$f_w = \gamma\cos\theta. \tag{10.12}$$

Using the electro-capillary equivalence, and the BLY law, we find

$$f_w = \gamma\cos\theta = \gamma\cos\theta_0 + \frac{C}{2}V^2. \tag{10.13}$$

The difference of contact angles between the "hydrophilic" part of the triple line and the "hydrophobic" part results in a net capillary force. Let us sketch a droplet placed on EWOD electrodes in the drawing of figure 10.6. The electro-capillary force acting on the droplet is readily deduced from (10.13)

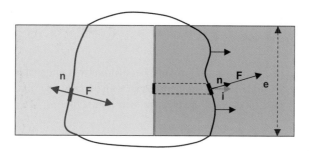

Figure 10.6 Sketch of the contact of a drop with the substrate (we have sketched the usual situation where the contact angle on the non-actuated electrode (left) is larger than 90°, and that with the actuated electrode (right) is smaller than 90°).

From (10.12), the capillary force in the direction x (unit vector $\vec{i}$) on the hydrophilic electrode is given by

$$F_x = \int_L \gamma\cos\theta\, \vec{n}\cdot\vec{i}\,dl, \tag{10.14}$$

$\vec{n}$ being the unit vector normal to the surface, and dl a unit element of the contour line. Equation (10.13) can be integrated to [3]

$$F_x = \gamma\cos\theta \int_L \vec{n}\cdot\vec{i}\,dl = e\gamma\cos\theta, \tag{10.15}$$

where e is the width of the electrode. Note that the real shape of the triple line has no importance. Figure 10.7 shows different forms of experimental contact lines. Noting that the x-direction force on the triple line outside the electrodes vanishes, it is concluded that the x-direction capillary force on the droplet, whatever its shape, is

$$F_x = e\gamma(\cos\theta - \cos\theta_0) = e\frac{CV^2}{2}. \tag{10.16}$$

10.2.3   Limitations of EWOD – Saturation, Dielectric Breakdown and Hysteresis

There are three limitations of EWOD: (1) the saturation effect, (2) dielectric breakdown, and (3) hysteresis.

Figure 10.7 Different shapes of droplets on EWOD electrodes: left, water droplet; right, ionic liquid droplet. (photo courtesy Ph. Clementz).

10.2.3.1   Saturation Limit

If the electric potential is increased, equation (10.2) predicts a zero contact angle for some value of the electric potential. In fact, above a certain value of the potential, the apparent contact angle does not decrease anymore. It is the saturation effect. Different explanations have been proposed for this effect [3,15-17]. It is very convenient, although approximate, to use the zero surface-liquid energy limit model pioneered by Ralston *et al.* [16,17]. We shall refer to this model by the abbreviation PQRS. In this approach, the effective solid-liquid surface tension decreases with the voltage according to equation

$$\gamma_{SL}^{eff}(V) = \gamma_{SL} - \frac{1}{2}CV^2,\tag{10.17}$$

and the voltage-dependant Young's law can be cast in the form

$$\cos\theta(V) = \frac{\gamma_{SG} - \gamma_{SL}^{eff}}{\gamma_{LG}}.\tag{10.18}$$

The effect of relation (10.18) is sketched in figure 10.8 for different values of the applied voltage. For a zero voltage, the force balance is that defined by the classical Young's law. As the voltage increases, the effective solid-liquid surface tension decreases, and the contact angle decreases according to the BLY equation. The lower limit for the effective solid-liquid effective surface tension is zero; when this value is reached, the minimum contact angle is obtained. This minimum value is the saturation contact angle θ_{sat}.

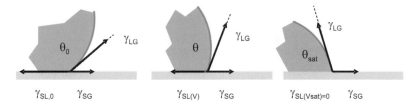

Figure 10.8 Sketch of the different contact angles depending on the applied voltage. Left: at zero potential, the contact angle is determined by the classical Young law; middle: the contact angle decreases when the applied voltage increases; right: the lower limit of the contact angle is obtained when the solid-liquid surface tension vanishes.

According to equation (10.18), at saturation $\gamma_{SL}^{eff} = 0$ and

$$\cos\theta_{sat} = \cos\theta(V_{sat}) = \frac{\gamma_{SG}}{\gamma_{LG}}. \tag{10.19}$$

Using equation (10.17) and Young's law, the saturation potential is

$$V_{sat} = \left[\frac{2\gamma_{SL,0}}{C}\right]^{\frac{1}{2}} = \left[\frac{2(\gamma_{SG} - \gamma_{LG}\cos\theta_0)}{C}\right]^{\frac{1}{2}}. \tag{10.20}$$

In order to take into account the saturation effect, Berthier has proposed using a modified BLY law [18]

$$\frac{\cos\theta - \cos\theta_0}{\cos\theta_S - \cos\theta_0} = La\left(\frac{CV^2}{2\gamma_{LG}(\cos\theta - \cos\theta_0)}\right), \tag{10.21}$$

where La is Langevin's function $La = \coth(3X) - \frac{1}{3X}$, and θ_S is the saturation angle. The Langevin law is well adapted to such a situation because it closely follows the quadratic law for small voltage and has an asymptotic behavior at large voltages. Below the saturation limit, equation (10.21) reduces to the usual BLY law. A comparison between experimental results and (10.21) is shown in figure 10.9.

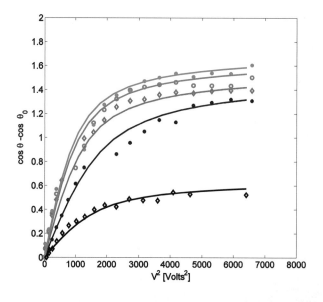

Figure 10.9 Modified BLY law using Langevin's function: comparison with experiments (continuous lines correspond to the Langevin function and dots to the experimental values).

10.2.3.2   Dielectric Breakdown

EWOD electrodes are coated with a thin dielectric layer in order to avoid electrolysis of the conducting fluid. Above this dielectric layer, a very thin hydrophobic layer is usually deposited to enhance the electrowetting effect by increasing the value of θ_0 above $90°$ (fig. 10.10).

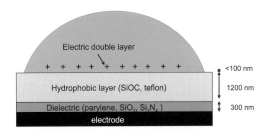

Figure 10.10 Schematic cross section of an EWOD digital element.

When the electric field is larger than a threshold – called the critical electric field, denoted EBD – the dielectric is disrupted, i.e. cracks suddenly form and the dielectric integrity is lost (fig. 10.11). This threshold is also called the theoretical dielectric strength of the material. It is an intrinsic property of the bulk material. The detailed physical explanation of dielectric breakdown is not the subject of this chapter and is well documented in the literature [19,20]. For a dielectric of thickness d, the critical electric field EBD is related to the dielectric breakdown voltage VBD by

$$V_{BD} = dE_{BD}. \tag{10.22}$$

Indications of the value of the critical electric field are given in table 10.1 for some common materials. In the preceding section, we have developed the notion of saturation potential. Typically, saturation potentials are of the order of 80 V for a substrate of capacitance $C \approx 2.210^{-5}$ F/m^2, obtained with a total dielectric/insulating layer of 1.5 μm thickness approximately. Thus the electric field at saturation is of the order of 55 V/μm. Note that this value is just below the breakdown value of Teflon. In other words, chips are usually designed to function at the saturation potential.

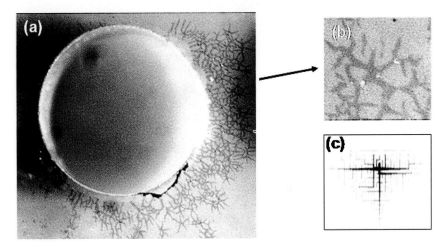

Figure 10.11 (a) Dielectric breakdown at the vicinity of a spherical object; the breakdown is materialized by the formation of cracks having the shape of tree branches (photo CEA-Leti); (b) close-up on the "tree effect"; (c) electrostatic breakdown model showing the growth of "failure tree" due to electron avalanche. Reprinted with permission from [54]; ©IOP 2003.

Table 10.1 Values of the breakdown voltage for some common materials (units V/μm. Note that the values indicated here are given for perfect and ideal materials. Real values are usually less than these.

Paraffin	Teflon	Oil	Glass	Mica
10	15	59	100	197

However, dielectric breakdown is sometimes observed at lower values of the potential. Breakdown frequently occurs when there are defects in the substrate surface or when objects such as cells or proteins adhere to the substrate. A possible explanation could be the anomalous value of the electric field in the vicinity of geometrical inhomogeneities. The contact of an object with the substrate is sketched in figure 10.12. Assuming that the liquid is perfectly conductive and that the object is insulating or at least its envelope is insulating, and if ε_1 and ε_d denote the relative permittivity of the object and the solid dielectric respectively, the electric potential is given by the Laplacian equation

$$\varepsilon_1 \nabla^2 V_1 = \varepsilon_d \nabla^2 V_d. \tag{10.23}$$

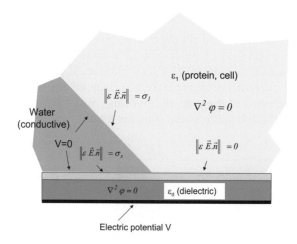

Figure 10.12 Electric scheme of the contact of an object with the substrate.

From a theoretical point of view, the electric potential can be obtained through the Schwarz-Christoffel conformal mapping. If the angle is sharp, there is a pole of the transform at the tip of the wedge (fig. 10.13). This anomalous, localized value of the electric field may increase the voltage above the dielectric breakdown voltage [3]. This could be the explanation why dielectric breakdown is sometimes observed when cells or proteins adhere to the substrate, or when the surface get crackled after a long period of use. Work is currently under way to reinforce the level of the dielectric breakdown voltage without raising the value of the capacitance.

10.2.3.3 Hysteresis

As a general rule, advancing and receding contact angles are different. This is due to the roughness of the substrate, and the difference exists even for apparently smooth surfaces. Elec-

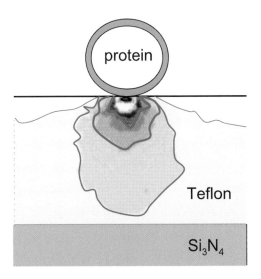

Figure 10.13 Contour plot of the magnitude of the electric field below a cell sticking to the surface (Comsol software).

trowetting is not an exception; the contact angle is not the same when the potential is gradually increased or gradually decreased (fig. 10.14). The contact angle difference between advancing and receding contact angles can be approximated by a constant angle α. This hysteresis angle depends on the quality of the substrate and is reduced for superhydrophobic substrates. It is usually in the range $1° - 30°$. Besides the contact angle difference, it is observed that a droplet of conductive liquid does not move from one electrode to the next as soon as an electric actuation is applied. A minimum voltage is required to have the triple line moving. It has been shown that the minimum voltage required to obtain interface motion is linked to the hysteresis contact angle α [3,21]:

$$V_{min} = 2\sqrt{\frac{\gamma_{LG}\alpha\sin\theta_0}{C}}.$$

(10.24)

10.2.3.4   Minimum and Maximum Actuation Voltages – Working Range

Taking into account saturation limit, dielectric breakdown and hysteresis, an approximate working range can be determined [21]. On one hand, a minimum actuation voltage based on hysteresis has been derived, on the other hand a maximum voltage linked to saturation and dielectric breakdown has been found. Hence, the domain for EWOD workability is then given by the following relation:

$$\alpha\gamma_{LG}\sin\theta_0 < \frac{CV^2}{2} < \gamma_{SL} - \gamma_{LG}\cos\theta_0.$$

(10.25)

At this stage, we remark that the interval $[V_{min}, V_{max}]$ depends on the capacitance C. Microdevices with thinner layers of dielectric have a larger capacitance and require lower level of actuation. There is a clear advantage in trying to increase the specific conductance of the dielectric layer: the electric potentials needed to actuate the droplets will then be lower. Equation (10.25) can be translated in terms of line force by

$$f_{min} = \alpha\gamma_{LG}\sin\theta_0 < f_{EWOD} < \gamma_{SL} - \gamma_{LG}\cos\theta_0 = f_{max},$$

(10.26)

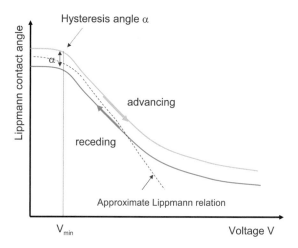

Figure 10.14 Hysteresis of electrowetting phenomenon and advancing and receding apparent contact angle. Insert: experimental observation of hysteresis (substrate SiOC with very low hysteresis level).

showing the range of electrowetting force. Using a typical value for the specific capacitance $C \approx 0.03$ mF/m^2, relation (10.25) shows that the minimum electro-capillary forces lies in the range 1-20 μN, and the maximum forces in the range 15-40 μN.

10.3   Droplet Manipulation with EWOD

In order to be effective, a droplet microfluidic technique should handle some basic operations. These operations are: droplet motion, division and merging, and dispensing. In the following sections we investigate how these operations are performed on EWOD systems and we show that a special design, called "covered" EWOD – as opposed to "open" EWOD systems, facilitates these operations. Indeed, while droplet motion is possible on both open and covered EWOD systems, the other operations require the covered EWOD geometry. First let us present typical architectures of "open" and "covered" EWOD systems.

10.3.1   Open vs. Covered EWOD System

In an "open" EWOD system, droplets are just deposited on the substrate, which is paved with electrodes, and isolated by a dielectric microlayer, covered by a hydrophobic coating (fig. 10.15).

Covered EWOD systems are just open EWOD system with a cover plate. In such systems, the substrate which supports the droplet is similar to that of open EWOD systems described previously. But, in this case, a top plate covers the droplet [22,23]. This top plate consists of an electrode usually made of ITO (Indium-Tin oxide) coated with a thin layer of Teflon. Thus, without electric actuation, the contact is hydrophobic with both plates, as shown in fig. 10.16.

In a covered EWOD system, the droplet follows closely the shape of the electrodes (fig. 10.17). In such a configuration, the vertical gap is very small compared to the horizontal dimension of the droplet, and the surface area of the free interface (liquid-air) is very small compared to that of the solid-liquid interface. Consequently, the energy of the liquid-air interface is much

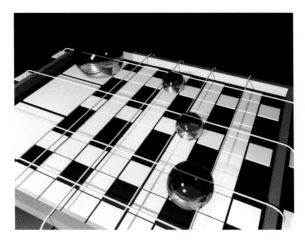

Figure 10.15 Photo of an "open" EWOD microsytem. The grounded electrodes are the catena wires running parallel to the chip. (Courtesy Y. Fouillet, CEA-Leti).

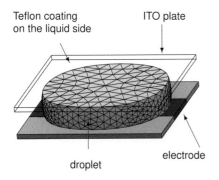

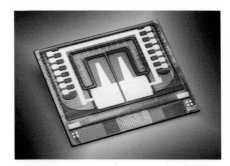

Figure 10.16 Schematic view of a covered EWOD microsystem; Left: droplet squished between two plates. Right: covered EWOD microsystem fabricated by the LETI (photo CEA-LETI).

smaller than that of the solid-liquid interface and the droplet adopts the shape of the underlying electrode. This is why a droplet is nearly square when the electrodes are square.

10.3.2   Droplet Motion

Because the friction on the substrate is quite small (even when a cover plate is present), the electrowetting force is usually sufficient to move a liquid droplet. Water, water with surfactants, and biologic fluids can be moved by EWOD. Using the capillary equivalence, it is comparable to the motion of a droplet under a gradient of wettability [24,25]. Figure 10.18 shows the Evolver simulated motion of the droplet. Note that Evolver does not include the dynamic and viscous forces; hence it can only indicate if the motion occurs. However, because the surface tension forces are usually dominant – as soon as the surface tension is sufficiently important – there is a similarity between the real motion and the simulated motion. Estimates of the Weber, Capillary and Ohnesorge non-dimensional numbers assess that the surface tension forces are dominant

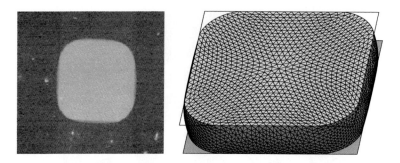

Figure 10.17 Left: photograph of a "square" droplet in a covered EWOD microchip; right: the droplet simulated with the Surface Evolver).

[26]. The Weber number is the ratio of the inertial and surface tension forces,

$$We = \frac{\rho V^2 R}{\gamma} = \frac{\rho V^2}{\frac{\gamma}{R}}, \tag{10.27}$$

where V is the liquid velocity, ρ the liquid density, and R the droplet radius. Usually for EWOD microsystems the Weber number is smaller than 0.01. On the other hand, the ratio of the viscous and the surface tension forces is given by the Capillary number,

$$Ca = \frac{\eta V}{\gamma}, \tag{10.28}$$

where η is the liquid viscosity. Usually also the Capillary number is less than 0.01. This is equally the case for the Ohnesorge number,

$$Oh = \frac{\eta}{\sqrt{\rho \gamma R}} = \frac{\sqrt{We}}{Re}, \tag{10.29}$$

where Re is the Reynolds number. It is concluded that the surface tension forces are dominant during droplet motion.

Let us analyze the forces on a droplet located on the boundary between an actuated (ON) and non-actuated (OFF) electrode. The line force is $f_a = \gamma \cos \theta_a$ on the actuated side and $f_0 = \gamma \cos \theta_0$ on the non-actuated side. The resulting electrowetting force is

$$F_w = e(\gamma \cos \theta_a - \gamma \cos \theta_0), \tag{10.30}$$

where e is the length of the junction between the electrodes. Because $\cos \theta_0$ is usually negative, two terms on the right side of the preceding equation add together.

In figure 10.19, it is shown that the calculated motion is very similar to the real motion.

10.3.3 Moving Droplet Velocity

In digital microfluidic systems, a droplet moves from one location on the chip to the next incrementally. Because inertia forces are small, the droplet accelerates very quickly, reaches a nearly steady velocity, and stops abruptly. In this section we focus on the "nearly steady state" velocity of the droplet.

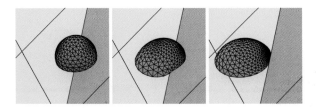

Figure 10.18 View of a droplet of ionic liquid moving to the left from a hydrophobic substrate to a hydrophilic substrate (Surface Evolver). Note that the "hydrophilic" forces applied on the front part of the triple line and the "hydrophobic" forces applies on the rear part of the triple line contribute together to the motion. This is why a hydrophobic substrate is preferred in EWOD devices.

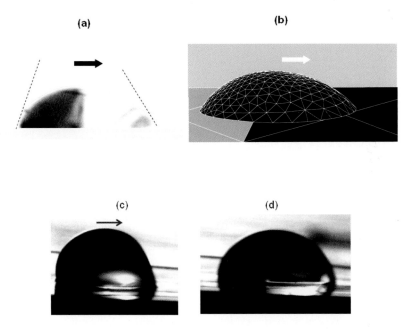

Figure 10.19 (a) Side view of a moving droplet (ionic liquid); (b) Evolver simulation of the same droplet; (c) photograph of a water droplet during its motion; (d) end of motion (courtesy CEA/LETI). One sees quite well the asymmetric shape of the droplet during its motion.

Velocity depends on the strength of the electro-capillary forces. It is then expected that velocity of the droplet is directly related to this force. In reality, the motion of micro-drops is a complex physical phenomenon involving surface tension, viscosity, dynamic contact angles, internal motion, etc., and is still a subject of research. In the following section, we present a very simple model proposed by Chen *et al.* [27] to make an estimate of the velocity of micro-drops during their motion in EWOD microsystems.

10.3.3.1 Model

We consider a droplet of volume *Vol* either sandwiched between two plates (covered EWOD) or sitting on a flat surface (open EWOD), as sketched in figure 10.20.

The following simplified analysis can be done: if the velocity of motion is constant, the

Figure 10.20 (a) Droplet moving in a covered EWOD configuration. (b) Droplet moving in an open EWOD configuration.

electrowetting force balances the friction force due to viscous dissipation at portions of the substrate in contact with the drop, assuming that the other resisting forces, such as contact-line resistance, are negligible compared to the viscous dissipation. Then, in the case of the covered system, the contact surface is assumed circular (which is not true in reality),

$$F_{viscous} \approx 2\pi r^2 \tau_w \approx \frac{12\mu Vol}{\delta^2} v_{cov},$$ (10.31)

where v_{cov} is the global velocity and τ_w is the substrate friction, approximated with the Poiseuille profile

$$\tau_w \approx \frac{6\mu v_{cov}}{\delta}.$$ (10.32)

Using the capillary equivalence, we can express the electrowetting force by

$$F_{elec} = 2r\gamma_{LG}(\cos\theta_a - \cos\theta_r)$$ (10.33)

and by equating equations (10.31) and (10.33), we deduce the droplet velocity of motion,

$$v_{cov} \approx \frac{\delta\gamma_{LG}}{6\pi\mu r}(\cos\theta_a - \cos\theta_r).$$ (10.34)

If the voltage is smaller than the saturation voltage, we can use the Lippmann-Young relation, and the velocity can be estimated by

$$v_{cov} \approx \frac{\delta}{12\pi\mu r}CV^2.$$ (10.35)

Equation (10.35) shows that the velocity of motion is proportional to the square of the electric potential.

Similarly, in the case of the open EWOD system,

$$F_{viscous} \approx \pi a^2 \tau_w.$$ (10.36)

The shear stress τ_w can be approximately expressed by

$$\tau_w \approx \frac{5\mu}{2h}v_{open},$$ (10.37)

based on an approximate peak velocity located at 2/5 of the drop height, according to numerical simulations. Then

$$F_{viscous} \approx \frac{5\pi a^2\mu}{2h}v_{open}.$$ (10.38)

On the other hand, the electrowetting force is

$$F_{elec} = 2a\gamma_{LG}(\cos\theta_a - \cos\theta_r),$$ (10.39)

and the droplet velocity is obtained by combining equations (10.38) and (10.39):

$$v_{open} = \frac{4h}{5\pi\mu a}\gamma_{LG}(\cos\theta_a - \cos\theta_r).$$ (10.40)

In terms of voltage (far from the saturation limit), we have

$$v_{open} = \frac{2h}{5\pi\mu a}CV^2.$$ (10.41)

It is easy to see that the velocity of a droplet is larger in an open EWOD system by forming the ratio v_{cov}/v_{open}:

$$\frac{v_{cov}}{v_{open}} = \frac{5}{24}\frac{\delta}{r}\frac{a}{h}.$$ (10.42)

In (10.42), the term a/h is of the order of 1, and $\delta \ll r$, showing that the velocity of motion is much larger in an open configuration. We note that this model assumes that the two dominant forces are the electrowetting and the friction forces, and supposes an established Poiseuille flow inside the droplet. It neglects internal recirculation, the effect of the triple line, the deformation of the droplet during motion, etc. Much is still to be done to understand the motion of droplets under electrowetting actuation. Lets us mention that the first numerical simulations of Dolatabati et al. [28] agree on some points with Chen's model, but disagree about the effect of the vertical gap δ. However, it is a general observation, according to equations (10.35) and (10.42), that the velocity of motion is proportional to the electrowetting force and inversely proportional to the viscosity.

10.3.3.2 Experimental Results

The variation of the velocity with the square of the voltage predicted by the preceding model has been checked by Pollack et al. [29] for low to moderate values of the applied voltage. There is a very interesting point in Pollack's publication: the velocity is the same when the dimensions are reduced homothetically, i.e., when the ratio δ/r is kept constant. This is exactly what is predicted by equations (10.35) and (10.42).

At large voltages, the electrowetting force saturates and in consequence the velocity of motion is limited by an asymptote (fig. 10.21).

Using the modified Lippmann law based on the Langevin function, equation (10.35) can be recast in the form

$$v_{cov} \approx \frac{\delta}{6\pi\mu r}\gamma_{LG}(\cos\theta_s - \cos\theta_0)La\left(\frac{CV^2}{2\gamma_{LG}(\cos\theta_s - \cos\theta_0)}\right),$$ (10.43)

and equation (10.42) in the form

$$v_{open} = \frac{4h}{5\pi\mu a}\gamma_{LG}(\cos\theta_s - \cos\theta_0)La\left(\frac{CV^2}{2\gamma_{LG}(\cos\theta_s - \cos\theta_0)}\right),$$ (10.44)

where La is the Langevin function, and θ_s the saturation angle. These relations between the droplet velocity and the voltage then have the shape shown in figure 10.21. There is an asymptote at high voltages that represents the upper bound of the velocity. In conclusion, velocity

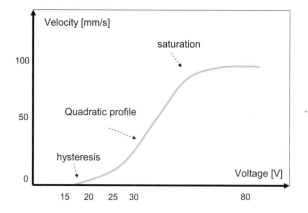

Figure 10.21 Schematic relation between the velocity of motion and the voltage, showing saturation for large actuation potentials.

of motion in electrowetting devices is limited by the saturation of the electrowetting force and very high velocities cannot be expected. This upper limit on drop velocity is typically 15 cm/s.

Experimental observations confirm that the droplet velocity decreases when the viscosity of the liquid increases (fig. 10.22). One expects that there is a viscous limit above which liquids cannot be moved in EWOD microsystems. In biology, viscous (even non-Newtonian) liquids are sometimes used, like polysaccharides, blood, or alginates. Their viscosity increases with the concentration of monomers, dimers, trimers, etc. [30]. For instance, in chemical applications, it has been observed that ionic liquids can be displaced on EWOD chips, but at a small velocity (bottom curve of figure 10.22).

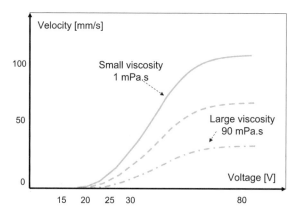

Figure 10.22 Droplet velocity is decreased by increasing the viscosity of the liquid.

Let us finally analyze the effect of surfactants. Remember that in biology and biotechnology, surfactants are currently used for many reasons, the most important being to disperse aggregates and to reduce adhesion on the solid walls. A comparison of the droplet velocity between de-ionized water and biologic buffer is shown in figure 10.23. We observe that hysteretic effects are reduced when surfactants are present in the solution, but the droplet motion is much slower. The reduction of the hysteresis there should be in the case of biologic buffers can be attributed to the

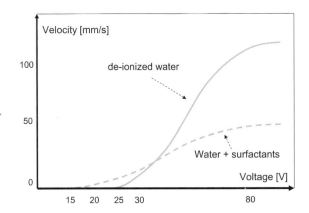

Figure 10.23 Observed velocities of a droplet as a function of applied voltage: velocities are higher when the surface tension is large, except at very low voltage.

reduction of the quantity $\gamma_{LG} \sin \theta_0$ (from equation 10.25), and the reduction of the velocity is due to a smaller value of the solid-liquid surface tension γ_{SL} (from equations 10.33 and 10.39).

10.3.4   Droplet Merging and Division

The merging of two droplets is straightforward. It suffices to move two droplets from opposite sides onto the same electrode. However, the division of one droplet into two is much more complicated. The principle of division is shown in figure 10.24. The droplet is stretched by the electrowetting forces along the electrode line, and pinched on the hydrophobic non-actuated electrode. The surface tension force of the free surface opposes these forces. The balance of these forces determines if the droplet can be split in two.

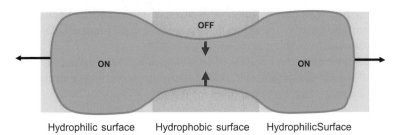

Hydrophilic surface   Hydrophobic surface   HydrophilicSurface

Figure 10.24 Principle of droplet division: division is obtained by a pinching in the central area and a stretching from the two ends.

Numerical investigations with Surface Evolver show that the division of a water droplet is not possible in open EWOD systems, but only in covered EWOD system, if the vertical distance between the two plates is not too large [31]. In the open EWOD system the pinching and stretching forces are not sufficient to dominate the surface tension forces and the droplet escapes laterally from the pinching region, randomly on one side or the other (fig. 10.25). This can be viewed as an example of the double-bubble instability described in Chapter 2.

This is not the case of covered systems, where the droplet can be efficiently separated in two equal daughter droplets – if the vertical distance is not too large. Let us define a characteristic

(a)   (b)

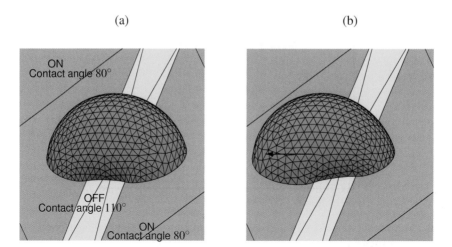

Figure 10.25 (a) at the start of the actuation, a pinching occurs in the middle of the water droplet; (b) droplet escapes randomly on one side.

dimension R representing the droplet volume Vol by

$$Vol = \pi R^2 \delta, \tag{10.45}$$

where δ is the vertical distance between the plates, and define the scaling number λ by

$$\lambda = \frac{\delta}{R} = \delta^{\frac{3}{2}} \left(\frac{\pi}{Vol} \right)^{\frac{1}{2}}. \tag{10.46}$$

It is numerically shown that if λ is sufficiently small, the droplet can be cut into two daughter droplets (fig. 10.26). The critical value of λ depends on the actuated/non-actuated contact angles and the surface tension. The key is that the splitting must happen faster than the droplet can respond to the double-bubble instability.

Let us investigate further the question of droplet splitting. Let us place ourselves in a more general case and ask the question: can a droplet placed between two horizontal parallel planes be cut into two daughter droplets by capillary means?

10.3.4.1 Droplet Stretching

The first idea that comes to mind is to stretch the droplet by capillary forces, in order to elongate a filament that will eventually break. Then the question is reduced to: can a droplet be cut by exerting capillary traction forces on two ends?

Let us start with the artificial situation where a partial splitting has already occurred, in order to facilitate the splitting. The droplet is then supposed divided in two regions linked by a large thread (Fig. 10.27). Because the liquid wets the two substrates (bottom and top plates), capillary forces are exerting traction on the wetted perimeter of the droplet, as sketched in figure 10.28.

Let us follow first a theoretical approach which makes use of the free energy: The total surface energy is

$$E = \gamma_{SL} S_{SL} + \gamma_{LG} S_{LG}, \tag{10.47}$$

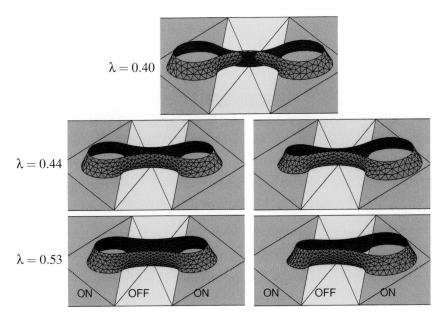

Figure 10.26 Division of a droplet is possible in a covered EWOD device if the ratio of the vertical gap and a characteristic dimension of the droplet (representing its volume) is small enough (the liquid is water, and the contact angles are $80°$, $120°$ and $110°$ respectively for the actuated, non-actuated electrodes and cover plate).

Figure 10.27 Sketch of a droplet placed between wetting walls, "artificially" stretched in two regions. The cover plane has been dematerialized for visualization.

where L denotes the liquid, G the gas and S the solid. Let us make the approximation that the free interfaces are flat. We have with the notations of figure 10.28 that

$$S_{SL} = 2(2\pi R^2 + Lw) \tag{10.48}$$

and

$$S_{LG} = 2h(2\pi R + L). \tag{10.49}$$

Moreover, the liquid volume is

$$V_L = hS_{SL}. \tag{10.50}$$

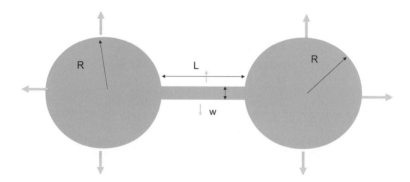

Figure 10.28 Schematic of the initial droplet.

If there is no evaporation, (10.48) implies that

$$dV_L = hdS_{SL} = 0. \tag{10.51}$$

Then, using (10.47)

$$dE = \gamma_{LG}dS_{LG}, \tag{10.52}$$

Using (10.49), and remarking that L is constant, we find

$$dE = 4\pi\gamma_{LG}hdR, \tag{10.53}$$

and

$$\frac{dE}{dR} = 4\pi\gamma_{LG}h > 0. \tag{10.54}$$

The droplet evolves to find its minimum energy morphology. Relation (10.54) indicates that the liquid filament cannot shrink – i.e. R cannot increase – because an increase of R would increase the energy. Hence, the droplet cutting is not possible by exerting capillary forces on the two ends.

We have checked this analysis numerically with Surface Evolver. The starting point is a droplet split in two circular regions, and linked by a thread (fig. 10.29). Using many different values of the parameters h, L, w, and R, cutting was never obtained.

10.3.4.2 Stretching and Pinching

Stretching the droplet is not sufficient to achieve splitting. Let us consider now an additional capillary force which is the pinching of the droplet. The question is then: can a droplet be cut by exerting capillary traction forces on two ends and a pinching in the middle? Figure 10.30 represents a droplet placed between two horizontal planes, with two hydrophilic (lyophilic) regions at both ends and a hydrophobic (lyophobic) region in the middle. In EWOD systems, such an action is obtained by actuating the two end electrodes and not actuating the middle electrode.

The sketch of the capillary forces is shown in figure 10.31. The hydrophilic regions tend to stretch the droplet while the hydrophobic region pinches the middle of the droplet.

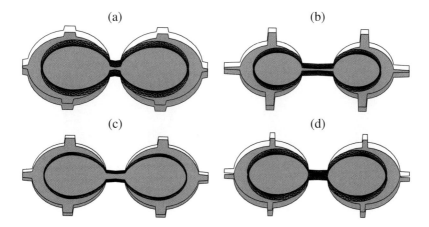

Figure 10.29 Splitting cannot be obtained by varying the different parameters: (a) small L; (b) longer L; (c) small h; (d) small w.

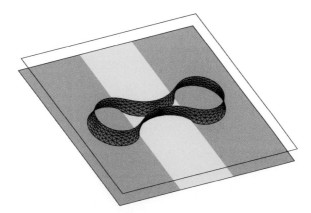

Figure 10.30 Simulation of a droplet pinched in the middle and stretched by the two ends. The cover plane has been dematerialized for visualization. Note that the cover plate is neutral ($\theta = 90°$), hence only the capillary forces on the bottom plate contribute to the splitting.

Let us follow a theoretical approach based on the free energy minimization. The total surface energy is

$$E = \gamma_{S1L}S_{S1L} + \gamma_{S2L}S_{S2L} + \gamma_{LG}S_{LG}, \tag{10.55}$$

using the same notations as above, and where the indices 1 and 2 denote respectively the pinched and rounded regions of the liquid. In the case of EWOD, the actuated solid-liquid surface tension is $\gamma_{S2L} = \gamma_{S2L}^{eff}$. Let us make the approximation that the free interfaces are flat. We have, with the notations of figure 10.31

$$S_{S1L} = 2Lw, \tag{10.56}$$

and

$$S_{S2L} = 2(2\pi R^2). \tag{10.57}$$

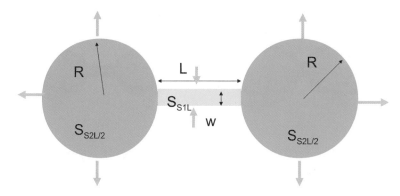

Figure 10.31 Schematic view of the droplet with the pinching in the thread region.

On the other hand, the liquid-gas surface area is

$$S_{LG} = 2h(2\pi R + L).$$

$$(10.58)$$

The liquid volume is

$$V_L = hS_{SL} = h(S_{S1L} + S_{S2L})$$

$$(10.59)$$

If there is no evaporation, (10.59) implies that

$$dV_L = hdS_{SL} = h(dS_{S1L} + dS_{S2L}) = 0.$$

$$(10.60)$$

Differentiating (10.55) yields

$$dE = \gamma_{S1L}S_{S1L} + \gamma_{S2L}dS_{S2L} + \gamma_{LG}dS_{LG},$$

$$(10.61)$$

Using (10.60), we find

$$dE = (\gamma_{S2L} - \gamma_{S1L})dS_{S2L} + \gamma_{LG}dS_{LG},$$

$$(10.62)$$

and noting that L is constant, but w and R vary,

$$dE = (\gamma_{S2L} - \gamma_{S1L})8\pi RdR + \gamma_{LG}4\pi hdR.$$

$$(10.63)$$

Finally

$$\frac{dE}{dR} = 4\pi[(\gamma_{S2L} - \gamma_{S1L})R + \gamma_{LG}h].$$

$$(10.64)$$

The comparison to (10.54) shows that there is now an additional term on the right hand side of (10.64). This time, the sign of dE/dR can be positive or negative depending on the values of the parameters. In order to have splitting, the free energy E must decrease when the radius R increases: $dE/dR < 0$. The condition for splitting is then

$$h < \frac{2R(\gamma_{S1L} - \gamma_{S2L})}{\gamma_{LG}}.$$

$$(10.65)$$

Lipmann's law states that electrowetting changes the value of the solid-liquid surface tension $\gamma_{S2L} = \gamma_{S2L}^{eff}$ (equation 10.1). Using the capillary equivalence for electrowetting, with regions 2 actuated and region 1 not actuated,

$$\cos\theta_0 = \frac{\gamma_{S1G} - \gamma_{S1L}}{\gamma_{LG}}, \tag{10.66}$$

$$\cos\theta_a = \frac{\gamma_{S2G} - \gamma_{S2L}^{eff}}{\gamma_{LG}}. \tag{10.67}$$

By difference, and using the equality $\gamma_{S1G} = \gamma_{S2G}$, we obtain

$$\cos\theta_a - \cos\theta_0 = \frac{\gamma_{S1L} - \gamma_{S2L}^{eff}}{\gamma_{LG}}. \tag{10.68}$$

Substitution in the Berge-Lippmann-Young (BLY) equation yields the relation

$$\gamma_{S1L} - \gamma_{S2L}^{eff} = C\frac{V^2}{2}, \tag{10.69}$$

where C is the specific capacitance, and V the tension. Substituting back in (10.65) yields

$$h < 2R\frac{CV^2}{2\gamma_{LG}} = 2R\xi, \tag{10.70}$$

where ξ is the electrowetting number. Relation (10.70) shows that splitting can be achieved if the vertical gap is sufficiently small and/or the electrowetting number sufficiently high, and/or the radius R sufficiently large. The radius R being approximately of the order of the electrode dimension, EWOD systems with small vertical gaps and/or large electrodes require a smaller voltage to cut the droplet. The fact that a high value of the electrowetting number facilitates the cutting is obvious: the more the droplet is stretched at both ends while it remains pinched in the middle, the easier is the cutting.

Note that relation (10.70) can be rewritten as

$$\frac{h}{R} < 2\frac{CV^2}{2\gamma_{LG}} = 2\xi, \tag{10.71}$$

showing that the electrowetting number – in our case the voltage – can be progressively relaxed as R increases during the splitting. In other words, once the thinning of the liquid junction has started, the splitting is inevitable and goes to its end.

The numerical analysis confirms this theoretical approach (Fig. 10.32). A droplet can be cut in two if the vertical gap between the two plates is sufficiently small. When the vertical gap h is large, no shrinking in the middle of the droplet can be achieved.

10.3.5 Droplet Dispensing

Dispensing denotes the operation consisting of extracting liquid from a reservoir in order to form "digital" droplets. Let us analyze first dispensing in an open EWOD configuration.

(a)   (b)

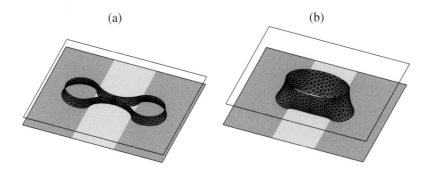

Figure 10.32 Splitting is achieved when h is small (a), but not when h is sufficiently large (b).

10.3.5.1   Open EWOD

It is intuitive that if the width of the door – i.e. the opening in the vertical wall – is small compared to the volume of liquid in the reservoir, it will be impossible to extract a droplet through the door. Let us first characterize the volume of the droplet by the length scale R defined by

$$Vol = \frac{4}{3}\pi R^3,\tag{10.72}$$

where Vol is the volume of the liquid in the reservoir, and let us define the scaling number λ as

$$\lambda = \frac{\delta}{R} = \delta\left(\frac{4\pi}{3Vol}\right)^{1/3},\tag{10.73}$$

where δ is the width of the door. Numerical calculation shows that when $\lambda < 1.05$ it is impossible to dispense any droplet from the reservoir (fig. 10.33). Hence open EWOD systems cannot be used for dispensing because the ratio λ is much smaller than 1 (when the reservoir is filled).

The threshold limit $\lambda = 1$ slightly depends on the liquid. Ionic liquid with a surface tension of 30 mN/m and an actuated contact angle of $50°$ can be dispensed for $\lambda > 0.95$. We reach the same conclusion: open EWOD systems cannot dispense droplets.

10.3.5.2   Covered EWOD

An experimental view of droplet dispensing in a covered EWOD system is shown in figure 10.34. First, a liquid tongue is pulled out of the reservoir on the electrode line by actuating the electrodes (1,2,3 in the figure). Once the tongue is extended, electrode 2 is shut off, resulting in the pinching of the liquid. Finally electrode 0 is actuated to perform a "back pumping," resulting in the detachment of the droplet.

Note that the extrusion of the liquid out of the reservoir is possible in the covered geometry, contrary to the case of an open geometry. A numerical simulation with Surface Evolver shows that the liquid progresses first to the boundary of the first working electrode, then after the actuation of the second electrode, it progresses to the limit of the second working electrode (fig.10.35).

The process leading to dispensing can be analyzed by a numerical approach (fig. 10.36). First, extrusion can be obtained by switching OFF the reservoir electrode and ON the working

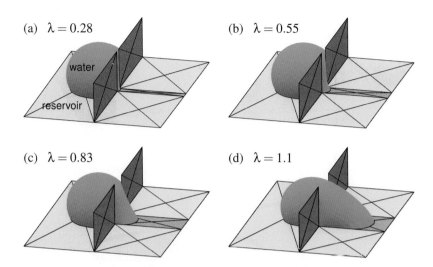

Figure 10.33 Numerical simulation of dispensing water droplets in an open EWOD geometry: dispensing is impossible when the volume of water in the reservoir is too large compared to the width of the "door". The contact angles are 120° in the reservoir, 120° with the Ordyl wall and 60° on the actuated working electrode.

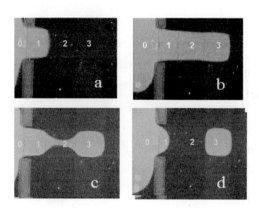

Figure 10.34 Dispensing of a water droplet in a covered EWOD system. (a) electrode 0 is OFF while electrodes 1, 2, and 3 are ON; (b) the liquid has progressed to the extremity of electrode 3; (c) electrode 2 is switched OFF for pinching; (d) electrodes 0 and 1 are set OFF for back pumping and a droplet isolated on electrode 3. (photo CEA-LETI).

electrodes. A liquid "tongue" progresses onto the first non-ctuated electrode. Then, pinching is triggered by switching OFF the cutting electrode and ON the reservoir electrode in order to achieve a back-pumping effect that lowers the liquid pressure and facilitates the pinching. In a final step, the liquid bridge on the cutting electrode gets thinner and eventually breaks, and a droplet is dispensed.

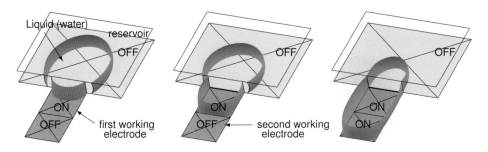

Figure 10.35 Dispensing from a reservoir in the covered EWOD geometry: the liquid invades first the first working electrode. After the second working electrode is turned ON, the liquid progresses further.

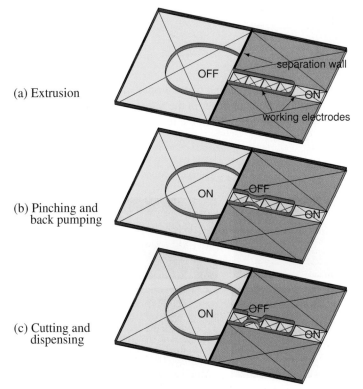

Figure 10.36 The three main steps for dispensing: (a) extrusion by switching off the reservoir electrode and on the working electrodes; (b) pinching and back pumping by switching on the reservoir electrode, and off the cutting electrode; (c) when cutting is obtained, the droplet is moved on the working electrodes away from the reservoir.

10.3.6   Coupling Between Covered and Open EWOD Systems

In light of the preceding sections, we can analyze the advantage of each type of EWOD system: open systems move droplets faster, which is advantageous when using the droplet as a carrier fluid; besides, they are directly accessible from the top, which facilitates manual or robotic intervention if needed. On the other hand, covered systems are required for the dispensing and division of droplets. Coupling the two systems by designing an EWOD chip partly open and partly covered has been thought of [32]. Figure 10.37 shows such a coupled EWOD system, with the open region at the left and the covered region at the right of the picture. In the covered region, the ground electrode is the ITO cover (transparent), and in the open region, the ground electrode is a catena running across the entire device.

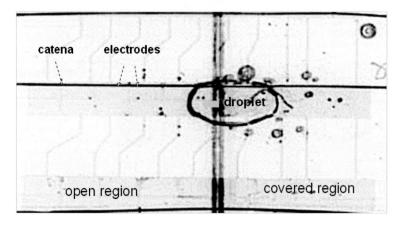

Figure 10.37 View of a coupled EWOD system: the open region is at the left and the covered region at the right of the picture. A catena runs all the way across the two regions (the catena is the zero potential electrode for the open EWOD system).

This concept relies on the fact that motion between a covered and an open region – and backwards – is possible under electrowetting actuation. We analyze numerically the possibilities of such a motion with Surface Evolver. It is emphasized again that this approach assumes that the capillary and electrowetting forces are dominant over inertial and viscous forces (i.e., the Weber number and the Ohnesorge and Capillary numbers are small). Then, using the Evolver results, and using the Laplace law, a very simple condition for droplet motion is derived.

Figure 10.38 shows the motion of a water droplet from a covered to an open region and vice-versa.

The numerical model shows that, in the case of a droplet of water in air, motion from an open to a covered region is possible provided that the vertical gap in the covered region is not too small, and motion from a covered to an open region is possible if the vertical gap is not too large. The contact angle with the upper plate θ_t is an important parameter of the motion; below $90°$ the droplet will have difficulties exiting the covered region towards the open region due to hydrophilic grip on the upper plate; above $120°$, the motion towards the covered region will be increasingly difficult due to hydrophobic repulsion on the upper plate. Another condition is that the electrode size be adapted to the liquid volume.

Figure 10.39 shows a diagram of the general pressure evolution in the droplet during a cycle. A cycle is defined by motion from the open region (noted 3D/open) to the covered region (noted 2D/covered) and back. Suppose the droplet starts from the 3D/open region (top left, figure 10.39). Electrodes in the 3D/open region are not actuated, whereas the electrodes in the 2D/covered region are actuated. The droplet then moves towards the covered region (step 1 in the figure). When the droplet has crossed the boundary and is located in the 2D/covered region (bottom right), the actuation is switched off and the droplet internal pressure suddenly increases (step 2). The actuation in the 3D/open region is then switched on and the droplet moves back

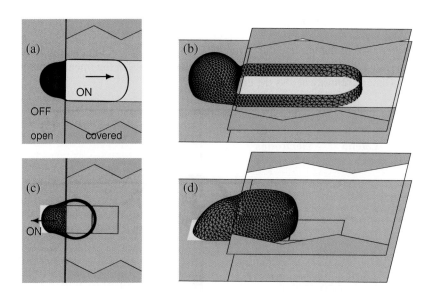

Figure 10.38 (a) Droplet moving from the open to the covered region; (b) close-up on the droplet shape; (c) droplet moving from the covered to the open region; (d) close-up on the droplet shape (the cover plate has been partly dematerialized for better visualization).

to this latter region (step 3). When the droplet is entirely located in the 3D/open region (bottom right), the actuation is switched off and the droplet recovers its initial conditions (step 4).

This analysis leads to the conclusion that a motion from one region to the other is accompanied by a monotone decrease in pressure. The condition for the motion from open to covered region is

$$P_{open,unactuated} > P_{covered,actuated}, \tag{10.74}$$

and conversely, for the motion from covered to open region

$$P_{covered,unactuated} > P_{open,actuated}. \tag{10.75}$$

Droplet pressure in each region can be calculated by using the Laplace law. For a drop of volume V, confined between two horizontal plates separated by a distance δ, internal pressure is given by

$$P_c = \gamma \left(\frac{-\cos\theta_t - \cos\theta_b}{\delta} + \sqrt{\frac{\pi\delta}{V}} \right), \tag{10.76}$$

where θ_t and θ_b are the contact angles with the top and bottom plates. In (10.50), the first term of the right hand side corresponds to the vertical curvature, and the second term to the horizontal curvature. For a sessile drop of the same volume (3D/open configuration), we obtain, again using the Laplace law,

$$P_0 = 2\gamma_0 \left(\frac{3V}{\pi(2 - 3\cos\theta + \cos^3\theta)} \right)^{-1/3}, \tag{10.77}$$

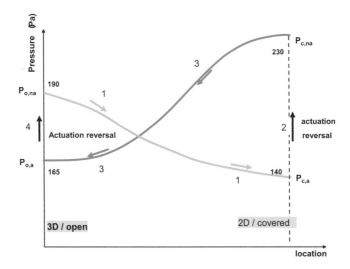

Figure 10.39 Droplet internal pressure during a cycle: the motion of the droplet corresponds to a decrease in pressure. Increase in pressure is obtained by suppressing the electric actuation when the droplet is on either side of the device.

where θ is the contact angle with the substrate. We are then left with two conditions derived from (10.76) and (10.77); the first one concerns the motion from the open to the covered region $P_0(\theta = \theta_0) > P_c(\theta_b = \theta_a)$, so that

$$2 \left(\frac{3V}{\pi(2 - 3\cos\theta_0 + \cos^3\theta_0)} \right)^{-1/3} > \left(\frac{-\cos\theta_a - \cos\theta_t}{\delta} + \sqrt{\frac{\pi\delta}{V}} \right). \qquad (10.78)$$

The second one is for the opposite motion from the covered to the open region $P_c(\theta_b = \theta_0) > P_0(\theta = \theta_a)$, so that

$$2 \left(\frac{3V}{\pi(2 - 3\cos\theta_a + \cos^3\theta_a)} \right)^{-1/3} < \left(\frac{-\cos\theta_0 - \cos\theta_t}{\delta} + \sqrt{\frac{\pi\delta}{V}} \right), \qquad (10.79)$$

where θ_0 is the non-actuated contact angle with the solid substrate and θ_a the actuated contact angle.

From this analysis, it has been shown that the motion from an open to a covered region of an EWOD micro-device – and conversely – is closely related to the difference of drop internal pressure between the departing and arriving regions.

10.3.7   Special Electrodes – Jagged Electrodes and Star-shaped Electrodes

10.3.7.1   Crenellated (Jagged) Electrodes

In electrowetting microsystems, micro-drops are displaced "digitally" on rows of electrodes. We have seen that the electrowetting force can move a droplet from one electrode to the next one. But this applies only if there is an overlap of the liquid on the next electrode. Micro-fabrication imposes a gap separating the electrodes. This gap is usually on the order of 5 to 30 μm, depending on the precision of the lithography process, compared to an electrode size of the order of 400 to 800 μm. This gap creates a permanent hydrophobic region between two neighboring electrodes. If the droplet has a volume such that it is "contained" by the boundaries of the electrode, then it cannot move to the next electrode when the latter is actuated.

In order to remedy the problem, jagged or crenellated electrodes have been designed as shown in figure 10.40. The idea behind such a design is that the droplet contact line on the teeth of one electrode overlaps the teeth of the next electrode. As soon as the next electrode is actuated, electro-capillary forces act to produce the motion of the droplet. Such jagged electrodes require more complicated micro-fabrication, but are very efficient for droplet motion, provided the geometry and dimension of the teeth are correctly designed.

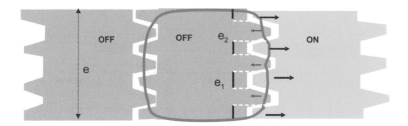

Figure 10.40 Sketch of the crenellated electrodes and position of the droplet. The force exerted by the neighboring electrode at the beginning of the displacement on the advancing contact line is symbolized by arrows.

We show that, at the very beginning of the motion, the electrowetting force on a droplet is proportional to the length of the contact line located on the neighboring actuated electrode. This result proves that the electrode's jagged boundary plays a key role in electrowetting actuation, and that its design must be carefully considered for drop motion to commence. We then investigate the position of the contact line on a jagged electrode. From a theoretical standpoint, this position is related to the theory of partial/total pinning on wettability boundaries, pioneered by de Gennes [32] and Ondarçuhu [33]. We first review the theory and then show how a criterion for determining the size of the teeth can be derived.

We analyze the electrowetting force on the droplet at the very beginning of the motion. The electrowetting force is due to the Maxwell stress tensor and, because the system is static at the onset of motion, can be translated as a capillary effect using the Lippmann-Young law. It has been established [31] that the electrowetting force during full motion from one electrode to the next is given by

$$F_{motion} = e\gamma(\cos\theta_a - \cos\theta_{na}), \tag{10.80}$$

where e is the width of the electrode, and θ_a and $\cos\theta_{na}$ are the actuated and non-actuated contact angles. In figure 10.40, we show the position of the contact line at the very beginning of the motion. In this case, the electrowetting force is

$$F_{start} = e_1\gamma\cos\theta_a + e_2\gamma\cos\theta_{na} - e\gamma\cos\theta_{na}, \tag{10.81}$$

where e_1 and e_2 are the total cross dimensions corresponding to the part of the contact line on the actuated and non-actuated electrodes. The first term on the right of (10.81) is the pulling force exerted on the liquid by the actuated neighbor electrode, and the second term is the force on that part of the contact line located on the non-actuated electrode (usually $\cos\theta_{na} < 0$). The last term of (10.81) corresponds to the receding contact line. Note that

$$e = e_1 + e_2. \tag{10.82}$$

After substitution in (10.81), we find

$$F_{start} = e_1\gamma(\cos\theta_a - \cos\theta_{na}). \tag{10.83}$$

Comparison of (10.80) and (10.83) shows that at the beginning of the motion, the force on the droplet is only

$$F_{start} = \frac{e_1}{e}F_{motion}. \tag{10.84}$$

We verify that if $e_1 = 0$, i.e. there is no part of the contact line overlapping on the neighboring electrode, then no electrowetting force acts on the droplet. Relation (10.84) shows that it is important to have at rest a significant part of the contact line overlapping the neighboring electrode, for motion to occur upon actuating it.

A comparison between experiment and model is shown in figure 10.41, where a droplet of water is placed on a asymmetrical electrode of a covered EWOD system.

(a)   (b)

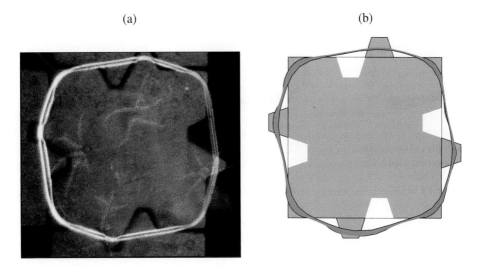

Figure 10.41 Comparison between experiment (a) and calculation (b) for a water droplet in a covered EWOD system: the interface is stretched out on the actuated electrodes and pushed inwards on the non-actuated electrodes.

The notion of elasticity of the triple line has been introduced by de Gennes and Ondarçuhu [32,33]. It is related to the surface tension and estimates the aptitude of an interface to follow a jagged contour (fig. 10.42).

The motion of a droplet overlapping a jagged electrode is shown in figure 10.43. It illustrates the preceding reasoning that states that the "starting" force is only a fraction of the full force during the motion.

(a)   **(b)**

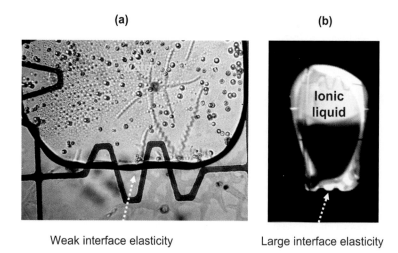

Weak interface elasticity   Large interface elasticity

Figure 10.42 Water has a relatively rigid interface (a), whereas an ionic liquid shows a more elastic interface (b).

(a)   (b)   (c)

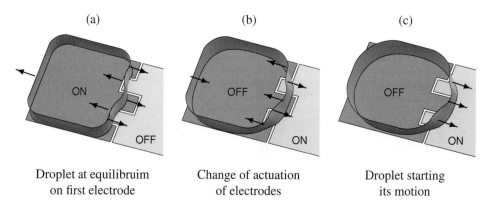

Droplet at equilibruim   Change of actuation   Droplet starting
on first electrode   of electrodes   its motion

Figure 10.43 Numerical modeling of the motion of a droplet in a covered EWOD system with crenellated electrode. The top cover has been dematerialized for clarity. (a) The droplet is at equilibrium, maintained on the first electrode. (b) The actuation has been shifted, and the resulting force on the droplet pulls the right triple line towards the second electrode. (c) The droplet accelerates its motion as soon a large part of the triple line has reached the actuated electrode.

10.3.7.2 Star-shaped Electrodes

In a general manner, star-shaped electrodes are used to maintain a droplet in a precise position. Solid substrates of EWOD microsystems are micro-fabricated using extra care to make them as smooth as possible. Surface defects can lead to unwanted pinning, resulting in the malfunctioning of the micro-chip. But a consequence of the smoothness of the surface is that micro-drops, if not anchored by a boundary line, may not always be positioned at the same location on the surface. This is especially the case in the reservoir of an EWOD device. The droplet may drift until it finds an anchored position by pinning to a singular point or to a boundary line. Hence,

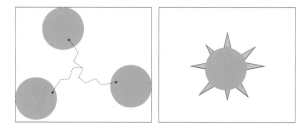

Figure 10.44 Left: on a smooth surface, a droplet may drift until it pins on a boundary; the free energy of the droplet is the same for any of the shown locations; right: a star-shaped electrode centers the droplet.

star-shaped electrodes have been designed to maintain a droplet at a given location (fig. 10.44).

The principle of drop centering by star-shaped electrodes is intuitive; it is similar to that of the preceding section where opposite electro-capillary forces are created by using crenellated electrodes. When the star-shaped electrode is actuated, the resultant of the electro-capillary forces is zero provided that the droplet is positioned at the center of the electrode (fig. 10.45).

A numerical simulation using the Surface Evolver confirms this analysis (fig. 10.46). A sessile droplet located on the plane is displaced towards the center by the actuation of the star-shaped electrode. After release of the actuation, it regains the same spherical shape, but is now positioned at the center of the star-shaped electrode.

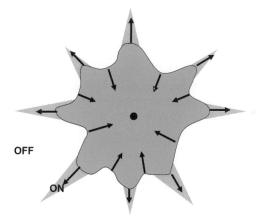

Figure 10.45 Droplet artificially pinned on a central position by the use of a star-shaped electrode. Any deformation of the triple contact line increases the free energy of the system.

10.3.7.3 Spike-shaped Electrodes

In the reservoir of an EWOD microsystem, the volume of liquid decreases each time a microdrop is dispensed into the system. Thus, the liquid volume of the reservoir progressively decreases. It is essential that the liquid be positioned right in front of the first "working" electrode of the microsystem, or else dispensing liquid may be stalled. In the geometry of droplet dispensing, a spike-shaped electrode can be used to "center" the droplet in front of the "opening

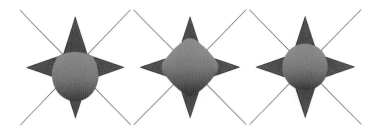

Figure 10.46 Surface Evolver simulation of a droplet on a star-shaped electrode. Left: the droplet is initially positioned arbitrarily on the plane; middle: when the star-shaped electrode is actuated, the droplet moves in such a way that it maximizes its contact surface on the actuated electrode; right: after the actuation is turned off, the droplet regains its spherical shape and is positioned at the center of the star-shaped electrode.

gate" in the vertical wall separating the reservoir from the "working" electrodes. In figure 10.47, we have numerically simulated how a droplet at first not positioned at the right location can be re-centered by successively switching on and off a spike electrode.

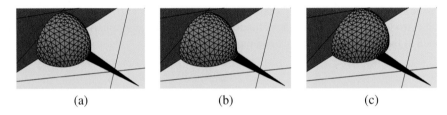

(a)   (b)   (c)

Figure 10.47 Simulation of a droplet being moved to the desired location by the successive actuation/de-actuation of a spike-shape electrode: (a) Droplet "de-centered," (b) actuation of the spike-electrode brings the droplet towartoward the center, (c) after de-actuation of the electrode, the droplet regains its original shape. Operation (a), (b), (c) are equivalent to a translation of the droplet along the vertical wall.

From a physical standpoint, the effect of the spike-shaped electrode is schematized in figure 10.48. When the droplet has an asymmetrical position relative to the electrode, and the electrowetting actuation is turned on, there is a resultant of the capillary forces on the triple line parallel to the vertical wall. This force translates the droplet until its position becomes symmetrical and the resultant vanishes.

10.3.7.4   Half Star-shaped Electrodes

Spike-shaped electrodes might not be sufficient for adequate dispensing from a large reservoir. The droplet may have escaped from the region above the spiked electrode. Half star-shaped electrodes then constitute a good solution (fig. 10.49). Even if, for some reason, the droplet has shifted away from the dispensing gate, such an electrode brings back the droplet into position.

Figure 10.50 shows the liquid in the reservoir positioned on the star-shaped electrode (after it has been de-actuated). The liquid is correctly placed for the dispensing of the next droplet through the opening gate at the top.

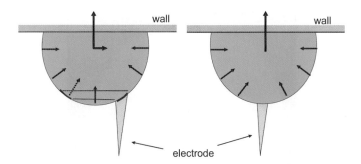

Figure 10.48 Sketch of the forces on the contact line: in an asymmetrical position relative to the electrode, there is a resultant capillary force parallel to the vertical wall. In the symmetrical position, this resultant vanishes and the droplet is at equilibrium.

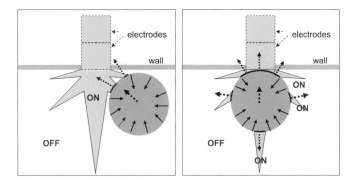

Figure 10.49 Sketch of the resultant of the forces when the drop in the reservoir has moved away from the opening gate. The star-shaped electrode brings back the liquid towards the opening gate.

10.3.7.5   Reservoir Electrodes

If no precaution is taken, the liquid in the reservoir can move away from the opening gate after the dispensing of a droplet. In this case, it is impossible to dispense the next droplet. This situation is similar to a pump that is drained. This effect is due to the curvature of the triple contact line between the corners of the gate (fig. 10.51). This analysis is confirmed by an Evolver calculation. Figure 10.52 shows the location of the droplet interfaces in the two cases (reservoir electrode ON or OFF). The bulging out of the interface in the OFF regime is extremely small.

Different solutions have been proposed to avoid this drawback. Usually an auxiliary electrode and an overlap of the first working electrode inside the reservoir are used. The basis for the design of the auxiliary reservoir electrode is similar to the design of half star-shaped electrodes. Also, in order for the liquid in the reservoir to overlap on the first working electrode, the first working electrode is crenellated as shown in figure 10.53.

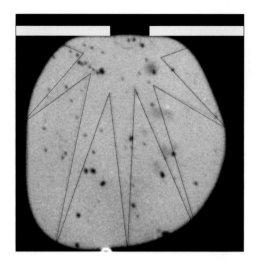

Figure 10.50 Verification of the role of the star-shaped electrode: the droplet has been pulled back towards the Ordyl wall by actuation of the star-shaped electrode; then the star-shaped electrode is switched off and the system is ready for dispensing a new droplet. (Courtesy CEA-LETI).

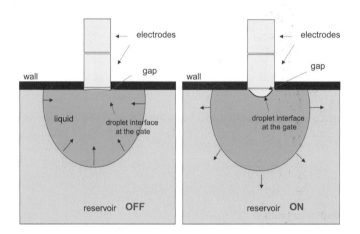

Figure 10.51 Sketch of a poorly designed reservoir: the reservoir drop cannot be dispensed on the working electrodes: when the reservoir electrode is OFF, the droplet internal pressure makes the droplet slightly bulge at the "door", but the overlap with the first "working" electrode is very small or even not existing (if there is a small gap). When the electrode is turned ON, the internal pressure decreases and the interface at the door bulges inside.

10.3.8   General EWOD Architecture

10.3.8.1   System Architecture

In the preceding sections we have investigated the basic manipulations of droplets in EWOD systems. These manipulations involved only a few electrodes. In this section, we give some insights into the more complex architecture of complete EWOD micro-devices. Usually EWOD

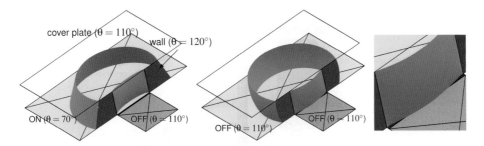

Figure 10.52 Simulation of the shape of the droplet in the reservoir. Left, the bottom electrode of the reservoir is actuated and the interface at the "door" bulges inside. Middle, the bottom electrode is switched off, and there is a very slight bulging of the interface at the door. Right, a close-up of the bulge, showing the bottom contact line slightly moving into the outside electrode. The maximum penetration of the interface depends on the dimension of the door and the surface tension of the liquid (here, for water and a door of 300 μm, the maximum bulging is 3 μm).

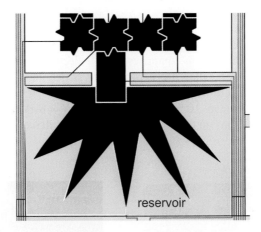

Figure 10.53 Design of an efficient dispensing system.

biochips include a few reservoirs containing the different reagents and buffers, and a sufficient number of electrodes to perform the required biological and chemical operations.

Figure 10.54 shows a typical biologic EWOD chip with its main, secondary, and dump reservoirs, and the electric wiring. The figure especially shows the complexity of the wiring. The electrodes are most of the time individually addressed. However, some electrodes are often difficult to connect to an electric line; it is also very costly in terms of microfabrication to have two levels of wiring in the substrate. Hence, a technique called "multiplexing" is used to address groups of electrodes without losing their individual functionality. We discuss this technique in the following section.

10.3.8.2   Multiplexing

EWOD chips require a paving of the substrate with electrodes. Some EWOD applications require a large number of electrodes and reservoirs. Addressing each one of the electrodes individually is not always possible if the wires are all in the same plane inside the substrate.

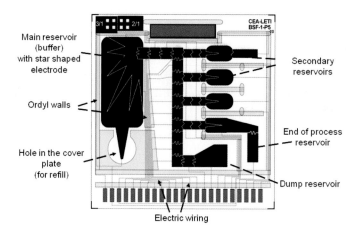

Figure 10.54 Covered EWOD biochip from LETI.

However, depending on the travel scheme of the droplets on the chip, some electrodes can be put on the same electric line [34-36]. Consider the design of an open EWOD system with catenaries (fig. 10.55). The system comprises $M = 5$ electrodes in a row and $N = 3$ rows. Droplets can be individually handled if it is possible to switch the catenaries to the voltage 0 or V and the electrodes to 0 or V. This would mean having $N * M = 15$ independent electric wires. The multiplexing of figure 10.55 requires only $M + N = 8$ electric switches. Of course, multiplexing brings limitations on the sequence of operations that can be realized at the same time.

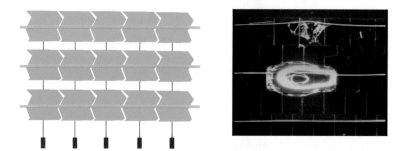

Figure 10.55 Left: Multiplexing scheme for an open EWOD system with catenae: number of electrodes in a row is $M = 5$, number of rows $N = 3$, number of electrical lines $M + N = 8$, total number of electrodes N*M=15. Right: EWOD system with multiplexing. Photograph Ph. Dubois CEA-LETI.

Multiplexing is very efficient for systems doing droplet manipulations in parallel. An example of such a microsystem is that of Moon *et al.* [37]. It realizes the generation of multiple droplets from a reservoir and parallel in-line sample purification of a carrier fluid containing proteins.

10.4   Examples of EWOD in Biotechnology – Cell Manipulation

Cells – if not too adherent to the substrate – can be moved together with the droplet. Applications of digital microfluidic applications have been done for cell separation [38,39] and cell lysis followed by PCR amplification for DNA recognition [40].

To perform the biologic protocol, EWOD is often combined with another technique for the specific transport of cells, such as dielectrophoresis (DEP) [38], optoelectronic tweezers OET [39] or magnetic beads [40,41]. The EWOD protocol is programmed to displace and manipulate the carrier fluid, while the adjoined technique is aimed at manipulating each cell. In the following, we give the examples of combined EWOD/DEP, EWOD/OET and EWOD/magnetic beads manipulation.

10.4.1   DEP and EWOD

DEP and EWOD techniques can be combined to concentrate particles and cells. The schematic of the process is shown in fig. 10.56. The method described here has been developed by Fan and colleagues [38]. In a covered EWOD configuration – because droplet division will be required later in the process – the central electrode is replaced by an array of parallel thin electrodes; the dispersed cells are successively pulled by attractive dielectrophoresis from one electrode to the other, towards one end of the droplet. After that, droplet separation is triggered by EWOD, and a concentrated solution is obtained.

Let us recall that the dielectrophoretic force is given by the expression [42]

$$F_{DEP} = 2\pi R_H^3 \varepsilon_f Re(f_{CM}) \nabla^2 |E_{RMS}|^2, \tag{10.85}$$

where R_H is the (hydraulic) radius of the particle or cell, E the electric field (E_{RMS} is the root mean square electric field), ε_f the liquid permittivity and f_{CM} the Clausius-Mossotti factor. This factor depends on the frequency according to

$$f_{CM} = \frac{\varepsilon_p * - \varepsilon_f *}{\varepsilon_p * + 2\varepsilon_f *}, \tag{10.86}$$

where the frequency-dependant electric permittivities are given by

$$\varepsilon_p * = \varepsilon_0 \varepsilon_p - j \frac{\sigma_p}{2\pi f}, \tag{10.87}$$

$$\varepsilon_f * = \varepsilon_0 \varepsilon_f - j \frac{\sigma_f}{2\pi f}, \tag{10.88}$$

where σ_p and σ_f are the electric conductivities of the particles and fluid respectively, f is the field frequency and j is the square root of -1. When the real part of f_{CM}, or $Re(f_{CM})$, is greater than zero, the dielectrophoretic force F_{DEP} attracts particles toward high field strength regions, which is referred to as positive – or attractive – DEP. Conversely, a negative $Re(f_{CM})$ generates negative – or repulsive – DEP, which repels particles from high field strength regions. Very often (for cells for example) there is a change between negative and positive DEP with the frequency.

10.4.2   EWOD and OET

Shah and colleagues [39] have demonstrated that EWOD can also be combined with OET, to develop a special OET, called LOET (for lateral-field OET). Again EWOD is used for manipulating the carrier fluid. In contrast to the preceding application using DEP, cells are individually

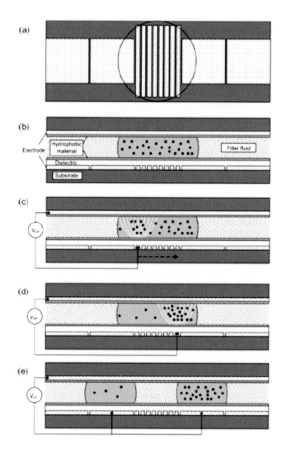

Figure 10.56 Manipulation of cells using EWOD and DEP: Cells are progressively pulled into the right part of the droplet, and concentrated by droplet division. Reprinted with permission from [38]; ©RSC 2008.

manipulated by OET. The advantage is that a single cell can be isolated in a droplet. The principle is shown in figure 10.57. The LOET uses interdigitated electrodes overlapping the EWOD actuation electrodes. It has been demonstrated that HeLa cells have been isolated in separated droplets with this method.

10.4.3   EWOD and Magnetic Beads

EWOD and magnetic force can also be combined. A demonstration has been made by Yizhong Wang and colleagues [41]. Recently, the principle of gene analysis in EWOD systems using magnetic beads has been established [40]: first, cell lysis is realized in a sessile droplet in an open EWOD system; second functionalized super-paramagnetic beads are added to capture the specific mRNA; then the targets are extracted by magnetic force out of the "mother" droplet. Finally a PCR is done to amplify the mRNA, allowing for detection.

The demonstration of extracting beads and targets from the droplet is shown in figure 10.58. The figure shows an open EWOD device (with catena) and a droplet containing magnetic beads aggregated by a mini-magnet placed below the substrate. If the electrowetting forces and the

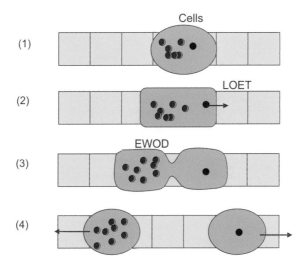

Figure 10.57 Principle of cell manipulation combining EWOD and OET [38] (LOET stands for lateral-field OET).

magnetic forces are sufficient, the magnetic aggregate separates from the droplet.

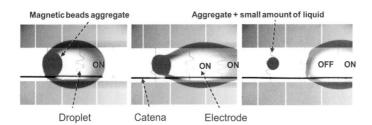

Figure 10.58 Combination of magnetic forces exerted on an aggregate of magnetic beads and electrowetting forces exerted on a conductive liquid droplet. In this case, the magnetic forces are sufficiently strong to pin the aggregate and the electrowetting forces are sufficiently large to move the droplet, leaving behind the aggregate with a small amount of liquid (photo CEA-LETI).

10.5   Examples of Electrowetting for Optics – Tunable Lenses and Electrofluidic Display

Electrowetting is a technique well adapted to optical applications [43-45]. In the following sections, we indicate applications of electrowetting for tunable liquid lenses and for screen display.

10.5.1 Tunable Lens

Electrowetting can be used to build tunable lenses. Berge, one of the founders of EWOD with the eponymous BLY equation, is at the origin of the development of such lenses (www.varioptic.com). The principle is directly based on the electrowetting principle (fig. 10.59) with an aqueous, conductive phase and an organic, nonconductive phase. When not actuated, because the electrodes surface is treated hydrophobically, the oil phase partially wets the electrodes and the interface between oil and water is concave (relative to the oil phase). When actuated, the contact angle of the water with the electrodes decreases, and the interface shape changes progressively with the magnitude of the actuation voltage. Above a certain voltage, the interface becomes convex (relative to the oil phase), and the focal point can be adjusted by tuning the magnitude of the applied voltage.

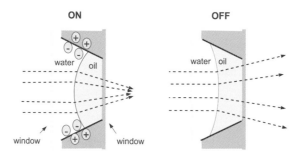

Figure 10.59 Schematic of a tunable lens actuation: left, the electrodes are actuated and the interface is convex (respectively to the oil phase); right, the electrodes are not actuated and the interface is concave (respectively to the oil phase).

10.5.2 Electrowetting Display

Electronic display can be designed using electrowetting [46]. The principle is shown in figure 10.60: a screen is composed of rectangular windows separated by a solid grid. Each window is a very/super-hydrophobic EWOD electrode which, when switched off, is covered by a fluorescent oil film. Once an electrode actuated, the oil film is repelled in the corners, and replaced by water. A digital screen is thus realized.

10.6   Conclusion

In this chapter, the principle of electrowetting and especially EWOD has been presented. It has been shown that such a principle could move droplets digitally on an electrode-paved substrate, and realizes the basic operations required to manipulate droplets. Extremely small volumes of liquids can be used and a low electric power is sufficient.

Applications of such a technique are twofold: biotechnology and optofluidics. In biotechnology, fast PCRs have been realized showing the same sensitivity as the conventional PCRs. New fields of application are sought in cellular biomicrofluidics. In optofluidics, many applications are emerging, from electronic paper [47] to optical switches, and to electrowetting and electrofluidic displays.

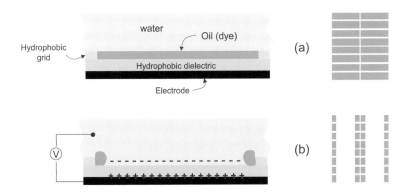

Figure 10.60 Principle of electrowetting display: (a) fluorescent oil totally wets the hydrophobic electrode (electrode off); (b) upon actuation, oil is repelled in the corners, and the water wets the electrode.

10.7 References

[1] A. Wixforth, J.P.Kotthaus, and G. Weimann, "Quantum Oscillations in the Surface-Acoustic-Wave Attenuation Caused by a Two-Dimensional Electron System," *Phys. Rev. Lett.* **56**, p. 2104, 1986.

[2] http://www.advalytix.com/advalytix/applications_1329.htm.

[3] J. Berthier. *Microdrops and Digital Microfluidics*. William-Andrew Publishing. 2008.

[4] G. Lippmann, "Relations entre les phénomènes électriques et capillaires," *Ann. Chim. Phys.* **5**, p. 494, 1875.

[5] B. Berge, "Electrocapillarity and wetting of insulator films by water," *Comptes rendus de l'Académie des Sciences, Séries II*, **317**, pp. 157–163, 1993.

[6] P. Perrot. *A to Z of Thermodynamics*. Oxford University Press, 1998.

[7] R.P. Feynman, R.B. Leighton, M. Sands, *The Feynman lectures on physics II,* Addison-Wesley, Section 8-2, 1977

[8] T.B. Jones, J.D. Fowler, Young Soo Chang, Chang-Jin Kim, "Frequency-based relationship of electrowetting and dielectrophoretic liquid microactuation," *Langmuir* **19**, pp. 7646–7651, 2003.

[9] T.B. Jones, "An electromechanical interpretation of electrowetting," *J. Micromech. Microeng.* **15**, pp. 1184–1187, 2005.

[10] Kwan Hyoung Kang, "How electrostatic fields change contact angle in electrowetting," *Langmuir* **18**, pp. 10318–10322, 2002.

[11] Jun Zeng, T. Korsmeyer, "Principles of droplet electrohydrodynamics for lab-on-a-chip," *Lab Chip* **4**, pp. 265–277, 2004.

[12] M. Vallet, B. Berge, "Limiting phenomena for the spreading of water on polymer films by electrowetting," *Eur. Phys. J. B* **11**, p 583, 1999.

[13] T.A. Driscoll, L.N. Trefethen, *Schwarz-Christoffel mapping,* Cambridge University Press, 2002.

[14] S. Thamida, H.-C. Chang, "Nonlinear electrokinetic ejection and entrainment due to polarization at nearly insulated wedges," *Phys. of Fluids* **14**, p. 4315, 2002.

[15] H.J.J. Verheijen, M.W.J. Prins, "Reversible electrowetting and trapping of charge: model and experiments," *Langmuir* **15**, p. 6616, 1999.

[16] V. Peykov, A. Quinn, J. Ralston, "Electrowetting: a model for contact angle saturation,"

J. Colloid Polym. Sci. **278**, pp. 789–793, 2000.

[17] A. Quinn, R. Sedev, J. Ralston, "Influence of the electrical double layer in electrowetting," *J. Phys. Chem. B.* **107**, pp. 1163–1168, 2003.

[18] J. Berthier, Y. Fouillet, Ph. Clementz, O. Raccurt, D. Jary, P. Claustre, and C. Peponnet, "An analytical model for the prediction of microdrop extraction and splitting in digital microfluidics systems," *Proceedings of the 2005 NSTI Nanotech Conference,* Anaheim, Ca, USA, May 8-12, 2005, pp. 664-667.

[19] S.J. Dodd, "A deterministic model for the growth of non-conducting electrical tree structures," *J. Phys. D: Appl. Phys.* **36**, pp. 129–141, 2003.

[20] S.M. Lebedev, O.S. Gefle, Y.P. Pokholkov, E. Gockenbach, H. Borsi, V. Wasserberg, N. Abedi, and J. Szczechowski, "Influence of high-permittivity barriers on PD activity in three-layer dielectrics," *J. Phys. D: Appl. Phys.* **37**, pp. 3155–3159, 2004.

[21] J. Berthier, Ph. Dubois, Ph. Clementz, P. Claustre, C. Peponnet, Y. Fouillet, "Actuation potentials and capillary forces in electrowetting based Microsystems," *Sensors and Actuators*, **134**(2), pp. 471–479, 2007.

[22] H. Moon, S. K. Cho, R.L. Garrell, and C-J Kim, "Low voltage electrowetting-on-dielectric," *Journal of Applied Physics* **92**(7), pp. 4080-4087, 2002.

[23] S.K. Cho, H. Moon, and C-J Kim, "Creating, transporting, and merging liquid droplets by electrowetting-based actuation for digital microfluidics circuits," *Journal of Micro-electromechanical systems* **12**(1), pp. 70–80, 2003.

[24] N. Moumen, R.S. Subramanian, J. McLMaughlin, "The Motion of a Drop on a Solid Surface due to a Wettability Gradient," *Langmuir* **21**, pp. 11844-11849, 2003.

[25] M.K. Chaudhury, G.M. Whitesides, "How to make water run uphill," *Science* **256**, pp. 1539–1541, 1992.

[26] J. Berthier, P. Silberzan. *Microfluidics for Biotechnology*. Second Edition, Artech House, 2010.

[27] Chao-Yi Chen, E. F. Fabrizio, A. Nadim, J. D. Sterling, "Electrowetting-based microfluidics devices: design issues," *2003 Summer Bioengineering Conference*, Sonesta Beach Resort in Key Biscayne, Florida, USA, June 25-29, 2003, pp. 1241-1242.

[28] A. Dolatabadi, K. Mosheni, A. Arzpeyma, "Behavior of a moving droplet under electrowetting actuation: numerical simulation," *Canadian Journal of Chemical Engineering* **84**, pp. 17–21, February 2006.

[29] M.G. Pollack, A.D. Shenderov, R.B. Fair, "Electrowetting-based actuation of droplets for integrated microfluidics," *Lab Chip* **2**, pp. 96–101, 2002.

[30] T.P. Lodge, "Reconciliation of the molecular weight dependence of diffusion and viscosity in entangled polymers," *Physical Review Letters*, **83**(16), pp. 3218-3221, 1999.

[31] J. Berthier, Ph. Clementz, O. Raccurt, D. Jary, P. Claustre, C. Peponnet, and Y. Fouillet, "Computer aided design of an EWOD microdevice," *Sensors and Actuators A: Physical* **127**, pp. 283–294, 2006.

[32] J. Berthier, Ph. Clementz, J-M Roux, Y. Fouillet, C. Peponnet, "Modeling microdrop motion between covered and open regions of EWOD microsystems," *Proceedings of the 2006 Nanotech Conference*, Boston, USA, May 7-11, 2006.

[33] P-G de Gennes, F. Brochard-Wyart, D. Quèrè, *Drops, bubbles, pearls, waves*. Springer, New-York, 2004.

[34] T. Ondarçuhu, "Total or partial pinning of a droplet on a surface with chemical discontinuity," *J. Phys. II France* **5**, pp. 227–241, 1995.

[35] Fei Su, Krishnendu Chakrabarty, R.B. Fair, "Microfluidics-based biochips: Technology issues, implementation platforms and design-automation challenges," *IEEE Transactions*

on Computer-Aided Design and Integrated Circuits and Systems **25**(2), Feb. 2006, pp. 211-223.

[36] M. Armani, S. Walker, B. Shapiro, "Modeling and control of electrically actuated surface tension driven microfluidic systems," *Proceedings of the 13th Mediterranean Conference on Control and Automation*, Limassol, Cyprus, June 27-29, 2005, pp. 131-138.

[37] K. Chakrabarty and F. Su, *Digital Microfluidic Biochips: Synthesis, Testing, and Reconfiguration Techniques.* CRC Press, Boca Raton, FL, 2006.

[38] Hyejin Moon, A. R. Wheeler, R. L. Garrell, J. A. Loob, Chang-Jin "CJ" Kim, "An integrated digital microfluidic chip for multiplexed proteomic sample preparation and analysis by MALDI-MS," *Lab Chip* **6**, pp. 1213–1219, 2006.

[39] Shih-Kang Fan, Po-Wen Huang, Tsu-Te Wang and Yu-Hao Peng, "Cross-scale electric manipulations of cells and droplets by frequency-modulated dielectrophoresis and electrowetting," *Lab Chip* **8**, pp. 1325–1331, 2008.

[40] G. J. Shah, A. T. Ohta, E. P.-Y. Chiou, Ming C. Wuc and Chang-Jin "CJ" Kim, "EWOD-driven droplet microfluidic device integrated with optoelectronic tweezers as an automated platform for cellular isolation and analysis," *Lab Chip* **9**, pp. 1732–1739, 2009.

[41] A. Rival, C. Delattre, Y. Fouillet, C. Chabrol, F. Bottausci, G. Catellan, N. David, X. Gidrol, F. Vinet, D. Jary, "Towards Single Cell Gene Expression on EWOD Lab On Chip," *Proceedings of the 2009 Nanobio-Europe Conference*, Grenoble, France, 16-18 June 2009.

[42] YizhongWang, Yuejun Zhao and Sung Kwon Cho, "Efficient in-droplet separation of magnetic particles for digital microfluidics," *J. Micromech. Microeng.* **17**, pp. 2148–2156, 2007.

[43] T. B. Jones. *Electromechanics of particles.* Cambridge University Press, 1995.

[44] G. Beni and S. Hackwood, "Electro-wetting displays," *Applied Physics Letters* **38**, pp. 207–209, 1981.

[45] J. Heikenfeld, N. Smith, M. Dhindsa, Kaichang Zhou, Murali Kilaru, Linlin Hou, Jilin Zhang, E. Kreit and B. Raj, "Recent progress in arrayed electrowetting optics," *Optics and Photonics News* **21** (1), pp. 20-26, January 2009.

[46] http://documents.epfl.ch/groups/s/si/si-unit/www/nano-bio-sensing/Wu_EPFL1.pdf.

[47] J. Heikenfeld, K. Zhou, E. Kreit, B. Raj, S. Yang, B. Sun, A. Milarcik, L. Clapp, and R. Schwartz, "Electrofluidic displays using Young-Laplace transposition of brilliant pigment dispersions," *Nat Photon* **3**, pp. 292–296, 2009.

[48] R. A. Hayes, B. J. Feenstra, "Video-speed electronic paper based onelectrowetting," *Nature* **425**, 383–385, 2003.

11

Capillary Self-assembly for 3D Microelectronics

11.1 Abstract

Since the first microprocessor, fabricated in 1971, the performance of electronic chips has been rapidly increasing. The development speed has been following Moore's law [1]. More precisely, Moore's law describes a long-term trend in the history of computing hardware, in which the number of transistors that can be placed inexpensively on an integrated circuit has doubled approximately every two years. However, miniaturization limits – such as interconnect length and interference between packed electric circuitry – has recently brought limitations to the historical trend. Two axes of development are being investigated: the first one, called "more Moore," consists of pursuing integration efforts to miniaturize ever further integrated circuit technology. The second one, called "more than Moore," seeks to add functionalities to the integrated circuits, without further reductions in the dimensions.

In both cases, three-dimensional integration appears to be mandatory [2]. The conventional two-dimensional architecture presents horizontal wiring that slows down the transmission speed. Three-dimensional stacking of chips would result in shorter interconnects and consequently deliver a shorter transmission time. A required condition is a precise alignment of the chips – allowing for vertical connections – and a sufficiently strong direct bonding to vertically assemble the components. Another advantage of three-dimensional integration is the possibility of stacking integrated circuits of different origins, fabricated with different process, such as MEMS and NEMS.

In order to be industrialized, three-dimensional assembly methods must be at the same time fast and precise: many chips must be placed in a very short time and every chip must be precisely aligned. Currently, the "pick and place" technique has been developed. This technique makes use of robotics to place a chip on a wafer or on another chip [3,4]. It is found that such a robotic solution either builds assemblies with submicron precision at a relatively low rate, or builds assemblies at a high speed but with poor precision. New methods are being investigated, especially self-assembly methods. Self-assembly using evaporating droplet and capillary alignment is one of the most promising methods, and is the topic of this chapter.

11.1.1   Introduction to Self-assembly

From a general viewpoint, self-assembly is an increasingly used method in micro- and nano-technology. According to G.M. Whitesides, "self-assembly is a spontaneous and reversible process that brings together in a defined geometry randomly moving distinct bodies through selective bonding forces" [5]. In other words, objects or structures can be built from the arrangement of randomly moving elements. The arrangement or ordering is obtained by the use of local forces and interactions. Usually, self-assembly is achieved at the microscopic or nanometric scale. For example, in biotechnology, self-assembly has been used to form cellular membranes [6]; in material sciences, coating of solid walls with aligned micro-spheres also results from self-assembly mechanisms [7]. Examples at the macroscopic scale are fewer because random motion is infrequent at such a scale. However, the notion of self-assembly can be extended to larger scales by considering that macroscopic objects not precisely arranged, but initially not very far from their final location, can be brought into place by local forces.

11.1.2   Three-dimensional Assembly of Silicon Chips on a Wafer

In this chapter, we focus on the self-assembly of silicon chips (also called dies) on wafers using an evaporating droplet method, and we investigate whether capillary forces are capable of alignment. It is expected that capillary forces will align the die and evaporation of the liquid droplet eventually brings the chip into contact with the wafer; finally, direct bonding of the die on the substrate will be performed at low temperature ($100°C$) [8-15]. This technique theoretically allows for self-alignment and assembly without any intermediate layer (fig. 11.1). The bonding strength should be high enough so that the assembly can handle post processing, such as thinning down or through-via etching for interconnects.

11.2   Ideal Case: Total Pinning on the Chip and Pad Edges

In this section, we assume that the droplet is attached along each edge of the chip or die on the upper side, and along pad boundaries on the lower side. We shall see later, in section 11.3, that this assumption constitutes an ideal case and that the real situation is more complex. However, such an assumption is an useful first step in the understanding of the phenomena occurring during capillary alignment.

More specifically, in this first section, we analyze the mechanisms of self-alignment and investigate the stable and unstable displacement modes. Using the characteristic times for capillarity, inertia and evaporation, it is shown that the self-alignment mechanism is governed by the capillary force exerted by the liquid-air interface.

Initially, the die is gently dropped on a water droplet sitting on a square hydrophilic pad patterned on the wafer surface. It is assumed that the release of the die does not induce an important inertial motion, i.e. the die is carefully placed on top of the droplet. At this point, three physical phenomena govern the die motion: (i) water-air interfacial forces that are expected to bring the die in an aligned position above its planned final location on the wafer, (ii) evaporation of the water that progressively moves the die towards the wafer surface, and (iii) gravitational forces linked to the weight of the die and the liquid.

Let us first compare the characteristic times of the different mechanisms acting on the die: capillarity, inertia and evaporation. The Tomotika time – or capillary time – is the time taken by a distorted liquid-air interface to regain its equilibrium shape against the action of the viscosity

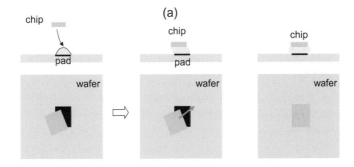

(1) Dispensing a droplet   (2) Rough positionning   (3) Accomplishing self-alignment

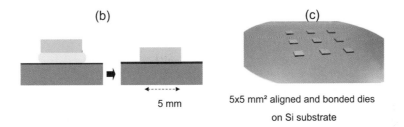

5 mm

5x5 mm² aligned and bonded dies
on Si substrate

Figure 11.1 Principle of die self-alignment: (a) sketch of the droplet and the die; (b) sketch of the die before and after evaporation; (c) view of aligned dies on a wafer.

[16]. It is given by the ratio

$$\tau_{capillary} = \frac{\eta R}{\gamma}, \qquad (11.1)$$

where η is the fluid viscosity, R a characteristic dimension of the interface (here the vertical gap h), and γ the surface tension. Using a water viscosity value of $\eta = 10^{-3}$ Pa.s, a vertical gap $h = 20\,\mu$m and a surface tension between water and air $\gamma = 72.10^{-3}$ N/m, we obtain $\tau_{capillary} < 10^{-6}$ seconds. At this scale, a distorted interface regains its equilibrium shape almost instantaneously.

An inertial time either based on the ratio between inertia and viscosity, or on the ratio between inertia and capillarity, can be defined. In the first case

$$\tau_{inertia} = \sqrt{\frac{\rho L^2 h}{\eta V}}, \qquad (11.2)$$

where h is the liquid thickness, L the dimension of the chip, ρ the density of the die and V the average velocity of the die during its motion. V is obtained from experiment and is of the order of $V \approx 0.1$ mm/s, and typically $h = 20\mu$m, $L = 5$ mm. In the second case

$$\tau_{inertia} = \sqrt{\frac{\rho L^2 h}{\gamma}}. \qquad (11.3)$$

The two inertia times are respectively of the order of 7 and 0.01 seconds.

Finally, we derive an evaporation time based on the diffusion of vapor [17]. The mass transfer equation is

$$\frac{dm}{dt} = D\frac{\delta c}{\delta n} 4hL,$$
(11.4)

where D is the diffusion coefficient, c the vapor concentration, n the normal unit vector and m the mass of liquid: $m = \rho_w hL^2$, ρ_w being the density of the liquid. An approximation of the vapor gradient of concentration is

$$\frac{\delta c}{\delta n} = \frac{c_{sat}}{L},$$
(11.5)

where c_{sat} is the saturation concentration (concentration at the interface). It can be shown that

$$\tau_{evap} = \frac{L^2}{4D}\frac{\rho_w}{c_{sat}}.$$
(11.6)

Physical tables indicate the approximate values $D = 21 \times 10^{-6}$ m/s^2 and $c_{sat} = 1.2$ kg/m^3. The evaporation time is then $\tau_{evap} \approx 250$ seconds. This value is in agreement with experiments showing complete evaporation after 4 to 5 minutes. Hence $\tau_{capillary} \ll \tau_{inertia} \ll \tau_{evap}$. The displacement of the die is then governed by capillary forces linked to the minimization of the liquid surface area along with the forces exerted by the liquid and die weight under the constraint of constant volume [18,19]. The surface energy of the droplet is given by

$$E_T = \gamma_{WA}S_{WA} + \gamma_{SW}S_{SW} + \gamma_{DW}S_{DW},$$
(11.7)

where S_{WA}, S_{SW} and S_{DW} are respectively the surface areas of the water-air interface, substrate-water interface and die-water interface, and γ_{WA}, γ_{SW} and γ_{DW} the surface tensions between water and air, substrate and water, and die and water. We make the simplifying hypothesis that the upper side of the droplet is pinned on the die edges, and on the the lower side of the droplet is pinned to the contour line of the hydrophilic pad. Hence, the contact angles do not intervene and the energy to be minimized is that of gravitational potential plus the water-air interface energy $E = \gamma_{WA}S_{WA}(x,y,z,\theta,\alpha,\phi,P,\gamma)$, where x,y,z are the coordinates of the mass center of the plate, θ, α, ϕ the directing angles of the plate, P the weight of the die and γ the surface tension between water and air. We use the Surface Evolver [20] to find the stable chip position which minimizes the total energy.

Four different displacement modes can be identified (figure 11.2) [18,21-23]: (1) the lift corresponding to a vertical motion of the plate, (2) the twist corresponding to a rotation of the plate in the horizontal plane, (3) the shift, which is a horizontal translation of the plate, and (4) the tilt or roll which are respectively a rotation around the horizontal y-axis or x-axis. Let us investigate successively these four modes and check if each of these modes is subject to self-alignment.

11.2.1   First Mode: Horizontal Displacement (Shift)

Let us denote by γ the surface tension, L the side length of the die, S the interfacial area, and V_l the liquid volume. The first displacement mode corresponds to a horizontal shift of the die (figure 11.3). It is obvious that the interfacial area is larger after a shift. Consequently the capillary forces will act to reduce the interfacial area by bringing back the die into alignment. A calculation with Surface Evolver shows that strong capillary forces pull the die back into

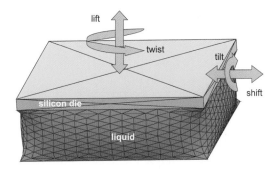

Figure 11.2 The four different modes (and possible reasons for misalignment): lift, twist, shift and tilt.

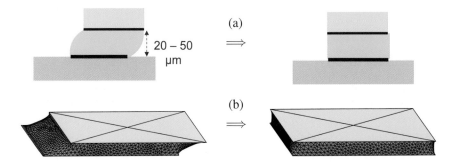

Figure 11.3 (a) After a horizontal displacement, the die is restored to alignment by capillary forces; (b) Evolver calculation of the pullback.

alignment. The force along the horizontal direction x is given by

$$F_x = -\frac{\delta(\gamma S)}{\delta x}. \tag{11.8}$$

Figure 11.4 shows the values of the restoring force for different liquid gaps. In the case of small shifts, this force is larger for small gaps because of the larger relative change of interfacial surface area.

This behavior can be checked against an approximate model where the interfaces would be planar (which is not exactly true). Under this assumption, it can be easily shown (Appendix A of this chapter) that the liquid-air surface energy is approximately

$$E \approx 2\gamma L h \left(1 + \sqrt{1 + \left(\frac{x}{h}\right)^2} \right), \tag{11.9}$$

where x is the horizontal shift, and the capillary force is

$$F_x \approx -2\gamma L \frac{1}{\sqrt{1 + \left(\frac{h}{x}\right)^2}}. \tag{11.10}$$

This formula compares well with Evolver results, as shown in figure 11.5.

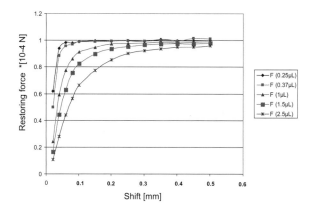

Figure 11.4 Restoring force vs. shift for different values of the droplet volume. (Courtesy F. Grossi, CEA-Leti.)

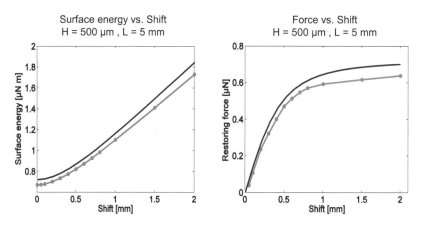

Figure 11.5 Comparison between approximate analytical model (plain line) and Evolver (dots). Left: interfacial energy; right: restoring force.

The energy curves in figure 11.5 show that the real interfacial area at zero shift – calculated with Evolver – is a little smaller than that corresponding to four planar interfaces ($S = 4hL$). This is linked to the shape of the interface, corresponding to a minimal surface area. For very small shifts, $x/h \ll 1$ and $F_x = -2\gamma Lx/h$; so the restoring force is a linear function of the shift. For large shifts $x/h \gg 1$, $F_x = -2\gamma L$, and the restoring force is constant. An important observation stemming from Eq.11.10 and from Evolver results is that the restoring force is larger for small vertical gaps h.

11.2.2   Second Mode: Rotation Around z-axis (Twist)

A twist is a rotation around the vertical z-axis. In this case too, a twist increases the interfacial area, and the die is pulled back to alignment, as shown in figure 11.6. As for the shift, the problem can be approached by using Surface Evolver and comparing the results to an approximate analytical model where the interfaces – initially planar – are twisted progressively (Appendix B).

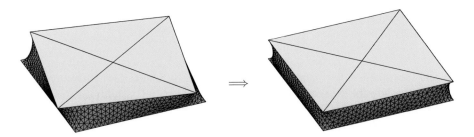

Figure 11.6 Die realigns after an initial twist under the action of a restoring torque.

Let us recall that the torque on the die is given by

$$T = -\frac{\delta(\gamma S)}{\delta\theta}.$$ (11.11)

Interfacial energy and torque are plotted in figure 11.7 as functions of the twist angle for three different liquid heights (100, 200 and 500 μm).

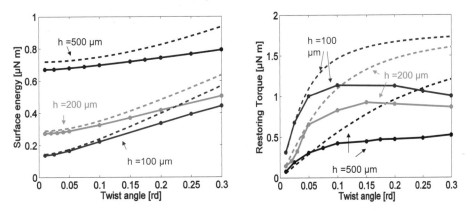

Figure 11.7 Energy (left) and torque (right) as functions of the twist angle, for three values of the gap. Dotted lines are Surface Evolver results and plain lines correspond to the analytical model.

The surface energy is obviously larger for a larger gap; but the derivative with respect to the twist angle is larger for small gaps, i.e. the variation of surface area with the twist angle is larger in the case of small gaps. Hence, the torque is larger for smaller liquid volumes. The analytical model produces a torque value given by

$$T \approx \gamma L^2 \left[\frac{a}{\sqrt{1+a^2}} + \frac{1}{a+\sqrt{1+a^2}} \left(\frac{1}{a} + \frac{1}{\sqrt{1+a^2}} \right) - \frac{\ln\left(a+\sqrt{1+a^2}\right)}{a^2} \right],$$ (11.12)

with $a = L\theta/2h$. For small twists, the analytical model compares well with Evolver, but not for large twists. In this latter case, the interfacial surface is considerably affected by surface tension forces and the hypothesis used by the analytical model is not physical. In such a case, a numerical approach, such as that of Evolver, is essential to predict the real shape of the interfaces and consequently deduce the values of the torque.

11.2.3   Third Mode: Vertical Displacement (Lift)

Because the vertical location of the die is a balance between gravity and capillary forces, it is expected that there is only one stable height for the die. This height depends on the liquid volume and die weight. This analysis recaps the work of Suzuki [24]. Evolver numerical simulation shows how the die is translated vertically after an initial push/lift and that the vertical gap depends on the weight of the die (fig. 11.8). Interfacial energy is shown in figure 11.9 and restoring force in figure 11.10. In each figure, a comparison has been made between Evolver results and a simplified analytical approach based on a circular curvature of the interfaces (see Appendix C).

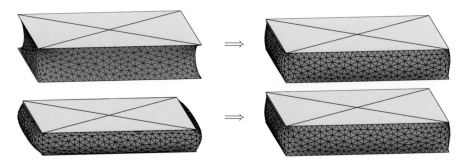

Figure 11.8 Die regains its stable position after a lift. Top: after an initial lift; bottom: after an initial compression.

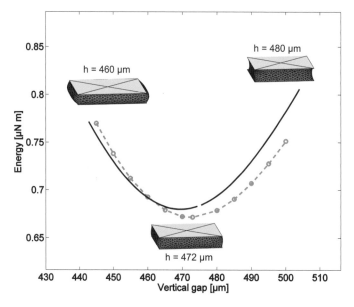

Figure 11.9 Surface energy as a function of the vertical gap: Continuous line corresponds to the analytical model and dotted line to Evolver results.

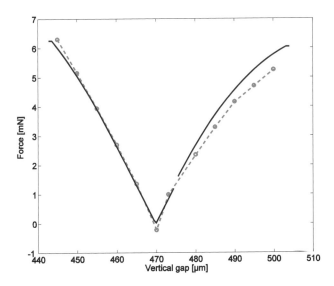

Figure 11.10 Restoring force vs. vertical gap. Continuous line: analytical model; dotted line: Evolver results.

11.2.4 Fourth Mode: Tilt (or Roll)

We will call tilt a rotation around the y-axis and roll a rotation around the x-axis. Basically, tilt and roll share the same behavior. In the case where the chip has a square shape, tilt and roll are exactly identical. Tilt is a complex phenomenon because the variation of the interfacial area is more difficult to intuitively predict. Recall from section 2.2.4 that for a circular die and circular pad without gravity, the minimum energy droplet has a spherical shape and the energy is independent of the tilt and roll. The very simple calculation depicted in figure 11.11 for square die and pad concludes that this shape problem is also indeterminate. For the same volume of the same liquid, assuming the simplest form of interfaces, the surface energy in a flat configuration is

$$E = \gamma S = \gamma (4Lh) = \gamma \left(4\frac{V_L}{L} \right), \tag{11.13}$$

where V_L is the liquid volume. In a dihedral morphology, the surface energy is

$$E = \gamma S = \gamma \left[2 \left(\frac{\alpha L^2}{2} \right) + \alpha L^2 \right] = \gamma \left(4\frac{V_L}{L} \right), \tag{11.14}$$

where α is the dihedral angle. Hence, it is the distortion of the interfaces that can make the difference and pinpoint the stable position. It is a second order problem. This remark leads to serious complications: from a numerical standpoint, the meshing of the surface needs to be sufficiently fine to produce a precise value of the energy. In return, the computation time is long. From a physical standpoint, the roles of the parameters – like the weight of the die, or the surface tension of the liquid – are difficult to predict.

  The numerical model shows that the dihedral position ((b) and (d) in figure 11.12) is the stable position whatever the initial depth of the liquid gap ((a) and (c) in figure 11.12).

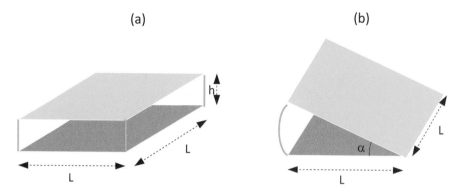

Figure 11.11 Two morphologies of the liquid having the same surface energy.

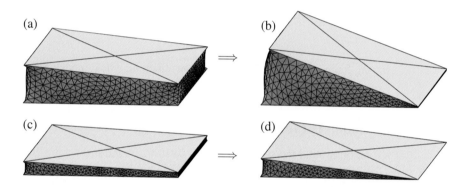

Figure 11.12 Die tilts to form a dihedral: (a) and (b) large liquid volume; (c) and (d) small liquid volume.

Interfacial energy and torque obtained numerically with the Evolver are plotted in figures 11.13 and 11.14. The horizontal lines correspond to the analytical approximation where the energy is constant.

The slightly smaller interfacial area for the dihedral morphology is due to small distortions of the interface in the die corners and along the die sides. The interface is concave in the first regions, and convex in the others (fig. 11.15). Parallel die-wafer morphology has 4 concave and 4 convex distortions, whereas the dihedral die-wafer morphology has only 2 concave and 3 convex distortions.

A remark about chip sliding: In the case considered here, the ratio h/L of the vertical to horizontal characteristic dimensions of the system is very small, of the order of 10^{-3}. For completeness, we extend our investigation to large volumes of liquid – i.e. the liquid thickness is of the order or larger than the chip dimensions ($h/L > 0.5$). Note that such a large liquid thickness usually results in overflowing, which will be studied later. Tilt associated to large liquid thickness triggers the sliding of the chip, as shown in figure 11.16. In such a case, a translational misalignment occurs after evaporation. It is also observed that sliding is enhanced by the chip weight.

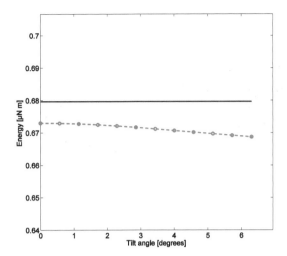

Figure 11.13 Comparison between the Evolver and analytical model: in case of tilt, the interfacial energy varies extremely slowly.

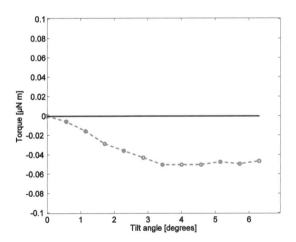

Figure 11.14 Tilt torque vs. tilt angle: continuous line indicates a zero torque, and the Evolver results (dotted line) show a small tilting torque.

11.2.5   Coupled Modes

From the previous analysis, it is expected that only coupled modes including roll and tilt will be unstable. This can be verified with Surface Evolver (fig. 11.17). An initial tilt and roll (fig. 11.17a) is progressively increased until one corner and then an edge of the chip contacts the substrate. The stable position is a wedge. In the case of an initial tilt and twist (fig. 11.17b), the twist vanishes quickly (it is a stable mode) but the tilt slowly increases. Finally, for an initial shift and twist – which are two stable modes – the shift vanishes first, then the twist (fig. 11.17c).

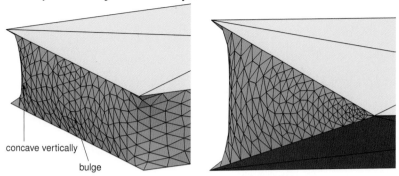

Figure 11.15 Bulging out/in shape of the interface in the two cases: left, parallel plates; right, contacting edges.

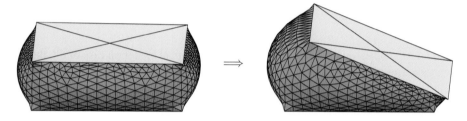

Figure 11.16 The chip slides horizontally after a tilt in the case of a large volume of liquid (and a sufficient chip weight).

11.3   Real Case: Spreading and Wetting

In the preceding section, we have focused on the alignment in a quasi-steady state, assuming that the liquid droplet has spread and is pinned along the edges of the chip and pad. It has been shown that this fully wetted state produces a precise alignment, even with the slight instability of the tilt/roll mode. However, a perfect wetting of the pad and chip has been assumed. In this section, we focus on the progressive spreading and wetting of the two surfaces and analyze the consequences on the alignment (fig. 11.18). In order to facilitate the pinning and alignment,

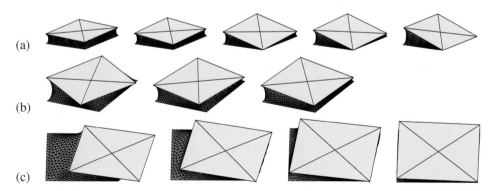

Figure 11.17 Coupled modes: (a): tilt and roll; (b) tilt and twist; (c) shift and twist. Unstable modes must include a tilt and/or roll; the other modes are automatically corrected by the capillary forces.

relief elements have been added on the pad and chip.

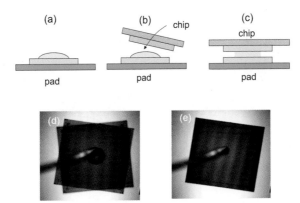

Figure 11.18 Top: sketch of the chip gently dropped on the droplet; bottom: experimental view of the alignment after the pipette suction has been deactuated and the liquid has aligned the chip.

The starting point is the moment in time where the chip is gently deposited on the droplet. The physical analysis is summarized in figure 11.19.

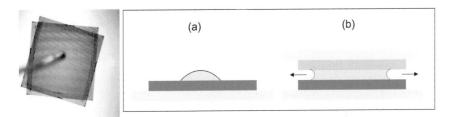

Figure 11.19 Left: Experimental view of the deposition of the chip on the pad; right: sketch of the chip and pad. Note that there is a relief on the chip and on the pad (the green and yellow on the right). This relief is aimed at improving the pinning on the edges, as will be detailed in the next section.

In the first stage, the droplet spreads under the action of the capillary forces and the weight of the chip. During this stage, there is no horizontal motion of the chip relative to the pad, as shown in figure 11.20 (vertically, the chip descends as the liquid spreads). This stage ends when the triple lines reach the first edges, either that of the chip, or that of the pad. In the second stage, the analysis of the forces on the chip edges show that there is a translational resultant that tends to zero out the shift (fig. 11.21 and 11.22). In the third stage, there is a resulting torque that brings the chip to alignment [25].

Although Surface Evolver cannot model the exact dynamics of the spreading, the model becomes relevant when the interface reaches the pad and chip edges. In such a topology, the surface tension forces are by far dominant. From a numerical standpoint, the difficulty in such a case is to specify constraints that are following the moving edges. It is shown that the preceding physical analysis is correct, i.e. the shift is first reduced, then the twist (fig. 11.23).

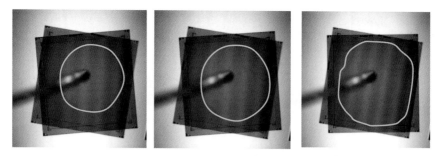

Figure 11.20 Experimental view of the spreading. The contact lines have been pasted in white for better visualization. The chip does not move until the contact line has reached the edges of chip and/or pad (courtesy S. Mermoz and L. Di Cioccio).

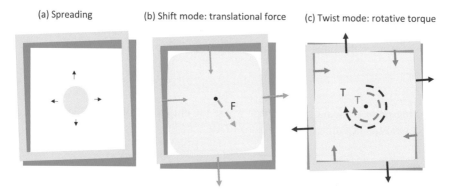

Figure 11.21 Sketch of the spreading. The chip only appears through its frame, for easier visualization. (a) the initial droplet starts spreading; (b) the liquid reaches edges (either those of the chip, or the pad), inducing a correction of the shift; (c) after the shift has been canceled, a restoring torque aligns the chip with the pad.

When looking closely at the corner, it is observed that wetting cannot be complete: a curvature radius always remains in the corners, even for extremely hydrophilic cases (fig. 11.24a). These small, symmetrical dewetted areas do not prevent alignment. However, a real wetting defect generally results in an incomplete alignment, especially when this defect is not

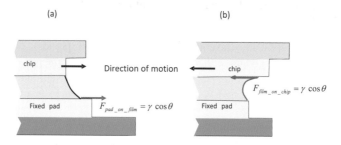

Figure 11.22 The two different situations of pinning: left, pinning on the lower (pad) edge; right, pinning on the upper (chip) edge.

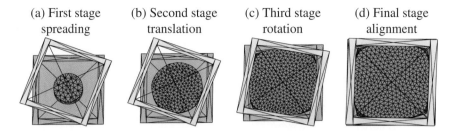

(a) First stage
spreading

(b) Second stage
translation

(c) Third stage
rotation

(d) Final stage
alignment

Figure 11.23 Evolution of water spreading and wetting on the two surfaces (Evolver). Note that the chip has been dematerialized for easier visualization.

symmetrical relative to the center of the pad or chip. In the case of figure 11.24b, a spreading defect has been set up on the pad, in its right lower corner. The liquid spreads to reach the edges, but cannot wet the corner, leaving remaining shift and twist.

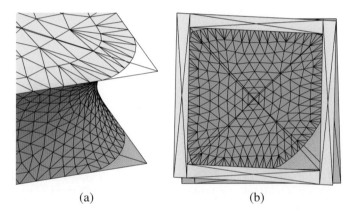

(a)   (b)

Figure 11.24 (a) Small dewetted zones in the corners do not prevent alignment. (b) Wetting defect in the lower right corner producesk a remaining shift and twist.

An analysis of the consequences of the presence of a partial dewetted region is shown in the sketches of figure 11.25. If the dewetted region is located in a corner, there is an unresolved shift; besides, if the dewetted region is asymmetrical (in respect to the diagonal) an additional twist superposes on the shift. This analysis is confirmed by the results of a simulation with Evolver (fig. 11.26).

In conclusion, self alignment is achieved when the spreading is total on the two surfaces – with the exception of the extremity of the corners which can never be totally wetted. It appears that the condition of spreading is not sufficient: a complete pinning has also been assumed to conclude that alignment was reached. In the following section we analyze the importance of the pinning on the chip and pad boundaries.

11.4   The Importance of Pinning and Confinement

An incomplete pinning on the boundaries results in an overspreading. Figure 11.27 shows experimentally observed and calculated overspread resulting in alignment defect. It is deduced

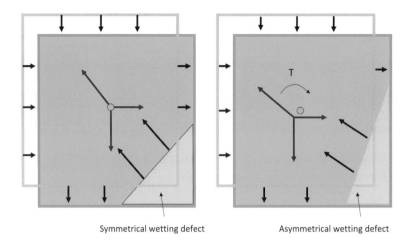

Symmetrical wetting defect   Asymmetrical wetting defect

Figure 11.25 A symmetrical dewetted region blocks the restoring of the shift; right an asymmetrical dewetted region blocks the restoring shift and creates an additional twist.

(a)	(b)	(c)
left shift	diagonal shift	shift and twist

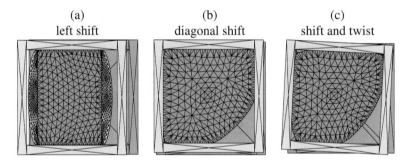

Figure 11.26 Evolver analysis of the consequences of a wetting defect. (a) Non-wetted band on right on bottom; top of liquid made transparent, so the white surfaces are the insides of the free surfaces between pad and chip. (b) Non-wetted corner. (c) Asymmetric non-wetted corner.

that poor pinning resulting in overspreading was the major difficulty of this alignment technique. The pinning of the interface on the pad and chip boundaries is then essential. A small overflow like that shown in figure 11.28 ruins the alignment. Different solutions are currently being explored.

There are two different approaches: the first one uses a planar wafer and a superhydrophilic pad; while the second uses a wafer with a relief, e.g. the pad is the top of a micro-pillar. In the first case, pinning is realized by a strong wetting contrast between the superhydrophilic pad (contact angle $2° - 5°$) and the less hydrophilic wafer (contact angle larger than $60°$). In the second case, the pinning is realized on the edges of the pad.

As we have seen in chapter 3, the canthotaxis limit determines the pinning. In the first case, the canthotaxis limit is given by the maximum possible bulging of the interface – which is the Young angle of the wafer – whereas in the second case, it is given by the Young angle on the vertical side of the pillar plus the pillar angle – usually $90°$ (fig. 11.29). Note that in the latter case, the pillar edge must be sharp or else the canthotaxis limit decreases [26].

The largest possible bulging angle of the liquid is difficult to determine. It depends on the

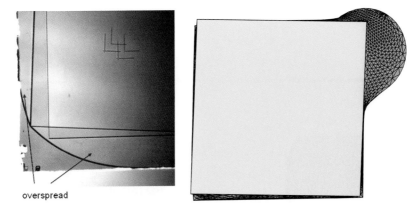

overspread

Figure 11.27 Left: misalignment due to water spreading on the wafer outside the hydrophilic pad; right: numerical simulation of an overspread, leading to an alignment defect.

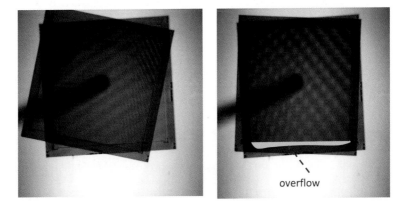

overflow

Figure 11.28 An overflow leads to a misalignment (courtesy S. Mermoz and L. Di Cioccio).

weight of the die, and also on the way the die is placed on top of the droplet. In this case, inertia – which has been so far neglected – could have an important effect. Hence, the limit depends on the protocol and should be determined experimentally.

11.5   Conclusion

In the first part of this chapter, we studied the capillary forces of the chip/pad system assuming that the liquid is attached to the chip and pad boundaries. In the second part we focused on the spreading of the liquid and the early motion of the chip when the liquid has not yet totally wetted the two surfaces. In both cases, it has been shown that restoring forces and torques were sufficient to bring the chip into alignment, under two conditions: the wetting of the pad and chip surfaces should be total (except for symmetric very small areas in the corners) and no overspreading linked to depinning should occur.

In order to avoid liquid overspreading, a sharp wetting contrast must be established between the pad and the rest of the wafer. Two different approaches can be made to reinforce the wetting contrast: first a hydrophobic treatment of the wafer, second the use of relief. Finally, the

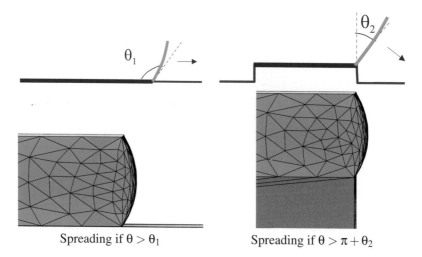

Spreading if θ > θ₁   Spreading if θ > π + θ₂

Figure 11.29 Left: canthotaxis limits for planar wafer; right: canthotaxis limit for a wafer with a relief (mesa).

canthotaxis limit has shown that one has to be especially cautious at the moment when the die is released on top of the droplet. An inertial motion of the die could be transmitted to the liquid and could trigger overspreading.

11.6   Appendix A: Shift Energy and Restoring Force

In this appendix, planar interfaces are assumed. The shift is then sketched in the figure 11.30.

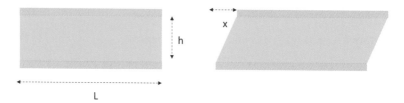

Figure 11.30 Sketch of the shift: left, at alignment; right, after a shift.

The volume of the liquid is then

$$V_L = hL^2, \tag{11.15}$$

and the interfacial area at alignment is given by

$$S_1 = 4\frac{V_L}{L}. \tag{11.16}$$

With the same assumption of planar interfaces, the interfacial area with a shift x is

$$S_2 = 2\frac{V_L}{L}\left(1 + \sqrt{1 + \left(\frac{x}{h}\right)^2}\right) = 2hL\left(1 + \sqrt{1 + \left(\frac{x}{h}\right)^2}\right). \tag{11.17}$$

Note that if $x \to 0$, then $E \to 4hL$. The surface energy difference is

$$\Delta E = \gamma(S_2 - S_1) = 2\frac{V_L}{L}\sqrt{1 + \left(\frac{x}{h}\right)^2}. \tag{11.18}$$

The restoring force is then

$$F_x \approx -2\gamma L\frac{1}{\sqrt{1 + \left(\frac{h}{x}\right)^2}}. \tag{11.19}$$

11.7   Appendix B: Twist Energy and Restoring Torque

In this appendix, the four interfaces are supposed to be planar at alignment, having square angles between each other. Recall that a twist around the z axis of angle θ corresponds to the rotation matrix

$$R = \left\{\begin{matrix} \cos\theta & -\sin\theta \\ \sin\theta & \cos\theta \end{matrix}\right\}. \tag{11.20}$$

The twist can be sketched as in figures 11.31 and 11.32. In order to obtain a twist of angle θ, we suppose that the interfaces are progressively twisted from $z = 0$ to $z = h$, with an angle of $(z/h)\theta$.

For a fixed y, the current point $M(s, y, 0)$ describes a spiral curve defined by

$$\tilde{M} = \left\{\begin{matrix} s\cos(\frac{z}{h}\theta) - y\sin(\frac{z}{h}\theta) \\ s\sin(\frac{z}{h}\theta) + y\cos(\frac{z}{h}\theta) \\ z \end{matrix}\right\}. \tag{11.21}$$

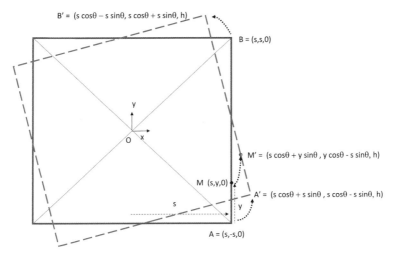

Figure 11.31 Sketch of the twist. The point M is describing AB while M' is describing A'B'. Note that $s = L/2$.

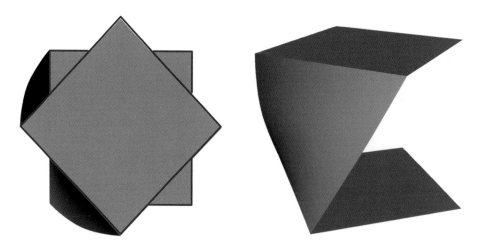

Figure 11.32 Three dimensional view of the twisted interface (only one twisted interface is shown in the figure), courtesy R. Berthier.

Let us consider the parametric function f such that

$$f : [-s,s] \times [0,h] \longrightarrow \Re^3, \tag{11.22}$$

$$(y,z) \longrightarrow \left\{ \begin{array}{c} s\cos(\frac{z}{h}\theta) - y\sin(\frac{z}{h}\theta) \\ s\sin(\frac{z}{h}\theta) + y\cos(\frac{z}{h}\theta) \\ z \end{array} \right\}. \tag{11.23}$$

This function produces the coordinates of $\tilde{M}$. The infinitesimal surface ds is given by

$$ds = \left\| \frac{\delta f}{\delta y} \times \frac{\delta f}{\delta z} \right\| dydz. \tag{11.24}$$

Using the calculated values

$$\frac{\delta f}{\delta y} = \left\{ \begin{array}{c} -\sin\frac{z}{h}\theta \\ \cos\frac{z}{h}\theta \\ 0 \end{array} \right\}, \tag{11.25}$$

$$\frac{\delta f}{\delta z} = \left\{ \begin{array}{c} -\frac{\theta}{h}[s\sin(\frac{z}{h}\theta) + y\cos(\frac{z}{h}\theta)] \\ \frac{\theta}{h}[s\cos(\frac{z}{h}\theta) - y\sin(\frac{z}{h}\theta)] \\ 1 \end{array} \right\}, \tag{11.26}$$

the vector product $\frac{\delta f}{\delta y} \times \frac{\delta f}{\delta z}$ is then

$$\frac{\delta f}{\delta y} \times \frac{\delta f}{\delta z} = \left\{ \begin{array}{c} \cos(\frac{z}{h}\theta) \\ \sin(\frac{z}{h}\theta) \\ \frac{\theta}{h}[-\sin(\frac{z}{h}\theta)(s\cos(\frac{z}{h}\theta) - y\sin(\frac{z}{h}\theta)) + \cos(\frac{z}{h}\theta)(s\sin(\frac{z}{h}\theta) + y\cos(\frac{z}{h}\theta))] \end{array} \right\} \tag{11.27}$$

which can be simplified to

$$\frac{\delta f}{\delta y} \times \frac{\delta f}{\delta z} = \left\{ \begin{array}{c} \cos(\frac{z}{h}\theta) \\ \sin(\frac{z}{h}\theta) \\ \frac{\theta}{h}y \end{array} \right\}. \tag{11.28}$$

The infinitesimal surface ds is finally

$$ds = \left\| \frac{\delta f}{\delta y} \times \frac{\delta f}{\delta z} \right\| dydz = \sqrt{1 + \left(\frac{\theta}{h}y\right)^2} \, dydz. \tag{11.29}$$

And the interfacial area (for one interface) is

$$A = \int_{z=0}^{z=h} \int_{y=-s}^{y=s} \sqrt{1 + \left(\frac{\theta}{h}y\right)^2} \, dydz = h \int_{y=-s}^{y=s} \sqrt{1 + \left(\frac{\theta}{h}y\right)^2} \, dy. \tag{11.30}$$

Integrating, we find a total interfacial energy of

$$E = 4\gamma \left[sh\sqrt{1+a^2} + \frac{h^2}{\theta} \ln\left(a + \sqrt{1+a^2}\right) \right] \tag{11.31}$$

$$\text{where} \quad a = \frac{s\theta}{h}. \tag{11.32}$$

Note that if $\theta \to 0$, then $a \to 0$ and $E \to 4\gamma Lh$. The torque T is given by the derivative of the energy with respect to the angle θ:

$$T \approx \gamma s^2 \left[\frac{a}{\sqrt{1+a^2}} + \frac{1}{a+\sqrt{1+a^2}} \left(\frac{1}{a} + \frac{1}{\sqrt{1+a^2}}\right) - \frac{\ln(a+\sqrt{1+a^2})}{a^2} \right]. \tag{11.33}$$

11.8   Appendix C: Lift Energy and Restoring Force

It is assumed that the chip is lifted up or down from its equilibrium position. Let us calculate the interfacial area using the simplifying hypothesis that each of the four interfaces has a circular profile as shown in figure 11.33.

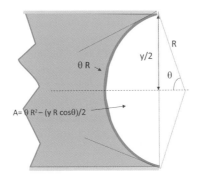

Figure 11.33 Sketch of the interface.

The total interfacial area is then

$$S = 4L(2\theta R), \tag{11.34}$$

and the (constant) droplet volume is

$$V = L^2 h_0 = L^2 y - 4L\left(\theta R^2 - \frac{y}{2} R\cos\theta\right). \tag{11.35}$$

Using the relation

$$R = \frac{y}{2\sin\theta},$$ (11.36)

we find

$$L^2 h_0 = L^2 y - L\frac{y^2}{\sin\theta}\left(\frac{\theta}{\sin\theta} - \cos\theta\right) \quad \text{if} \quad 0 < \theta \le \frac{\pi}{2},$$ (11.37)

$$L^2 h_0 = L^2 y + L\frac{y^2}{\sin\theta}\left(\frac{\theta}{\sin\theta} - \cos\theta\right) \quad \text{if} \quad \frac{\pi}{2} < \theta < \pi.$$ (11.38)

For a given angle θ, this last relation is a quadratic polynomial in y whose root is

$$y = \frac{1 - 1\sqrt{1 - 4\frac{2}{\sin\theta}\left(\frac{\theta}{\sin\theta} - \cos\theta\right)\frac{h_0}{L}}}{\frac{2}{\sin\theta}\left(\frac{\theta}{\sin\theta} - \cos\theta\right)} \quad \text{if} \quad 0 < \theta \le \frac{\pi}{2},$$ (11.39)

$$y = \frac{-1 + 1\sqrt{1 + 4\frac{2}{\sin\theta}\left(\frac{\theta}{\sin\theta} - \cos\theta\right)\frac{h_0}{L}}}{\frac{2}{\sin\theta}\left(\frac{\theta}{\sin\theta} - \cos\theta\right)} \quad \text{if} \quad \frac{\pi}{2} < \theta < \pi.$$ (11.40)

11.9   References

[1] G.E. Moore, "Cramming more components onto integrated circuits," *Electronics.* **38** (8), p. 4, April 1965.

[2] M. Koyanagi, T. Fukushima, T. Tanaka. "Three-dimensional integration technology and integrated systems," *Proceedings of the 2009 Conference on Asia and South Pacific Design and Automation*, Yokohama, Japon, 19-22 December 2009, pp. 409-415.

[3] http://www.datacon.at/.

[4] http://www.set-sas.fr/en/.

[5] G. M. Whitesides, B. Grzybowski, "Self Assembly at all scales," *Science* **295**(5564), pp. 2418–2421, 2002.

[6] Jieshu Qian, Meng Zhang, Ian Manners and Mitchell A. Winnik, "Nanofiber micelles from the self-assembly of block copolymers," Trends in *Biotechnology* **28**(2), pp. 84–92, 04 December 2009.

[7] T. Pinedo Rivera, O. Lecarme, J. Hartmann, E. Rossitto, K. Berton, and D. Peyrade, "Assisted convective-capillary force assembly of gold colloids in a microfluidic cell: Plasmonic properties of deterministic nanostructures," *J. Vac. Sci. Technol. B* **26**(6), pp. 2513–2519, 2008.

[8] T. Fukushima, E. Iwata, T. Konno, J.-C. Bea, K.-W. Lee, T. Tanaka, and M. Koyanagi, "Surface tension-driven chip self-assembly with load-free hydrogen fluoride-assisted direct bonding at room temperature for three-dimensional integrated circuits," *Appl. Phys. Lett.* **96**, 154105, 2010.

[9] T. Fukushima, T. Tanaka, M. Koyanagi. "3D System Integration Technology and 3D Systems," *Advanced Metallization Conference Proceedings*, pp. 479–485, 2009.

[10] W. Zheng, H.O. Jacobs. "Fabrication of multicomponent microsystems by directed three-dimensional self-assembly," *Advanced Functional Materials* **15**(5), pp. 732–738, 2005.

[11] H. Moriceau, B. Bataillou, C. Morales, A.M. Cartier, F. Rieutord. "Interest of a short plasma treatment to achieve high quality Si-SiO2-Si bonded structures," *7th Intl. Symp. on Semiconductor Wafer Bonding.* ECS Proceedings PV2003-19, p. 49 and p. 110, 2003.

[12] Q.Y. Tong, U. Gösele. *Semiconductor wafer bonding*. John Wiley and Sons, pp. 57-67, 1999.

[13] K. Sato, T. Seki, S. Hata, A. Shimokohbe. "Self-alignment of microparts using liquid surface tension - behavior of micropart and alignment characteristics," *Precision Engineering* **27**, pp. 42–50, 2003.

[14] F. Grossi, L. Di Cioccio, F. Rieutord, O. Renault, J. Berthier, J-C Barbé, F. De Crécy and L. Clavelier. "Self Assembly of Die to Wafer using Direct Bonding Methods and Capillary Forces," *Proceedings of the 2008 MRS Fall Meeting*, Boston, USA, 1-5 December 2008.

[15] F. Grossi, L. Di Cioccio, S. Vincent, M.D. Diop, L. Bally, N. Kernevez, F. Rieutord. "Self assembly of die to wafer using direct bonding methods and capillary techniques," *Proceedings of the 2007 Imaps Conference*, Scottsdale, USA, March 19-22, 2007.

[16] J. Berthier, P. Silberzan. *Microfluidics for Biotechnology*, Second Edition. Artech House Inc., 2010.

[17] K.S. Birdi, D.T. Vu, A. Winter. *J. Phys. Chem.* **93**, pp. 3702–3703, 1989.

[18] A. Greiner, J. Lienemann, J.G. Korvink, X. Xiong, Y. Hanein, K.F. Bohringer. "Capillary forces in micro-fluidic self-assembly," *International Conference on modeling and Simulation of Microsystems*, pp. 198–201, 2002.

[19] J. Berthier. *Microdrops and Digital Microfluidics*. William Andrew Publishing, February 2008.

[20] K. Brakke. "The Surface Evolver," *Exp. Math.* **1**, 1992.

[21] J. Berthier, K. Brakke, F. Grossi, L. Sanchez and L. Di Cioccio, "Self-alignment of silicon chips on wafers: A capillary approach," *JAP* **108**, 054905, 2010.

[22] J. Berthier, K. Brakke, F. Grossi, L. Sanchez, L. Di Cioccio, "Silicon die self-alignment on a wafer: stable and unstable modes," *Sensors and Transducers Journal* **115**(4), p. 135, 2010.

[23] P. Lambert, M. Mastrangeli, J.-B. Valsamis, G. Degrez, "Spectral analysis and experimental study of lateral capillary dynamics for flip-chip applications," *Microfluid Nanofluid* **9**, pp. 797–807, 2010.

[24] K. Suzuki. "Flow resistance of a liquid droplet confined between two hydrophobic surfaces," *Microsystem Technology* **11**, pp.1107–1114, 2005.

[25] J. Berthier, K. Brakke, L. Sanchez, L. Di Cioccio, "Self-alignment of silicon chips on wafers: the effect of spreading and wetting," *Nanotech NSTI Conference*, 13-16 June 2011, Boston, pp. 528-531.

[26] J. Berthier, F. Loe-Mie, V.-M. Tran, S. Schoumacker, F. Mittler, G. Marchand, N. Sarrut, "On the pinning of interfaces on micropillar edges," *J. of Colloid and Interface Science* **338**, pp. 296–303, 2009.

12

Epilogue

The knowledge of the behavior of droplets and interfaces is of utmost importance in microfluidic systems. Indeed, two- or multi-phase microflows are increasingly used in many different domains of microtechnology. In biotechnology for instance, droplets are commonly manipulated by droplet or digital microfluidics to achieve DNA detection and recognition, or encapsulation of cells. Capillarity is the driving mechanism for thread and paper microfluidics, producing fast and cheap solutions for the detection of biologic targets, like pathogens or bacteria. Fluids in contact with air can be moved by open and suspended microfluidics using capillary forces, enabling the fabrication of new microsystems for the investigation of cellular behavior, or totally passive compact fluidic systems, i.e. without requiring any external actuation. 3D microelectronics is searching for new solutions in capillary self-alignment, using interfaces to produce restoring forces and torques. Microfabrication is also investigating the same properties to align cover seals of microchips. In materials science, arrangement and ordering of nanoparticles is achieved by the progressive withdrawal of an interface, enabling for example the fabrication of optical fibers. Also in optics, tunable lenses can be achieved by changing on demand the curvature of a liquid interface, and new display screens of high quality use capillary arrangement of opaque electrolyte droplets in an oil phase. Finally, in mechatronics, capillary actuation has been demonstrated to be more efficient than conventional electromagnetic switches. These different applications are shown in figure 12.1.

Capillary and surface tension forces are the dominant forces in all these applications. Note that applications requiring electric actuation such asuch as digital microfluidics, optical displays, tunable lenses, etc., use simultaneously the theory of capillarity and the capillary equivalence. These considerations have led to the writing of this book. Many different situations occurring in microsystems have been investigated and the theoretical background as well as the numerical modeling has been developed in such a way that the reader can retrace the steps and develop their own application. A special effort has been made to present the two facets of the approach to interfaces and droplets behavior, namely the force approach and the energy approach. Their duality has been pointed out as much as possible. Not all situations can be treated in a single book. A focus has been made on the static or quasi-static behavior of droplets and interfaces, which is fundamental for a large panel of applications. A justification is the fact that, in microsystems, dynamic forces like viscous and inertia forces are usually small compared to capillary forces. Of course, many cases are still to be investigated, but it is the aim of this book to give the reader the potentiality to tackle these new problems by themself on the theoretical and numerical basis developed here.

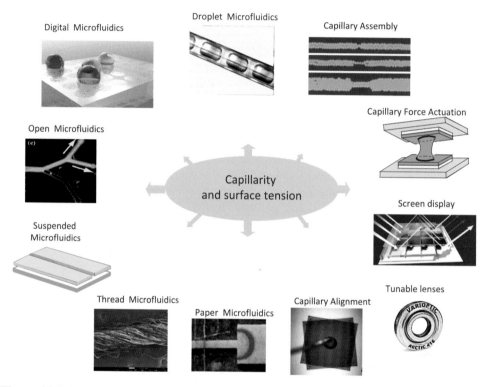

Figure 12.1 Capillary and surface tension forces are at the heart of many applications for microsystems. Note that for the applications based on electrowetting (digital microfluidics, tunable lenses, screen display and capillary force actuation) a conventional capillary theory and the electrowetting capillary equivalence intertwine. Reprinted with permissions: successively ©ACS, 2005; ©AIP, 2011; courtesy Liquavista; courtesy Varioptics; ©Springer 2010; ©AIP, 2011.

Index

Also of Interest

Check out these published and forthcoming related titles from Scrivener Publishing

The Physics of Micropdroplets
Jean Berthier and Kenneth Brakke
Published 2012. ISBN 978-0-470-93880-0

Integrated Biomaterials in Tissue Engineering
Edited by Murugan Ramalingam, Ziyad Haidar, Seeram Ramakrishna, Hisatoshi Kobayashi, and Youssef Haikel
Published 2012. ISBN 978-1-118-31198-1

Intelligent Nanomaterials
Processes, Properties, and Applications
Edited by Ashutosh Tiwari Ajay K. Mishra, Hisatoshi Kobayashi and Anthony P.F. Turner
Published 2012. ISBN 978-0-470-93879-9

Introduction to Surface Engineering and Functionally Engineered Materials
Peter Martin
Published 2011. ISBN 978-0-470-63927-6

Handbook of Bioplastics and Biocomposites Engineering Applications
Edited by Srikanth Pilla
Published 2011. ISBN 978-0-470-62607-8

Biopolymers: Biomedical and Environmental Applications
Edited by Susheel Kalia and Luc Avérous
Published 2011. ISBN 978-0-470-63923-8

Renewable Polymers
Synthesis, Processing, and Technology
Edited by Vikas Mittal
Published 2011. ISBN 978-0-470-93877-5

Plastics Sustainability
Towards a Peaceful Coexistence between Bio-based and Fossil fuel-based Plastics
Michael Tolinski
Published 2011. ISBN 978-0-470-93878-2

Miniemulsion Polymerization Technology edited by Vikas Mittal
Published 2010. ISBN 978-0-470-62596-5

Polymer Nanotube Nanocomposites
Synthesis, Properties, and Applications
Edited by Vikas Mittal
Published 2010. ISBN 978-0-470-62592-7

Handbook of Engineering and Specialty Thermoplastics
Part 1: Polyolefins and Styrenics by Johannes Karl Fink
Published 2010. ISBN 978-0-470-62483-5
Part 2: Water Soluble Polymers by Johannes Karl Fink
Published 2011. ISBN 978-1-118-06275-3
Part 3: Polyethers and Polyesters edited by Sabu Thomas and
Visakh P.M.
Published 2011. ISBN 978-0-470-63926-9
Part 4: Nylons edited by Sabu Thomas and Visakh P.M.
Published 2011. ISBN 978-0-470-63925-2

Biomedical Materials and Intelligent Medical Devices
Edited by Ashutosh Tiwari, Hisatoshi Kobayashi and Anthony P.F. Turner
Forthcoming August 2012